T0134676

Lecture Notes in Networks and Systems

Volume 181

The series "Lecture Notes in Networks and Systems" publishes the latest developments in Networks and Systems—quickly, informally and with high quality. Original research reported in proceedings and post-proceedings represents the core of LNNS.

Volumes published in LNNS embrace all aspects and subfields of, as well as new challenges in, Networks and Systems.

The series contains proceedings and edited volumes in systems and networks, spanning the areas of Cyber-Physical Systems, Autonomous Systems, Sensor Networks, Control Systems, Energy Systems, Automotive Systems, Biological Systems, Vehicular Networking and Connected Vehicles, Aerospace Systems, Automation, Manufacturing, Smart Grids, Nonlinear Systems, Power Systems, Robotics, Social Systems, Economic Systems and other. Of particular value to both the contributors and the readership are the short publication timeframe and the world-wide distribution and exposure which enable both a wide and rapid dissemination of research output.

The series covers the theory, applications, and perspectives on the state of the art and future developments relevant to systems and networks, decision making, control, complex processes and related areas, as embedded in the fields of interdisciplinary and applied sciences, engineering, computer science, physics, economics, social, and life sciences, as well as the paradigms and methodologies behind them.

Indexed by SCOPUS, INSPEC, WTI Frankfurt eG, zbMATH, SCImago.

All books published in the series are submitted for consideration in Web of Science.

More information about this series at http://www.springer.com/series/15179

Artem Antonyuk · Nikita Basov
Editors

Networks in the Global World V

Proceedings of NetGloW 2020

Editors
Artem Antonyuk
St. Petersburg State University
Centre for German and European Studies
St. Petersburg, Russia

Nikita Basov
St. Petersburg State University
Centre for German and European Studies
St. Petersburg, Russia

ISSN 2367-3370 ISSN 2367-3389 (electronic)
Lecture Notes in Networks and Systems
ISBN 978-3-030-64879-4 ISBN 978-3-030-64877-0 (eBook)
https://doi.org/10.1007/978-3-030-64877-0

This Springer imprint is published by the registered company Springer Nature Switzerland AG
The registered company address is: Gewerbestrasse 11, 6330 Cham, Switzerland

Preface

The primary goal of the 'Networks in the Global World' (NetGloW) conference series, organized by the Centre for German and European Studies of St. Petersburg State University and Bielefeld University, is to bring together network researchers from around the globe, to unite the efforts of various scientific disciplines in response to the key challenges faced by network scholarship today and to exchange original research results—thus enabling analyses of global social processes as well as theoretical and methodological advancements.

NetGloW series took off in 2012, with the conference subtitled 'Structural Transformations in Europe, the USA, and Russia,' and hosted researchers, political practitioners, and business representatives from all around the world to highlight the challenges catalyzed by the growing importance of networks in the world and to reflect on these societal shifts. The 2014 event mainly focused on linking theoretical and methodological developments in network analysis. NetGloW '16 thematically revolved around relations between diverse networks. In 2018, the main conference topic was devoted to the principles bringing networks to life, dealing with the logics and mechanisms that generate network structures in Europe and around the world.

The principal theme of the fifth NetGloW held on July 7–9, 2020 engaged with the notion of context. It is context that molds understanding of a specific relation and, thereby, of a particular network. We proposed conference participants to discuss different types of networks—would these be personal, symbolic, material, organizational, urban, regional, or state networks—in their situational, cultural, historical, institutional, temporal, spatial, and professional contexts. We invited the speakers to reflect on how networks are defined, interpreted, operated, and inhabited within and across these and many other types of contexts and, in turn, how networks affect and transform their contexts. On top of that, various kinds of networks comprise the contexts of each other, defining each other and co-evolving. For instance, cultural networks express the meaning of interpersonal ties which, in turn, mold cultural structures as individuals negotiate meanings. We invited papers on theoretical conceptualizations of the interplay between networks and their contexts, on the methods to incorporate peculiarities of contexts in network analysis—especially those drawing on interpretive approaches, mixed techniques, and subjective meanings of research participants—and on substantive applications examining networks in specific empirical contexts. Papers using data on European societies or dealing with issues topical for Europe were particularly welcomed.

The core subject areas of NetGloW in 2020 resembled those of the previous editions of the conference, as well as the overall approach: the focus on advances in network analysis combining different methods and data to address the challenges in studying diverse types of networks across cultures, societies, states, economies, and cities—with a primary focus on Europe. Like before, a particular emphasis was on linkages between theory, method, and applications, considering how theory-driven principles can be tested and which research settings are suitable for such investigations.

This volume comprises a selection of papers presented at the fifth edition of 'Networks in the Global World.' Following rigorous selection out of 217 initial conference applications, we invited 130 full papers and, finally, accepted 23 of 42 papers sent in response to our call, relying on single-blind peer reviews by at least two qualified reviewers for each paper. Thus, only as few as 11% of the conference's initial submissions have passed all the selection procedures and became the chapters of this book.

The papers included in the present volume cover five thematic areas: qualitative and mixed-method network analyses, computational methods and techniques of network analysis, online networks, political and social movement networks, and networks in education. The papers contribute to the study of the focal topic of this year's conference—networks and their contexts—as well as to the methodology of network analysis and its applications across various disciplines and subject areas.

We would like to thank all the authors for presenting at NetGloW '20 and for preparing chapters for this book. We also sincerely thank the reviewers for contributing their time and expertise to making the volume possible. We hope this collection of papers inspires both the authors and the readers to explore the numerous facets of network analysis and to push its frontiers further—both at the NetGloW conferences to follow and beyond.

Nikita Basov
Artem Antonyuk

Organization

Programme Committee

Nikita Basov	St. Petersburg State University, Russia
David Krackhardt	Carnegie Mellon University, USA
Artem Antonyuk	St. Petersburg State University, Russia
Camille Roth	Humboldt University Berlin, Germany
Iina Hellsten	University of Amsterdam, The Netherlands
Aleksandra Nenko	ITMO University, Russia
Peng Wang	Swinburne University of Technology, Australia
Svetlana Bodrunova	St. Petersburg State University, Russia

Reviewers

Alieva, Deniza	Management Development Institute of Singapore in Tashkent, Uzbekistan
Antonyuk, Artem	St. Petersburg State University, Russia
Bellotti, Elisa	University of Manchester, UK
Bodrunova, Svetlana	St. Petersburg State University, Russia
Burmistrova-Jennert, Elizaveta	St. Petersburg State University of Culture, Russia
Bykov, Ilia	St. Petersburg State University, Russia
Crossley, Nick	University of Manchester, UK
Des Marais, Eric	Okayama Prefectural University, Japan
Doreian, Patrick	University of Pittsburgh, USA/University of Ljubljana, Slovenia
Dudysheva, Elena	Shukshin Altai State University for Humanities and Pedagogy, Russia
Evmenova, Elizaveta	St. Petersburg State University, Russia
Fontes, Breno	Federal University of Pernambuco, Brazil

Fuhse, Jan	Humboldt University of Berlin, Germany
Giordano, Giuseppe	University of Salerno, Italy
Gradoselskaya, Galina	National Research University Higher School of Economics, Russia
Kacanski, Slobodan	Roskilde University, Denmark
Kazantseva, Tatiana	St. Petersburg State University, Russia
Keuchenius, Anna	University of Amsterdam, The Netherlands
Kirschbaum, Charles	Insper Institute of Education and Research, Brazil
Knyazeva, Irina	St. Petersburg State University, Russia
Korableva, Olga	St. Petersburg State University, Russia
Koseki, Shin-Alexandre	Ecole Polytechnique Fédérale de Lausanne, Switzerland
Kostyrev, Andrey	National University 'Chernihiv Collegium' named after T. G. Shevchenko, Ukraine
Kruglikova, Olga	St. Petersburg State University, Russia
Kubelskiy, Miroslav	St. Petersburg State University, Russia
Kurochkin, Alexander	St. Petersburg State University, Russia
Malai, Dumitru	University of Kassel, Germany
Marotta, Ilaria	University of Naples Federico II, Italy
Mikhailova, Oxana	National Research University Higher School of Economics, Russia
Moroz, Anna	National Research University Higher School of Economics, Russia
Nenko, Aleksandra	ITMO University, Russia
Oleskin, Alexander	Moscow State University, Russia
Platonov, Konstantin	St. Petersburg State University, Russia
Podkorytova, Maria	ITMO University, Russia
Poluboyarinova, Larisa	St. Petersburg State University, Russia
Porshnev, Alexander	National Research University Higher School of Economics, Russia
Roth, Camille	French National Center for Scientific Research (CNRS), France
Rubtcov, Aleksei	Saint Petersburg Institute of History of Russian Academy of Sciences, Russia
Shcheglova, Tamara	National Research University Higher School of Economics, Russia
Sherstobitov, Aleksandr	St. Petersburg State University, Russia
Smoliarova, Anna	St. Petersburg State University, Russia
Titkova, Vera	National Research University Higher School of Economics, Russia
Tolstukha, Ekaterina	University of Glasgow, UK
Tsenzharik, Maria	St. Petersburg State University, Russia
Zhukov, Andrei	St. Petersburg State University, Russia

Organizers

Partners

Deutscher Akademischer Austauschdienst
German Academic Exchange Service

Contents

Qualitative and Mixed Network Analyses

Content and Context in Social Network Analysis

Nick Crossley(✉)

University of Manchester, Manchester, UK
nick.crossley@manchester.ac.uk

Abstract. This paper discusses the importance of content and context in social network analysis. Arguing against the commonly held view that content and context are 'soft', add-on details whose analysis might supplement social network analysis proper I suggest that they are integral complexities which are removed in the process of simplification whereby networks are constructed as objects of scientific investigation but which must be reintroduced in many cases if analysis is to be meaningful and robust. This will often entail a mixed-method approach to social network analysis.

Keywords: Social network analysis · Content · Context · Mixed methods · Qualitative analysis · Abstraction

1 Introduction

In this paper I discuss the importance of content and context in social network analysis (SNA). I begin by exploring their relation to networks. There is a relatively common way of treating content and context which assumes them to be separate, ontologically, from networks. Content is conceived as something which can be taken in and out of a network without in any way affecting the network, as objects can be taken in and out of a box without affecting the box. Context, conversely, is conceived as something akin to a box which a network can be put in or lifted out of, again without appreciable effect. Content, network and context are conceived as three separate entities, one nested in the other like a Russian doll.

This conception is problematic. Networks and their content and context are not separate in reality; only in our thought about and representations of reality. We separate them out in the process whereby we construct networks as objects of scientific investigation. This point requires elaboration.

2 Constructing Networks

It is common today to acknowledge that objects of scientific inquiry are 'constructed'. This is as true of our object of enquiry, social networks, as of any scientific object. There is a real world which exists independently of our attempts to capture and analyse it and different measures and models are better or worse at doing so. However, reality, as it is

A. Antonyuk and N. Basov (Eds.): NetGloW 2020, LNNS 181, pp. 3–14, 2021.
https://doi.org/10.1007/978-3-030-64877-0_1

known to us empirically, through observation and experience, is complex and messy, and science, in the first instance, is a process of simplification and abstraction. We abstract certain aspects of the observed world, pulling them to the foreground of our thought and vision, and pushing others into the background; separating, in thought, qualities and properties which are inseparable in reality.

The concept of modelling to some extent captures this. Models are abstractions from and simplifications of the observed world, or some aspect of it. This is true of statistical models (e.g. exponential random graph models or stochastic actor-oriented models) and it belongs to the very rationale of blockmodeling; blocks being simplified representations of clusters of nodes occupying equivalent positions in a network, and identity matrices being simplified versions of adjacency matrices. However, even the 'observed networks' captured in our adjacency matrices are models: simplifications of a messy and complex reality.

We do not observe networks in the world as such. Rather, we employ coding schemata, algorithms and/or structured questionnaires in an effort to separate out and abstract a tiny fraction of detail from what is, in fact, a hurly burly of social activity; populating a matrix and thereby constructing a network. We reduce sophisticated social actors with varied biographies and dispositions to the status of nodes, with a standardised range of attributes. Likewise ties, which in reality are unique histories of interaction between social actors, carried forward and lived between them in the present, which we reduce to uniform 'types', corresponding to our own predetermined criteria, disregarding them when they fail to tick any of our boxes. We put boundaries around these sets of nodes and ties, making in/exclusion decisions which separate them from further potential 'nodes' and 'ties' which they are in other ways connected to, constituting them as discrete objects. And we abstract the patterns of connection that we observe by these means from the settings within which our nodes 'do' their relations, not separately but in the course of doing multiple other activities with varied goals, norms, dynamics and frameworks of meaning; activities which more often occupy the foreground of their thought and attention than the relations embedded in and embedding them, and which, in turn, are further shaped by actors' understandings of and responses to 'external events', near and far, which impinge upon them and upon their activities.

There is nothing wrong with this. I do not make these claims with the intention of criticising SNA. Simplification and abstraction are what science does; necessarily so. However, it is important to be reflexively aware that this is what we are doing and also of the possibilities for error it engenders. We may fail to simplify sufficiently, rendering our object overly complex and ourselves unable to see the wood for the trees. More often, however, we oversimplify, reducing our object to a point where it is unable to tell us very much of use or interest about those aspects of the world we are interested in and indeed, to a point where it is no longer possible to make reasonable interpretations of our measures and models.

Debates about content and context, properly conceived, concern these processes of simplification and abstraction. Content and context are the details removed in the process of constructing a network, available to be 'added' to analysis because they were previously 'subtracted', and the question is how much we can ignore them before we begin to undermine our efforts to understand and explain whatever aspects of the social world it is that we are seeking to understand and explain.

There is no single formula which will resolve such debates. The importance of details will always depend upon the nature of our projects and research questions. We use SNA to answer very different types of question and pursue very different types of research interest, and these variations affect the types and levels of detail which are important to us. The idiographic concerns of the anthropologist researching the social structure of a particular society are different to those of the nomothetically inclined social scientists interested in, for example, social capital or social influence. And the methodologist puzzling over the effectiveness of their algorithm will require different details again. However, the question is always relevant and must always be addressed anew in every project.

3 Qualitative Complexity

In what follows, I seek to expand upon these opening remarks in further detail. Before I do, however, it is necessary to make a brief note on mixed-method approaches and qualitative research in particular. Qualitative methods, which are often associated with 'content' and 'context' in our discussions, are typically belittled as 'soft' in these discussions. 'Soft' implies less rigour and in some cases that may be justified. However, the rigour of quantitative models and methods is often only achieved, in social science at least, at the cost of the oversimplification referred to above, and qualitative analysis is a means of capturing the detail and complexity which 'harder' models and methods are unable to accommodate and tend therefore to exclude. Mixing methods, in this sense, is a matter of quantifying that which can be plausibly and meaningfully quantified whilst simultaneously recognising that much that is analytically important in the social world studied by social network analysts, not least meaning itself, cannot always be satisfactorily reduced to a number such that qualitative data and analysis are necessary too, and such that dialogue and integration between quantitative and qualitative aspects of analysis are necessary. Qualitative analysis, soft or not, (re)captures some of the real-world complexity which the simplifying assumptions of quantitative research, because unable to process such complexities, screens out.

A full discussion of details and complexities typically excluded by quantitative approaches which qualitative research (re)admit would take far more space than I have. I will limit myself here to two illustrative points.

Firstly, we typically standardise in quantitative approaches; asking the same questions of all nodes/ties, recording them in a uniform fashion, as numbers in a matrix, and then processing all in a common procedure. This is good science and the reliability of measures and models depends upon it. However, any ethnographic or historical researcher knows that specific nodes, ties and relational events sometimes have important idiosyncrasies which influence events but which, as idiosyncrasies, cannot be uniformly measured across a network, or even if they can, are important in ways which defy such measurement. In my work on UK punk and London's early punk network in particular, for example, I noted that the relational dynamic between two particular nodes, Malcolm McLaren and Bernard Rhodes, which involved a mix of competition and cooperation, played a disproportionately important role in mobilising key events and developments [1]. For a time during the period I analysed (though, again adding complexity, only some and not all of the time), they competitively sought out bands to manage and promoted their key band (*the Sex Pistols* in McLaren's case and *the Clash* in Rhodes'), in an effort to claim the top position in their music world. And in doing so they made a huge contribution to the generation of London's emerging punk world and its constitutive network. This became apparent from my qualitative reading of archival and secondary sources and it is not clear that it could have come to light by quantitative means. Moreover, I could not further analyse it mathematically or statistically, beyond, for example, measuring the centrality of McLaren and Rhodes (it was high across several measures), because it was the dynamic of their particular relationship rather than, for example, their type of relation or relations between nodes of their type, that made the difference. It was crucial to an understanding of the network, however, as was amenable to qualitative analysis.

This is an example of details which are relevant to only a subset of nodes or ties within our network and in some cases only one, and which for this reason escape the standardising procedures of the quantitative approach. However, in other respects quantitative methods require difference. An attribute which all nodes share would not be much use for many, if any quantitative purposes. But it could be important. To take an example I will return to, much of what most of us are doing in our networks currently is affected by Covid-19, or rather by our knowledge and understanding of it, but we all have that knowledge and understanding, such that it would not make a particularly interesting or useful variable. We are all affected (in our behaviour) by our awareness of the virus; this needs to be taken into account if our current interactions and relations are to be properly understood; but the awareness is not easily rendered as a variable because, at a certain level at least, it is invariant, a constant. Everybody has it.

The list of important facets of social life squeezed out by the demands of SNA could go on but the above hopefully suffices to demonstrate the basic point. Our models and methods are powerful tools but they presuppose a reduction of our worldly observation to matrices, and not everything that is important, even about networks, let alone the hurly burly of social interaction from which our networks are selectively abstracted, can be reduced in this way. We need 'content' and 'context' and we need qualitative methods to capture and analyse them.

4 Vertices and Edges

I will begin my further elaboration of these remarks by considering an extreme example of decontextualisation (Fig. 1). We can derive a large number of measures for the network in Fig. 1. We can detect cliques and other such structures. We could block or statistically model it. But what would that tell us? What would we learn? How would we make sense of our measures and models? Learning that node nine is the most degree central node in the network or that the clustering co-efficient of the network is 0.77 tells us very little in and of itself, in the absence of information regarding vertices and edges. Who and what are the nodes in Fig. 1? And in what way are they connected? Does this graph represent nations bonded by trade agreements? Cities connected by flows of commuters? Children who play together in a school playground? Measures and models are evidence but what they are evidence of is unclear in the absence of details about who is connected and how they are connected. That is to say, in the absence of context and content.

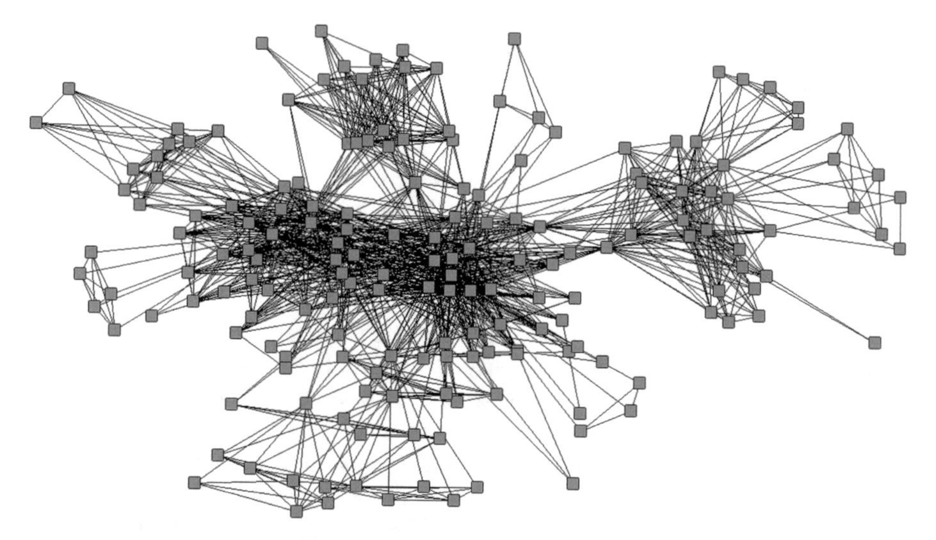

Fig. 1. A decontextualised network

The measures and models of SNA are highly formal. We often make some assumptions about nodes and ties; for example, that each node is capable, in principle, of forming the tie we are observing with any of the others (unless we have a two-mode or multi-modal network). But the algorithms are otherwise largely indifferent to the attributes of either nodes or ties. Cities connected by migration flows and children connected through friendship are equivalent structures from the point of view of our algorithms. This is a strength. It makes SNA versatile. However, it also generates the potential for serious error and requires of us, the reflexive researchers who use its algorithms, that we do take account of the attributes of nodes and ties. The basic attributes of our nodes determine what they can do in the network and how they can be affected, and this is no less affected by tie attributes; the nature of the relations between them.

Borgatti offers a rigorous demonstration of this in relation to tie types and centrality. What different centrality measures can tell us, he shows, depends upon what, if anything, is flowing through the network, and more generally upon the type of tie involved [2]. The same is true of other measures and analytic routines, such as core-periphery and block models. What we can infer from the existence of, for example, a core-periphery divide and the location of particular nodes in one camp or the other depends upon the relation involved. The same, moreover, is true of basic node attributes. Whether a node is a human being or an organisation, for example, makes a difference to how it is likely to be affected by and respond to its location in a network. Organisations can, for some purposes and in some cases, be regarded as 'social actors' [3], as human beings typically are, but there are clear differences between these two types of actor. Organisations do not respond emotionally to their position in a network, for example.

Milgram touches upon this point in his celebrated work on small worlds [4]. Referring to the 'six degree' finding, which he renders as 'five removes', and reflecting upon the proximity it is often taken to signify, he warns:

> … this is in large measure misleading, a confusion of two entirely different frames of reference. If two persons are five removes apart, they are far apart indeed. Almost anyone in the United States is but a few removes from the President …. But this is true only in terms of a particular mathematical point of view, and does not, in any practical sense, integrate our lives with that of [… the President…] We should think of the two points as being not five persons apart but "five circles of acquaintances" apart – five "structures" apart. This helps us to set it in its proper perspective. [4: 67]

Milgram's identification of 'frames of reference', 'proper perspective' and 'a particular mathematical point of view' is effectively a call for us to put graph theoretical measures in context; to recognise that in the human world or perhaps rather the human social world, six degrees can be (even if it is not always) a long way. A virus, such as Covid-19, which every country on earth is struggling with as I write, can traverse six degrees pretty quickly and easily, as our global pandemic testifies. Six degrees is not a long way for Covid-19, at least not initially. However, as responses to the pandemic show, human actors, in contrast to other potential hosts, can become reflexively aware of their vulnerability and potential for passing on the virus, and can adjust their behaviour by, for example, shifting to non-physical forms of contact and interaction, thereby slowing the movement of the virus through what is essentially the same network structure. We can interact at a two-metre distance or via technological devices, filtering potential virus flow out of our interactions and thereby making six degrees a much longer distance for the virus to travel.

What happens to a virus in a pandemic, when the human carriers of the virus become aware of the virus and their role in its transmission, happens all the time to other goods (and bads) flowing through human networks because, as writers such as Mead, Simmel and Goffman show [5–7], human actors almost always reflexively manage their interactions and relations, attempting to control the flow of information and other goods they might transmit. Moreover, because their alters and interactions have meaning for them and they orient to that meaning. A friend becomes a potential source of danger in the context of a pandemic and their stories of travel and meetings repel where previously they might have attracted. And many other such meanings regulate flows in human networks, including the circles of acquaintance to which Milgram refers. Personal information about the President can pass through six degrees with haste, as Bill Clinton learned to his cost, but much of it does not, not because his circle is structurally closed and potential channels of dissemination do not exist but rather because his confidants are aware of their status and code some of the information they receive and interactions they either participate in or observe as 'classified' – at least for the time being, all other things being equal. Channels which are open for some purposes are closed for others, and crypto-networks form within networks, demarcated not by an absence of relations and interactions but rather by meanings, identities and the reflexive awareness of their nodes; that is, by aspects of content and context which researchers need to be attuned to.

We might revise Milgram's warning slightly, in light of this discussion. 'Six degrees' can be either a long or a short distance and, as the slowing (and unfortunately newly accelerating) diffusion speed of Covid-19 suggests, its magnitude can vary over time, depending upon indigenous meanings and the reflexive awareness of nodes. His basic point remains, however; the significance of a measure, such as mean geodesic, depends upon content and context.

Milgram's contention pulls in the opposite direction to the claims of the small world pioneers of network science who have built upon his work [8, 9]. One of their key claims is that we observe small world networks and/or small world effects across a multitude of very different complex systems: from the neural network of the nematode worm, via birds in flight and swarming fireflies, to human transport infrastructures and the internet. Network science explores structures which are common across very different network domains, 'owned' by very different scientific disciplines. This is fascinating and important, but Milgram reminds us that the similarities only exist at a very high level of abstraction; a level which may sometimes be too high to be truly meaningful. That very different networks share common structures is revealing and deserves the attention it has generated but we should remember that *they are very different networks* and that it is often necessary to drill down into their differences and detail, their context and content, if we are to properly understand what is going on within and across them.

5 From Edges to Social Relations

All SNA worth its name stipulates node and tie types, using this information (often tacitly) to constrain and inform interpretation of measures, partitions and models. However, there are different ways of approaching node and tie types which probe the questions of content and context which they potentially beg to a greater and lesser extent and which, in that respect, vary in their scientific rigour and adequacy. Social ties of any particular type, for example, generally depend, for their existence, upon mutual recognition, agreement and shared meanings between those party to them. To be 'friends', for example, is to recognise and orient towards one another as friends, acknowledging and adhering as far as possible to norms of friendship which stipulate what friends do for one another and what they do not do to one another. Friends reflexively manage their friendship in accordance with the meaning which friendship has for them and the norms which attach to these meanings. Importantly, however, these norms and meanings may vary between network neighbourhoods (corresponding, in turn, perhaps to age groups or generational cohorts, ethnicity, sexuality and social class). If researchers are to make valid inferences from tie type to network effects, it follows, they cannot rely upon 'off the peg' definitions of those types but must rather investigate what 'friendship', 'collegiality', 'romance' or whatever type of relation they are interested in means in the particular network or network neighbourhood they are investigating.

Cross-cultural and historical studies bring this point into sharp relief because researchers are usually more aware that they cannot impose their own definitions of, for example, friendship or business relations and because they may encounter relations with which they are completely unfamiliar. Knowing where guanxi relations exist in a particular Chinese community and between whom is of little value to a Western network analyst, for example, unless they know what guanxi relations are; that is, unless they have researched the content and context of a tie type which, like all tie types, is defined by those party to it. This is only a dramatic illustration of a point which applies to all SNA, however, given both that all tie types are similarly subject to local, indigenous definition and that we can never presume to know of local variations in advance of some contextual and usually qualitative investigation of them. It may turn out that those we are researching define their relations pretty much as we would, orienting to similar norms, but we cannot know that in advance.

Certain key ideas which have informed the development of SNA have failed to take full account of context in this sense: Heider's structural balance theory, for example, which stipulates that if *i* has a positive tie with *j* and a negative tie with *k* then *j* too should have a negative tie with *k,* in order for *j* to maintain a sense of cognitive consistency [10]. This theory, at least as many network analysts use it, is posited in terms of abstract structural configurations which can be formally expressed and graphically illustrated. However, there is no compelling logic to it in purely formal terms. It only achieves any prima facie plausibility insofar as we assume, and assume that nodes assume, that positive ties entail an obligation to oppose one's friends' enemies; that is, if we assume a particular cultural content for ties, which is understood and reflexively observed by those party to them. Or perhaps if we assume conditions under which *i, j* and *k* meet simultaneously (in the same context), in circumstances where refereeing between conflicting partners is either not possible or undesirable. Clearly such conditions do not always hold and may, in fact, be relatively rare. In the contemporary Western context, where friendship is framed by norms of individualism and autonomy, the idea that it might entail an obligation to adopt alter's likes and dislikes does not ring true; friends often relish their differences, reserve the right to make their own choices and experience no sense of inconsistency in doing so.

Social structures, formally defined, cannot be balanced or imbalanced in and of themselves because they do not exist in and of themselves, independently of content and context. Social structures are 'made of' relations and interactions which, in turn, involve meanings, norms etc. and it is only in virtue of these meanings and norms, which direct human activity, that structures can be balanced or imbalanced.

We could make a similar case for transitivity and Granovetter's 'forbidden triad' [11]. There is a certain practical force pulling friends of friends together *in some contexts.* Their shared friend is akin to a network 'focus' [12], drawing them into a common orbit. As Simmel suggests, however, individual actors are sometimes the (only) point of intersection between the different social circles (e.g. family, neighbourhood, work, leisure-based groups) in which they are involved, and whether or not these social circles remain separate or merge via the 'focus' of their intersecting member depends in some part on the local cultures of the circles and, in particular, their orientation to outsiders [13]. In her classic kinship study, for example, Bott observed social class differences in the extent to which family and friendship relations were segregated [14]. Working class spouses typically observed a norm of segregation between spouse and friend relations, thereby discouraging transitivity. Open triads involving spouses and friends were encouraged and, if anything, closure was 'forbidden'. Middle class couples, by contrast, more often shared friends and considered it right to do so, thereby encouraging transitivity. What was 'forbidden' was dependent upon both local class culture (which was itself reproduced in networks of course (see below)) and the type of relation in question. Likewise covert networks, whose ties and nodes are kept secret from the other circles in which their participants are involved [15]. Or to give a different example, geographically distanced and/or technologically mediated relations, whose mediation presents an obstacle to the meeting of friends of friends: I am unlikely to ever befriend your pen pal because they live miles away and your interactions with them are closed to me by a sealed envelope. Like structural balance, the tendency towards closure hypothesised by Granovetter is not purely structural but rather depends upon culture, meaning and the mediation of relations; that is, upon context and content.

Three qualifications are important at this point. Firstly, norms, obligations and other cultural aspects of ties are actively (re)negotiated by those party to them, albeit drawing upon wider reference points and groups. We must be careful not to reify them or externalise them from the interactions they shape. Furthermore, they often diverge in practice from the official line that a less savvy researcher might be fed.

Secondly, the diffusion and effectiveness of the norms, meanings etc. which shape networks and mediate network effects are themselves affected by network structure. As Coleman argued, network neighbourhoods are more likely to be able to define and enforce norms and definitions where they are relatively dense and closed [3, 16]. Bott's aforementioned work on kinship relations, for example, argued that working class communities held on to traditional patterns of spouse/friend segregation and were able to do so, within a wider context of change, because and to the extent that their networks were 'tight-knit' (i.e. dense) [14]; and Milroy made a similar case for traditional speech patterns [17].

Finally, an understanding of local norms and definitions does not absolve us of the need to test, in practice and using SNA, whether patterns of relations we might hypothesise to exist on the basis of identified norms and definitions actually exist. Particular norms might lead us to expect high levels of transitivity, for example, but that does not mean that we will find it, less still that we can assume it. In my experience qualitative understandings of context seldom line up neatly with SNA measures at first, prompting the researcher to return to the data (and perhaps to the field) to explore possible reasons for this. Perhaps the context was initially misunderstood? Perhaps a particular measure is being misinterpreted? Perhaps a survey of ties hasn't captured a relation as it is practiced in context? In many cases it is possible to identify the causes of the discrepancy and, with reflection and further work, to bring our understanding of context and our measurement of structure into closer alignment – an important analytic step in its own right. But even where this does not happen the key point remains that structural and contextual data are two sides of a single coin which we must allow to be mutually constraining (in a positive way) in our analyses.

6 From Vertices to Social Actors

I have focused largely upon details of tie types in the above, but detail regarding nodes is similarly useful. In one respect this is a matter of simple categorical attributes which we might expect to impact exogenously upon a network. We might expect to observe homophily effects, for example, or we might expect incumbents of a particular category to be more central or more marginal in a network. This might be an effect we are interested in or one we are hoping to discount as a potentially confounding factor in relation to something else that we are interested in. In either case, however, our thought and expectations arise from an understanding, perhaps tacit and potentially wrong, of the meaning which particular attributes take on in our network of interest. Newcomb makes an interesting claim in this respect in an early discussion of homophily:

> It is not a very useful notion … because it is indiscriminate. We have neither good reason nor good evidence for believing that persons of similar blood types, for example, or persons whose surnames have the same number of letters are especially attracted to one another. The answer to the question, Similarity with respect to what?, is enormously complex. [18: 577]

Newcomb is right. Homophily is often complex. But it is also contextual, relating to the categorical schemas which actors draw upon in their everyday lives, and the meanings and value which they attach to particular distinctions – which in themselves may vary between the activities and setting in which they are involved. Social science provides us with a standard battery of status distinctions to feed into our models (e.g. gender, social class and ethnicity) but these are often fairly generic and fail to capture niche distinctions (e.g. between migrants from different parts of what, to the researcher, is 'the same country') which may have profound effects within some networks. Furthermore, social actors often invent their own status distinctions: e.g. Elias and Scotson's 'established and outsiders' [19], Willis' 'lads' and 'ear'oles' [20], and street gang affinities. This is equally true of value homophily, which may hinge upon distinctions (e.g. 'this' style of jazz vs 'that') that the researcher is entirely oblivious to, unless they are themselves an insider to the social world they are studying and therefore familiar with its context.

Nodal attributes are not given. They are, as the word suggests, *attributed to* nodes or other objects within the very relations and interactions that we seek to capture in our adjacency matrices, and this process of attribution, which nodes apply both to self and other, draws upon lay categorical schemas devised and drawn upon by nodes, again in interaction, and their values: what matters to them. As Newcomb notes above there are multiple similarities and differences between nodes in a network which might potentially form a basis for homophily. Contextual investigation of lay schemata and values, and indeed of events (e.g. scandals) and gossip which might shape and/or activate them, is one way of exploring which are more likely to actually do so. And not only that; such analysis contributes to our understanding of homophily by illuminating the reasons which might underlie and motivate homophily: e.g. the beliefs that motivate avoidance of 'the other'.

Not that homophily is always subjectively motivated in this way. Nodes in a network are typically doing far more than their relations and, indeed, more than the relational processes (e.g. 'influence') that we are interested in, and what else they are doing affects their doing of relations in various ways. In the unlikely event that a local blood transfusion service was to start a recruitment drive, asking members of different blood groups to attend on different days, for example, this would create a network focus, in Feld's sense

[12], generating precisely the blood group homophily that Newcombe was sceptical about. Events may make blood group a salient selection variable – perhaps without nodes even being aware of the fact. The example is silly but the point more generally is that the ties which nodes form reflect the activities and events, both routine and exceptional, in which they are engaged, such that contextual knowledge of such activities and events is crucial to a proper understanding of networks.

Feld's focus concept [12] begins to tease this out by allowing us to envisage how nodes with similar interests might, without any express desire to meet likeminded others, nevertheless cross paths and thereby enjoy an increased likelihood of forming and maintaining ties. This is but one way, however, in which the events, activities, practices and spaces in which nodes participate and mingle, that is to say, context, might impact upon their networks.

7 Conclusion

Social life comprises interactions and relations. The social world is a network, and SNA is invaluable because it enables us to capture and analyse it as such. It affords unique purchase upon the social structures that comprise a key focus of our studies. Like any other method, however, it abstracts and simplifies. Abstraction and simplification can be entirely justified and are unavoidable to some extent, but how far and in what ways it is justified to abstract and simplify depends upon our project and must always be open to discussion. The simplification required to run a sophisticated model may be a high price to pay; gain in statistical sophistication being outweighed by reduced purchase on the complexity of whatever is under investigation. However, this can be compensated if we supplement sophisticated modelling with qualitative investigation which drills into the real-life complexity bracketed out by quantification and modelling. Mixing methods combines the strengths of these two different approaches and provides a corrective for their respective weaknesses.

References

1. Crossley, N.: Networks of Sound, Style and Subversion: The Punk and Post-Punks Musical Worlds of Manchester, London, Liverpool and Sheffield 1976–1980. Manchester University Press, Manchester (2015)
2. Borgatti, S.: Centrality and network flow. Soc. Netw. **27**, 55–71 (2005)
3. Coleman, J.: Foundations of Social Theory. Harvard, Belknap (1990)
4. Milgram, S.: The small world problem. Psychol. Today **1**, 61–67 (1967)
5. Mead, G.H.: Mind, Self and Society. Chicago University Press, Chicago (1967)
6. Simmel, G.: The sociology of secrecy and secret societies. Am. J. Sociol. **11**, 441–498 (1906)
7. Goffman, E.: The Presentation of Self in Everyday Life. Penguin, Harmondsworth (1959)
8. Watts, D.: Small Worlds. Princeton University Press, New Jersey (1999)
9. Newman, M., Barabási, L., Watts, D.: The Structure and Dynamics of Networks. Princeton University Press, New Jersey (2006)
10. Heider, F.: The Psychology of Interpersonal Relations. Wiley, London (1958)
11. Granovetter, M.: The strength of weak ties. Am. J. Sociol. **78**, 1360–1380 (1973)
12. Feld, S.: The focused organisation of social ties. Am. J. Sociol. **86**, 1015–1035 (1981)

13. Simmel, G.: Conflict and the Web of Group Affiliations. Free Press, New York (1955)
14. Bott, E.: Family and Social Networks. Free Press, New York (1957)
15. Crossley, N., Edwards, G., Harries, E., Stevenson, R.: Covert social movement networks and the secrecy-efficiency trade off: the case of the UK suffragettes (1906–1914). Soc. Netw. **34**, 634–644 (2012)
16. Coleman, J.: Free riders and zealots: the role of social networks. Sociol. Theory **6**, 52–57 (1988)
17. Milroy, L.: Language and Social Networks. Blackwell, Oxford (1980)
18. Newcomb, T.: Prediction of interpersonal attraction. Am. Psychol. **11**, 575–587 (1956)
19. Elias, N., Scotson, J.: The Established and the Outsiders. Sage, London (1994)
20. Willis, P.: Learning to Labour. Routledge, London (1977)

Mixed Methods in Egonet Analysis

Elisa Bellotti(✉)

Mitchell Centre for Social Network Analysis, School of Social Science, University of Manchester, Manchester, UK
Elisa.bellotti@manchester.ac.uk

Abstract. Social network analysis that uses mixed methods has gained momentum in the last ten to fifteen years. Although not new, as SNA can count on a long and illustrious tradition of qualitative and ethnographic studies, the approach that relies on qualitative methods in data collection, analysis and interpretation, and mixes it with more or less sophisticated statistical modelling, has been the centre of theoretical and methodological debates. Drawing on empirical examples that combine various quantitative and qualitative tools in collecting, contextualizing, analysing and interpreting social network data, this article illustrates how mixed methods can be used in egonets analysis. After discussing the ontological and epistemological foundations of SNA, the article illustrates how to build egonets typologies, model their outcomes, and test and reformulate social network theory. Without pretending to be exhaustive, the article offers a solid and rounded guide to mixed methods in egonets analysis.

Keywords: Mixed methods · Social network analysis · Egonet typologies · SNA models · Qualitative data

1 Introduction

The debate over the use of mixed methods in social network analysis has risen steadily and consistently over the past 15 years [1–4], in response to criticisms that see social networks as arid and reductive simplifications of the complex nature of personal relationships [5, 6]. The criticism was not completely unfounded: in social network studies we often see an overwhelming attention to the highly sophisticated modelling techniques that we use to observe mechanisms of formation and development of ties, at the expense of the theoretical conceptualization and measurement of the relationships used in the analysis. We sometimes draw conclusions about the role of family, friends, acquaintances, colleagues, over a huge variety of selection and influence mechanisms, but we rarely discuss the nature of these ties, why we expect them (and not others) to play a role in the behavioural practices we aim to observe, and when and where these ties may or may not activate network mechanisms. Likewise, sometimes we observe network structures becoming denser, sparser, more centralized, more polarized, without knowing what do these configurations imply in different contexts and how actors perceived them subjectively, therefore influencing the adapting behaviours accordingly.

A. Antonyuk and N. Basov (Eds.): NetGloW 2020, LNNS 181, pp. 15–33, 2021.
https://doi.org/10.1007/978-3-030-64877-0_2

The debate on how to address the problem of measuring social relationships in a quantifiable way, as Edwards [7] points out, has stimulated theoretically driven calls for mixed methods within sociology [8–10] and anthropology [11, 12]. In such theoretical calls, network formalizations are claimed to be limited in showing how intentional and creative action contributes to the constitution of the web of relationships that in turn shapes the opportunities for such subjective action [8]. The focus on the structural elements that define the relations among sets of positions, as theorized by White et al. [13], is supposed to hide the importance of the cultural context in which relationships and interactions take place [14].

In reality, social network analysis can count on a long and illustrious tradition of observing social networks in culturally specific settings, where the adoption of mixed methods comes in hands. A notable precursor is the famous study of an Italian slum in Boston [15], and ethnographic observations were also at the core of the studies of the Manchester School of Anthropology, which flourished in the late 60s under the influence of significant scholars like John Barnes, Clyde Mitchell, Elizabeth Both, and Bruce Kapferer [16, 17]. Since then, a solid and consistent stream of anthropological ethnographic studies in SNA thrived [18–23].

The tradition continued with Harrison White and Charles Tilly, who, during the 90s, played a crucial role in broadening the spectrum of analysis of network structures and engineering the emergence of what might be called "the New York School" of relational sociology [24]. In particular, White's work on identity and its narrative construction [25], which follows his equally fundamental work on formal measurement of positions in networks [13], can be considered at the core of this focus on cultural and contextual elements in social network analysis.

To date, the number of studies that pay attention to the subjective, contextual, and cultural nature of social relationships and the patterns they form is on the rise. To grasp the complexity of social fabrics, scholars adopt a wide variety of methods, both qualitative and quantitative. The combination of these methods requires, as in any good research design, robust planning and justification, as well as flexibility and sensibility. The highly formal statistical analysis that we can use to disentangle structural mechanisms of selection and influence within a social context is only the final step of an empirical process that starts with the identification of case studies, definition of the population to be investigated, contextualization of the study, conceptualization of the social relations to be observed, operationalization of the ties and mechanisms we believe are of importance in our study, and data collection protocols. These steps are not peculiar of social network analysis but are at the core of any empirical study. What is peculiar of social network analysis is that when looking at the fabrics of social interactions, we are not simply interested in the individuals involved in the processes, but also in the more or less durable relationships they interlace.

Using some exemplar empirical studies, this article aims to guide the reader through the ways in which we can mix qualitative and quantitative methods in observing the contextualized nature of social networks. The aim of these studies varies: some are set to observe how social networks are formed, some to look at the consequence that network structures may entail. Some intend to build descriptive network typologies, while others aspire to reinforce or reconfigure theories. The empirical material selected for the scope is not in any way exhaustive of the vast range of mixed methods studies in social networks, as this is not intended to be a review article. I simply chose pieces of research that, through the years, helped me to clarify and illustrate, to myself and my students, the process of designing robust social network analysis studies. Furthermore, I only focus on egonet studies, as an overview of mixed methods in whole network studies would require more space than it is available in a journal article.

In discussing the possible mixed methods designs, I follow Hollstein's classification [26: 12–18], which is grounded in the established literature on mixed methods [27]. Hollstein identifies various types of possible designs, which are not mutually exclusive but can be used together in the same research:

- sequential designs, in which one methodological strand follows the other in either exploratory (quantitative to follow qualitative) or explanatory (qualitative to follow quantitative) modes;
- parallel designs, in which qualitative and quantitative strands are employed more or less simultaneously, and therefore data collection on one strand does not rely on the other strand's results;
- fully integrated designs, in which qualitative and quantitative strands can be mixed, at various stages of the research process, both in parallel and sequentially, and dynamically inform each other in a continuous dialogue between different types of results and inductive/deductive logics;
- embedded (or nested) designs, in which one strand constitutes a small contribution to the overall study, for example, a structured questionnaire to validate a set of qualitative results, or ethnographic observations to contextualize quantitative results;
- conversion designs, which involve the transformation of data of one type into the other type (qualitative into quantitative or vice versa).

I begin with discussing, in the next section, the ontological and epistemological foundations of social networks, which gives the justification to the use of mixed methods. I then organize the empirical material according to the scope of the analysis: in the third section I provide a definition of egonets, and in the following subsections I introduce empirical studies that have used mixed methods to produce egonets typologies, studies that have modelled egonets formations and outcomes, and studies that have aimed at reformulating social network theories. Each empirical study is introduced in light of the above mixed methods classification, to clarify the strategies and sequences of data collection and analysis adopted by researchers.

2 The Ontological and Epistemological Foundations of Mixed Methods

Social networks are contextual, socially constructed, subjectively perceived and dynamically evolving entities. We build the networks we are embedded in within the social context in which we live: those contexts give us the possibilities, and increase the probabilities, of interacting with other individuals. Social ties, Feld reminds us, concentrates around "foci – social, psychological, legal or physical entities around which joint activities are organized" [28: 1016]: we are more likely to establish relationships with people who live geographically nearby, with whom we share interests or friends in commons, who work in similar or interdependent economic sectors, and the like.

When we interact, we utilize interpretative categories to make sense of the people around us. We expect familial relationships to be different from working relationships, young people to behave differently than elderlies, women to be more emphatic than men: we typify our relationships, and stereotypy people around us. At the same time, each concrete interaction is a unique negotiation of the nature of that specific relationship, in which we build a story that is distinctive to the person we relate to, using our past interactions to create future expectations [29, 30]: we may do it tacitly, without expressing every time what Schutz [31] called the "taken for granted" of a particular social setting, or we might do it overtly, for example, when we argue with a friend who we believe has broken the rules of our relationship [3]. In White's terminology, these negotiation efforts are forms of controls that individuals exercise over the concurrent definitions of the various situations in which they interact with other actors, who themselves try to impose their definitions. These forms of control constitute actors' unique identities, intended as the various styles of actions that individuals develop to handle the complexities of overlapping social circles [3: 49]. Because relationships are uniquely storied, we also end up interpreting them in a subjective way. What a friend, a brother, or any other type of personal relationship is for me will not be entirely the same of what it is for someone else: my own friends may have different ways of defining and perceiving friendship, young people may have different expectations than older people, women may have different interactive styles than men, and the term friend may mean something different in different cultures [32–35].

The relationships we establish with other individuals overlap and intertwine to create the social fabric of our everyday life. These fabrics are what we call social networks, webs of more or less formalized and durable interactions that create the social environment in which we live. Social networks can be short lived and extemporaneous, like the interactions that may happen between passengers who get stranded in an airport who end up sharing food, lending each other phones to call home, looking after each other's luggage. These interactions will likely end once they all embark on their journeys, but they still produce a temporary organized social structure. Social networks can also be more durable and prescriptive, like Universities' fraternities with initiation rituals, strict hierarchies, and codes of conduct. They develop their own system of enrolment, their own rules of interactions, their own social and cultural norms, their own symbols of recognition. Crossley [30, 36] describes these networks using Becker's concept of art worlds [37], as in a network of interactions demarcated by individuals' mutual involvement in a specific set of activities, driven by shared interests, conventions, and resources[1].

The fact that social networks are contextual, socially constructed, subjectively perceived and dynamically fluid does not mean that they do not ontologically exist in the empirical realm. When we conceptualize social class as a separate entity from individuals, which supposedly act upon them creating opportunities and imposing boundaries, we are reifying a social structure that does not exist outside social construction. But when we observe and measure the concrete relationships people establish with individuals who are more or less similar to them in terms of economic and cultural background, the activities they organize together, the tastes and values that they share, the collective identities that may emerge, the opportunities and constraints these interactions provide, and the practices of distinctions or conflicts against other sectors of societies that these relationships may foster [38], we are not reifying a social structure. We are observing how individuals, by virtue of relating to each other, create, maintain, and dissolve such social structures, whose boundaries span beyond superimposed social categories.

This aspect of social networks analysis, of rejecting superimposed social categories in the definition of social boundaries, has been at the core of the discipline since the 1960s. Reflecting upon the design and the planning of the East York study conducted in Toronto in 1968, one of the first egonet studies mapping the contours of a local neighbourhood, Wellman [39: 429] notes: "I realized that if we saw communities as networks, we no longer had to think of them as necessarily bounded by place (neighbourhood) or solidary groups (kinfolk)". Neighbourhoods, families, social classes, and institutions are not defined and bounded by geographical proximity, blood relations, similarities of socio-economic standings, affiliations to organizations, roles and functions: social network analysis opens the black box of social groups and reformulate them empirically as patterned fabrics of social interactions and relationships. The ontological nature of social networks lies in the consistency of such fabrics, the extension and durability of (partially overlapping) social relations.

Epistemologically, this means that the way in which social network analysis grasps the knowledge of structural configurations is by observing, measuring, and explaining the interconnectivity of the ties that link together different actors. When we look at social networks, we cannot simply aggregate individuals in groups according to pre-determined categories like class, gender, age, etc. We need to observe the patterned structure that such individuals build by virtue of interacting with each other's. The way in which we access the ontological consistency of such structures is by asking people to talk about their interactions and relationships, by observing them, and/or by reconstructing them

[1] White uses the concept of disciplines [25: 63], structural configurations that emerge out of blocks of equivalent positions in which actors developed collective identities. Within a discipline, actors share priorities and criteria for the acceptance of newcomers, have stable and recognised modes of actions, and establish local expressions of social control. Disciplines exhibit comprehensive and hierarchical schemes of comparisons and classifications, or *catnets*, that emerge from the struggle for control. I discussed the limits of structural equivalence as grounding terrain for *catnets* in Bellotti [3: 57–62].

from the more or less permanent traces of social relations that they leave behind. The key issue, in social network analysis but also in any other type of social research, is how we transform the messy and complex batch of raw data into formal units of measurement [10], being them themes in narratives, values in variables, numbers in networks.

Ultimately, the social network community is highly aware of the contextual nature of social networks. While searching for generalizable mechanisms (transitivity, homophily, clusterability) across networks, researchers are conscious that the same mechanism may be the outcome of very different processes, may assume very different meanings, and may imply very different consequences. Qualitative methods are an efficacious and valuable complement to the measurement of network structures, as they allow to contextualize them and grasp the significance of processes, positions, and outcomes [1].

3 Mixed Methods in Egonets

An egonet is the network which forms around a particular social actor, which normally includes ego, her alters and all relevant ties between alters, although sometimes this definition is relaxed to exclude information on ties between alters [40]. One advantage in using an egonet approach is that it allows observing what Simmel [41] calls 'intersecting social circles' and White [25] network domains or 'net-doms' [40]. As individuals, we are more likely to interact and establish relationships with people who frequent our same social environments, which in modern and contemporary worlds multiply in distinct social circles that may or may not overlap [41]. Egonet are usually constructed using name generators, which allow the collection of structural data (ego-alter ties, alter-alter ties, and often alters' characteristics) for a selected number of egos. This approach is also known as the relational approach [40], opposed to the resource approach that only looks at the stock of resources an individual can access via position or resource generators [42, 43].

Given that information is only collected from a focal actor, egonets are a subjective representation of local neighbourhoods. Individuals name significant relationships depending on the questions they are asked (i.e. 'Whom do you talk to?', 'Whom do you ask for advice?', 'Who are your friends?'), which can be interpreted differently by different people in different contexts. People may forget to name some significant alters (for example, whom did they ask for advice in specific occasions), or they may intentionally want to keep some relationships secret (for example, a personal friendship with a boss). Their responses may be biased by the activities they have recently done (it is more likely to name colleagues we recently interacted with), or by the perception they have of the interviewer (men may not talk in derogative terms about their female contacts in front of a female interviewer). People may not remember or know certain characteristics of their alters (where did a friend study, how old an uncle is), and they may not be aware of alter-alter ties. Seminal studies on the validity and reliability of egonet data collection are Bernard et al. [44] and Freeman and Romney [45]. As egonets are subjective representations of individuals' personal relationships, it is quite common to mix name generators with qualitative interviews, usually in-depth or semi-structured. By recording the subjective accounts of network structures, qualitative tools gain an insider view of the interactional processes which generate those structures [7].

The scope of egonet qualitative research is usually to build typologies of personal experiences, but empirical analysis can also extend to modelling network outcomes, and, in line with standard qualitative research, it can help to reinforce or reconfigure theoretical frameworks. In the following paragraphs I illustrate empirical examples that serve each of these purposes.

3.1 Building Typologies

Egonets studies can use a mixed method approach to build social typologies. As in every social network study, they may concentrate on observing the mechanisms that produce network ties and structures, or the outcomes that such structures entail (or a combination of both).

Of the first type is the seminal study of Bidart and Lavenu [46], which looks at the impact of different life trajectories such as entry into the labour market, geographical mobility, and family formation on the size and composition of the personal networks of 66 young people in Normandy (France). Bidart and Lavenu [46] conducted a longitudinal parallel mixed methods study, as both quantitative and qualitative data were collected at the same time. They sampled their interviewees based on age (between 17 and 23 at the time of the first wave), gender, and educational background. They reconstructed their personal networks every three years using contextual name generators, each of which concentrate on a specific life context (studies, work, leisure activities, family, neighbours, etc.). They collected sociographic characteristics of the alters and alter-alter ties, although this last information was not used in the paper discussed here, as its scope is to observe the changes in personal network sizes over time. The "quantitative" data collection that produced the numerical measurement of network size and composition was accompanied by in-depth "qualitative" interviews, in which personal and relational changes were discussed at length.

The analytical strategy consists in mixing the quantitative information about personal networks, namely the changes in their size, with the subjective narratives that recollect the mechanisms and meanings of reconfigurations. Bidart and Lavenu [46] identify four types of network trajectories purely based on the quantitative measurement of changes in size, within which they identified sub-typologies constructed with qualitative materials. For example, the first typology of trajectories groups together people whose network reduced its size at each wave. Within this typology, the qualitative materials allow to distinguish between two types of individuals: those whose network decreased because they quickly left the teenage style of sociability to fully enter the job market; and those who left school to face unemployment or early motherhood. The thick description of the mechanisms that produced the reduction of network size presents us with two opposite contextual explanation: networks reduce following the swift entry in the job market as a consequence of losing ties established at school, but equally reduce for those who are precociously excluded from the job market as a consequence of missing the opportunity of forge relationships in working environments.

If Bidart and Lavenu's study only concentrates on the mechanisms of network formations, Bernardi, Kleim and Klärner [47] focus the analysis to the consequences of network structures, looking at the influence of personal networks over fertility attitudes, intentions, and behaviours. In this study, Bernardi and colleagues compare two purposefully constructed samples of a population from an Eastern German and a Western German city, where rates of fertility are different. They interviewed 61 people, contacted via snowballing, and asked them to identify the informal relationships salient in fertility decision-making. In each interview they used a target of concentric circles [48] to measure ties' strength, and a network grid to measure alter-alter ties (Fig. 1). They then accompanied the name generators with a semi-structured interview to explore why people were named and placed in specific circles in the target, and the subjective meaning of each tie.

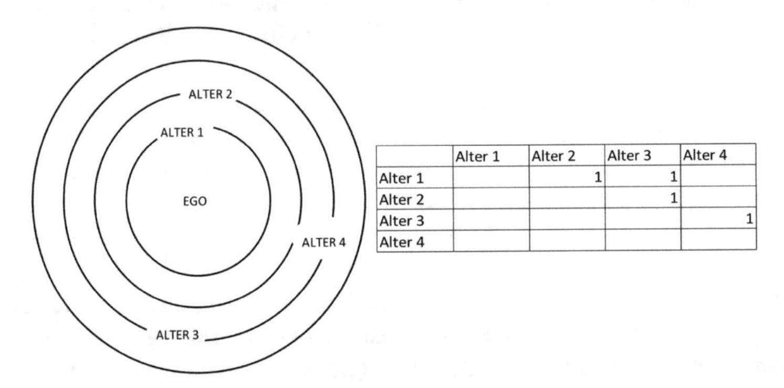

	Alter 1	Alter 2	Alter 3	Alter 4
Alter 1		1	1	
Alter 2			1	
Alter 3				1
Alter 4				

Fig. 1. The target of concentric circles and the grid of alter-alter ties

The analytical strategy starts with a pure qualitative analysis to identify different mechanisms of social network influence over fertility intentions and behaviours (for example, social support or social pressure), followed by the descriptive statistics of the ranking of friendship dyads, where family ties are more likely to be stronger than colleagues, neighbours and acquaintances. The combination of the qualitative and quantitative results allows to explore which type of influence mechanisms are exerted by which type of dyads, where findings suggest that while strong ties convey social pressure and social support, weak ties can be sources of social learning and contagion, as well as exercise contrasting influence. Children in the network, for example, can be perceived positively and motivate own childbearing intentions, and they can also be perceived negatively and prevent egos from having children.

These two examples combine descriptive quantitative and qualitative analysis to produce typologies of network trajectories in the first case, and of influence mechanisms in the second case. In the next section, I introduce two studies that go beyond pure description and offer means of explaining how personal networks and individual behaviours may be interwoven.

3.2 Modelling Networks and Outcomes

Explaining social mechanisms and trying to predict them, is always extremely difficult. Social scientists usually rely on statistical methods to model social phenomena, but classic inferential statistics have not given robust evidence to date [49]. More important, inferential statistics cannot be used in social network analysis, because its core assumptions, of independent observations, is violated by the dependencies of actors and dyads that are observed in social networks (for an extensive discussion of these aspects, see [3, 50]). The issue is not simply methodological but goes to the ontological and epistemological roots of social networks, which, as Carrington explains, "is fundamentally neither qualitative nor quantitative, nor a combination of the two. Rather, it is structural" [51: 35]. Social network analysis diverges from the epistemological ground of classic statistics in two ways. Descriptively, it departs from the logic of the more, the more/the more likely [52: 244]: the mathematics behind social network analysis – graph theory – does not represent "quantities" but "structures" [51: 35], as in patterns that emerge from interweaving ties. Inferentially, it diverges from the logic of representative sampling: these emerging patterns are systemic, in the sense that we cannot observe a portion of a network to infer the structure of the rest. They are also highly contextual, which means that network characteristics are not generalisable via design-based inferences [53].

Social network researchers have addressed the problem of dependencies extensively, developing families of statistical modelling that allow to observe both the formation mechanisms that induce network structures, or the outcomes that such structures may facilitate. Examples of these families are network disturbance models [54–56] and network effect models [57], in which the social structure (i.e. the network) influence the error terms – network disturbance – or an individual level outcome variable – network effect (for an extensive discussion of these models see [50]). Other popular families are exponential random graph models (ERGM), which infer the probability of some basic and foundational properties of network structures to be found in empirically observed networks [58, 59], and SIENA models that look at the dynamics of networks over time [60].

In this section I present two studies that have adopted a mixed method strategy with the aim of modelling either the formation of egonet structures [61], or the potential outcomes that these structures enhance [52]. Methodologically both studies use a parallel design, insomuch they collect numeric network data and qualitative material simultaneously; and a conversion design, because they both transform qualitative data into numerical values. Analytically, the two examples are interesting because they offer two radically different options for modelling network structures and outcomes. Both are valid alternatives and should be selected depending on the specific research questions and the type of available data.

The first study is Lubbers and colleagues' investigation of the changes in the personal networks of 25 Argentinian migrants in Catalonia [61]. The subjects of the empirical research were interviewed twice, in 2004–2006 and again in 2007–2008. The data collection process followed a parallel mixed methods design, in which interviewees were asked a series of questions about themselves; a name generator in which they had to mention at least 45 alters they know, had some contact within the previous two years and could be potentially contacted again; a series of alters' attributes; and alter-alter ties. The outcoming egonet visualization was then used to conduct a qualitative interview in both waves, with the scope of isolating important predictors of network change (involution, stability, and evolution). The research design resembles very closely the one of Bidart and Lavenu [46] described in the previous paragraph, but the analytical strategy goes beyond the simple description of personal network typologies. Once the potential predictors of change were identified with qualitative analysis, data were converted into numbers and added to statistical models to observe which factors (individual characteristics, life events, ego-alter characteristics, and/or network characteristics) play a role in explain and predict network changes.

Lubbers et al. [61] analytical strategy makes use of:

1. Multilevel logistic regression analysis to model the persistence of a relationship over time.
2. Multilevel linear regression analysis to model how persistent relationships change over time.
3. Multivariate regression analysis to model how networks' structure and composition changed over time.
4. SIENA – Stochastic Actor-Oriented Models – [62] to model the changes in relationships among alters.

The outcome of these combined models of analysis indicates how much a predictor may significantly contribute to changes in egonets.

In (1) and (2), which look at the presence or absence of a dyad over time, and the changes in its value, the authors overcome the problem of dependencies of ties by adopting a multilevel approach. The unit of analysis here are individual dyads, which are either absent or present (logistic regression) at time 2 compared to time 1, or they change in value over time (linear regression). But dyads are not independent observations: they are nested together under the focal actor who intentionally forms, ceases, or modifies them. The multilevel approach allows to express the probability that dyads that are generated by the same individual will be more similar to each other compared to the ones generated by someone else. Not only dyads are nested cross-sectionally, but we expect their values at time 2 to be dependent from their values at time 1: the authors solve the problem of temporal dependencies by using an auto-correlation approach, in which time 1 values are added as covariates. Further discussion of multilevel approaches in egonet analysis is available in Crossley et al. [40].

When modelling how egonet structures change over time (3) – for example, do they become more or less dense? More or less centralised? – a multivariate regression analysis is sufficient and does not violate the assumption of independency, because each egonet structure is observed independently from the others. The main issue in this type of analysis is related to the statistical power, by which the power of a study in detecting a real effect that is not caused by pure chance is determined by the number of observations in relation to the number of parameters estimated. A rule of thumb is that ordinary least squares need five cases per independent variable in the analysis, and maximum likelihood needs at least 10 cases per independent variable [63]. I will get back to this aspect when discussing Hollstein and Wagemann's paper [52], which offers a viable solution to the problem.

Finally, modelling the changes in alter-alter ties (4) requires statistical models specifically developed to account for the rich dependence structure that changes in embedded ties will create. SIENA [62] models such changes as Markov processes, where probabilities of tie changes are determined by the current state of a network, and each tie change modifies the state of the whole network which again shape the possibilities of further tie changes [64]. Lubbers et al. [61] offer the first example of SIENA application to egonet studies, which has subsequently been discussed in Crossley et al. [40].

The second study is Hollstein and Wagemann's examination of the conditions under which personal networks are effective in facilitating the transition to working life of 88 teenagers and young adults from Southern Germany [52]. The sample (45 men and 43 women) was interviewed three times: at the age of 17–19, at the end of vocational training, and two years after. Of these, a subgroup of 35 people experienced a relatively bad start to their working life, being registered as unemployed at the time of the first interview and enrolled in an employment assistance programme: the study concentrated on these, who at the time of the last interview either found or not a job. Data were collected, at each wave, using: a qualitative interview that focused on work/employment history, leisure time and family, including questions over the respondent's personal network and resources; a sociographic questionnaire of personal data, including career-related information; four name generators, collected with the visual aid of concentric circles [48]; a series of alters' attributes.

Hollstein and Wagemann's choice of analytical strategy departs completely from the logic of standard statistics and embraces the perspective of Configurational Comparative Methods [65, 66]. These methods analyse each case (for example, a respondent's interview) as a combination of potential factors that may produce a specific outcome. For example, a school dropout may report on lack of support from teachers, deprived school environments, a difficult family environment, while another may mention the same lack of support from teachers but also a higher investment in sport activities. The scope of the analysis is to detect causal mechanisms in a set of cases without assuming these mechanisms to be identical across different cases. The term CCM refers mainly to qualitative comparative analysis (QCA), a purely mixed method approach that code qualitative material in combinatory elements that can be categorised into crispy (0/1), multiple (0; 1; 2; 3 […]) or fuzzy (ordered or continuous) sets; aggregation of elements into configurations following Boolean logic – a form of algebra centred around the three operators "OR", "AND" and "NOT"; and the making of general statements about necessary or sufficient conditions (combinations of elements) to infer cause in cases' observed outcomes.

In doing so, QCA aims to discover the underlying mechanisms that may explain these configurations by relying on causal logic that is intrinsically different from "the more, the more/the more likely" of classic statistical analysis. In classic statistical analysis we may say that we see a correlation between unemployment and criminal convictions, as in 'people who have been convicted stand fewer chances of finding a job'. In QCA instead, we develop set-theoretic hypotheses constructed on the logic of 'if, then', which advance several forms of causation mechanisms. Equifinal causation indicates that conditions can be alternative to one other, for example people can compensate the disadvantage of being convicted in different ways (increasing educational skills OR volunteering), producing hypothesis in which neither one nor other condition is necessary, but at least one is sufficient to produce a condition which is in itself necessary but insufficient to produce an outcome – for example, increasing educational skills OR volunteering may facilitate in finding a job ONLY IF the conviction period has terminated – [52: 245, 67].

Conjunctural causation implies that causes may exert effects only in conjunction with other causes (increasing educational skills AND volunteering), producing hypothesis in which neither of these conditions are sufficient alone, but together are sufficient (although not necessary) to produce an outcome – for example, increasing educational skills AND volunteering can facilitate finding a job, but this can also be facilitated by having an employed mother AND receiving an endorsement by a teacher [52: 245, 68: 62].

One of the advantages of QCA is that it is not limited by statistical power, because it does not observe variance across cases. Indeed, QCA was first developed to confront the problems of comparative analysis, the main one being that research at a macro level (e.g., national comparison) very often has to deal with a limited number of cases (e.g., 27 EU countries) but a large number of variables [69]. It is therefore well suited for analysis in middle-range studies, which can be based on a continuum between a small number and a large number of cases.

However, the most important distinction between the foundational logic of the two studies discussed above is what the models can explain, and what assumptions they make on the social phenomena they do not explain. In Lubbers et al. [61] the scope of the analysis is to observe network formation mechanisms (as in forming, maintaining, changing or ceasing a tie), by looking at how these mechanisms depend on the person who exercises its own agency in deciding whom to relate to, on the characteristics of the dyads that are formed over time, and on the overall effects that networks exert on agents and dyads by virtue of embedding them into a structure. Conclusions that can be drawn from such analysis are, for example, that dyadic relationships are more persistent when contacts are frequent and alters are more central in personal networks; but also that these mechanisms interact with the type of alters, in which we see relationships with Spaniards to last more in recent migrants' networks than ties with Argentineans, while less recent migrants have more stable ties with Argentineans. What cannot be said, in such analysis, is why other ties are not equally maintained. Standard statistical analysis is modelled against the null hypothesis which states that there is no correlation between frequency of contacts, centrality in networks, and persistency of ties. We know, from Lubbers et al. [61] analysis, that frequent and central ties are more likely to last, but we do not know the reasons why other ties have not lasted.

In set-theoretic hypothesis, instead, we can assume asymmetric causation. Going back to Hollstein and Wagemann's example [52]: to verify the hypothesis that having being convicted for a crime may have repercussions in finding a job, we should not be able to find any case in which people convicted for a crime did not have difficulties in finding a job. But people without criminal convictions may also have problems in finding jobs for other type of reasons, and such cases do not falsify our hypothesis because these do not make assumptions in regards.

Ultimately, both analytical strategies of multilevel and SIENA analysis and of QCA, are excellent ways to say something of the potential causes that produce social phenomena: the choice between the two depends on the research questions that guide an empirical study. In the case of Lubbers and colleagues, the aim of the study is to calculate the probability that certain individual characteristics, dyadic properties, and network structure influence the evolution of personal networks. The tested hypothesis explores the contribution of these factors, without making any assumption about the relevance of other, concurrent factors. In the case of Hollstein and Wagemann, the aim is to observe all the possible combinations that emerge from the analysis of qualitative data, and evaluate which of these may be trivial (e.g.: you need to be born to find a job) and consistent (to what extent the conclusion that a condition is sufficient and/or necessary is based on empirical data, or what proportion of data may diverge from this conclusion). We may conclude that Lubbers et al. [61] method of analysis is more suitable for testing already existing hypotheses, while Hollstein and Wagemann [52] serves the purpose of exploring new theoretical possibilities, or in other words, that the former study has a predominant deductive structure, while the latter an inductive one. The contribution that mixed methods in social network analysis can give to theoretical advancement, and the dialectic between inductive and deductive processes, is the topic of the next section.

3.3 Reformulating Theory

Qualitative methods have always been considered as suitable approaches to explore novel theoretical frameworks, extend existing ones or challenge them altogether. By collecting subjective perceptions on social phenomena, qualitative methods allow to explore under-researched aspects or novel ways of explaining how certain phenomena come to be. They inform theories, because as Small says: "the soundness of a theoretical premise, in social sciences, turns on its ability to accurately capture how people think, interact, and make decisions" [70: 153].

Small's in-depth study [70] of a cohort of 38 first-year graduate students in three academic departments is an elegant example of how a small-scale qualitative study can capture essential elements of people's behaviours and use them to inform theoretical advancements. Small's research focuses on the composition of what the literature calls the core discussion network [71], a core network of confidants whom people turn to when they need to talk about something important for them. The focus of the research emerged inductively from a previous study on the support network mothers made through day-care centres [72]. Surprisingly, those women would happily leave their children with parents they barely knew. In other words, they would entrust strangers, or what we call in social network analysis, weak ties. Puzzled by the finding that diverges from the commonly accepted opinion that people trust their strong ties, Small started examining the structural theory of core discussion network closely. He reconstructed how the theory first came to be, by looking at the early 1960s attempts to measure personal networks [73–75] which informed the design of a name generator question to be included in the US General Social Survey [76, 77]. Burt's aim was to measure North Americans' personal networks, and he was convinced that asking whom people discuss personal matters with would elicit intimate and emotionally close relationships. He was not interested in knowing exactly whom people did talk to, but used it as a proxy for strong ties. This proxy, of asking discussion partners to elicit strong ties, eventually ended up being conceptually reverted, and strong ties have been considered the ones used to discuss important matters since.

Small followed, up to this point, a classic inductive research process: he started with some puzzling qualitative finding and went to search for the existing theories that may make sense of these odd results. As the theory of strong ties constituting core discussion networks did not seem to explain his results, he decided to put the theory to test. He tailored a small qualitative study set up to verify exactly the premises of the structural theory of core discussion networks. The scope of this study was twofold: on one hand, he wanted to test the validity of the GSS name generator in eliciting whom people talk to; on the other hand, he wanted to explore why people may not talk to the ones they consider very close, in case they did. Deductively, he wanted to test an existing theory; inductively, he wanted to go beyond it and build a new set of assumptions.

Methodologically, he interviewed 38 graduate students who entered a PhD programme in a large US institution. Interviews were conducted at the beginning of the first semester (few weeks in), then after 6 months, then after other 6 months, then a year later. He subministered the exact GSS questionnaire, asking whom they talked to when they had important matters to discuss in the previous 6 months, including alters' name interpreters. Then he conducted a mixture of semi-structured and unstructured interviews, where probes switch the question around. He asked to list the three most impelling concerns, recall when was the last time they talk about them, with whom, why, and why they did not talk about them with the people they nominated on the GSS generator (in case they did not).

Results indicated that 48% of intimate conversations were with someone not named as confidant, and that students talked to them either because they intentionally avoided strong ties; or because they were more relevant and emphatic about the discussed topic; or simply because they were there and available when they needed to talk. These finding gave confidence to his initial scepticism about the validity of the GSS name generator in eliciting the core discussion network and suggested that what the GSS reproduces is not the practice of confiding, but the network structure of close ties. Results gave Small a new set of assumptions that challenged the original theory: that people may have good reasons to avoid strong ties when confiding; that people may act spontaneously and without reflecting on consequences, so they may confide to someone without consciously planning the decisional process; and that foci of social activity [28] may affect whom people confide with, where mothers confide with each other just because they see themselves often [70: 17–19].

Armed with this new set of assumptions, Small then asked himself if his findings were only typical of a specific cohort of PhD students, or could be empirically generalised to a wider population. He thus designed and commissioned an online survey on core networks and important alters of 2010 adult respondents whose characteristics (gender, race, and employment status) matched the US population. When asked to report their core discussion networks, and then the people they are close to, in 45.3% of the case the people named did not match [78]. He followed up with a second survey on the reasons why students may talk to someone, which was administered online on a non-representative sample of 2211 college students from different ethnic background [79]. He found that the primary reason for talking to whom they did was alter availability, and that a third of the time the decision was not planned but spontaneous.

Small's complex, fully integrated design is exemplar of mixed method research. He combines inductive and deductive strategies, by which he gains insights of the limits of an established theory from results of a qualitative study (inductive), set up an empirical study to test his assumption (deductive), reformulate the theory with new evidence from his results (inductive) and test it again over a new set of collected data (deductive). His reasoning is counterintuitive, like scientific theorising usually is [80], and challenges well established assumptions not with the intent of diminishing their validity, but to refine their range and their portability. What he proposes is not "a comprehensive theory, but a set of related assumptions that by prioritising practice over structure, introduces a new and different set of questions" [70: 151]. The practices of interactions, and the contexts in which they happen, make these new assumptions portable to other contexts, allowing to explain why people, for example, may not talk about inner struggles, may not report violence, may not disclose secret behaviours. In line with the tradition of qualitative research, his findings are theoretically generalizable, in the sense that they can be transferred and potentially observed in comparable yet different contexts.

4 Conclusions

This article has discussed empirical examples of how social network analysis can be mixed with a variety of qualitative and quantitative methods in egonets research. As Carrington reminded us [51], social network analysis is not qualitative nor quantitative, but structural: it is a mathematical representation of social phenomena that looks at the emerging patterns of interweaving ties, and their outcomes. As such, it can be mixed with both qualitative and quantitative tools. When collecting data, tools can be purely qualitative and generate texts (semi-structured, in-depth, thematic interviews; ethnographic observations; secondary analysis of archival materials), or quantitative, and generate numbers (name generators, rosters, ego and alter attributes questionnaires, alter-alter ties grids). In conversion studies, qualitative data are then converted into quantitative information, or qualitative and qualitative data can be kept parallel. Analytical strategies can be equally qualitative (content analysis, thematic analysis, discourse analysis), and/or quantitative (descriptive statistics, inferential models, set-theoretic models), and interpretation of results takes advantage of the insights that both approaches offer.

What all these studies show is that quantitative analysis is incredibly valuable in detecting typologies and modelling network formations mechanisms and their outcomes, while qualitative analysis is invaluable in exploring the subjective and sometimes diametrically opposed trajectories that lead actors to the positions they occupy and the outcomes they enjoy.

Thus, in Bidart and Lavenu's example [46], a similar network dynamic (reduction) is the outcome of completely opposite trajectories, of people entering the job market when young, and of people falling into early unemployment. Likewise, in Bernardi et al.'s example [47], the same tie (having friends or relatives with children) may influence opposite choices (of having or not children of their own).

Lubbers and colleagues [61] and Hollstein and Wagemann [52] offer two radically different approaches to quantitative modelling of network mechanisms. In both cases data were gathered via qualitative and quantitative methods (interviews, questionnaires, name generators and name interpreters), but while Lubbers et al. tested causation mechanisms which contribute to the dynamics of network ties and structures with multilevel regression analysis and SIENA actor-oriented models, Hollstein and Wagemann depart from classic statistics to embrace set-theoretic causation models. As we discussed, these two strategies serve different purposes of deductively test or inductively generate theories, going beyond the mainstream assumption that qualitative and quantitative methods diverge in their inductive and deductive epistemological scopes. Mixed methods approaches can serve the purposes of inductively generating puzzling research questions as well as testing dominant theories, as Small [70] magisterially shows in his book.

Overall, mixed methods egonet studies serve the purpose of showing how cultural elements, as in attitudes, values, beliefs and behaviours, emerge out of and are negotiated within social interactions, as well as how these interactions are structured in local networks that reinforce or challenge emerging cultural elements. In this perspective, the decision of having a child is influenced by the opinions and behaviours of the people who matter for us [47], but also changes the opportunities we have to establish or maintain our social relationships [46]. Likewise, the struggle to find a job can cause the reduction in the number of meaningful relationships [46], but at the same time the very lack of such relationships may hinder our ability in finding a job [52]. We might subjectively believe that in case of need we will turn to our closest relationships for help, but everyday practices suggest that we are likely to consciously or unconsciously rely on the relational opportunities that our structural environment offer to us [70].

As such, mixed methods in social networks allow to go beyond the limited description of local mechanisms and typologies that are characteristics of much of qualitative research, offering opportunities to combine subjective narrative accounts with sophisticated analytical strategies that can test how cultural elements and structural features co-evolve. At the same time, mixed methods social network analysis allows to overcome the limits of classic inferential statistics, offering opportunities to evaluate the theoretical portability of structural and cultural mechanisms and formations, where actors' and objects' positions in the social structure may be explained by different, concurrent, and sometimes opposite individual trajectories. Mixed methods social network analysis can be a viable solution to move beyond the non-generalizability of qualitative data, where subjective accounts never reach the explanatory power of quantitative analysis; but can also offer a valid alternative to the rigid and limited simplifications of classic statistical linear models, that force data into the line that best summarizes the variance of cases, by minimizing the deviations of standard errors and by dropping outliers. When combining qualitative and quantitative methods in social networks, we can explore at the same time the emergence of structures that include both dominant and peripheral actors, and the individual trajectories and subjective stories that drive actors in their positions. The hope is that future research in social sciences will increasingly adopt such a powerful perspective.

References

1. Hollstein, B.: Qualitative approaches. In: Scott, J., Carrington, P.J. (eds.) The Sage handbook of Social Network Analysis. Sage, London (2011)
2. Dominguez, S., Hollstein, B. (eds.): Mixed-Methods Social Network Research. Cambridge University Press, New York (2014)
3. Bellotti, E.: Qualitative Networks. Mixed Methods in Sociological Analysis. Routledge, London (2015)
4. D'Angelo, A., Ryan, L.: Social network analysis: a mixed methods approach. In: Mckie, L., Ryan, L. (eds.) An End to the Crisis in Empirical Sociology? Routledge, London (2016)

5. Mønsted, M.: processes and structures of networks: reflections on methodology. Entrepreneurship Reg. Dev. **7**, 193–213 (1995)
6. Coviello, N.: Integrating qualitative and quantitative techniques in network analysis. Qual. Market Res. **8**(1), 39–60 (2005)
7. Edwards, G.: Mixed-Method Approaches to Social Network Analysis. Review paper. ESRC National Centre for Research Methods (2010)
8. Emirbayer, M., Goodwin, J.: Network analysis, culture and the problem of agency. Am. J. Sociol. **99**, 1411–1454 (1994)
9. Mische, A.: Cross-talk in movements. In: Diani, M., McAdam, D. (eds.) Social Movements and Networks. Oxford University Press, Oxford (2003)
10. Crossley, N.: The social world of the network: combining quantitative and qualitative elements in social network analysis. Sociologica 1 (2010)
11. Riles, A.: The Network Inside Out. University of Michigan Press, Ann Arbor (2001)
12. Knox, H., Savage, M., Harvey, P.: Social networks and the study of relations: networks as method, metaphor and form. Econ. Soc. **35**(1), 113–140 (2006)
13. White, H.C., Boorman, S., Breiger, R.: Social structure from multiple networks: blockmodels of roles and positions. Am. J. Sociol. **81**, 730–780 (1976)
14. Brint, S.: Hidden meanings: cultural content and context in Harrison Whites structural sociology. Sociol. Theory. **10**, 194–208 (1992)
15. Whyte, W.F.: Street Corner Society. The Social Structure of an Italian Slum. University of Chicago Press, Chicago (1943)
16. Mitchell, J.C. (ed.): Social Networks in Urban Situations. Manchester University Press, Manchester (1969)
17. Freeman, L.C.: The Development of Social Network Analysis. A Study in the Sociology of Science. Empirical Press, Vancouver (2004)
18. Bernard, R.H.: Greek sponge boats in Florida. Anthropol. Q. **38**, 41–54 (1965)
19. Bernard, R.H., Killworth, P.D.: On the social structure of an ocean-going research vessel and other important things. Soc. Sci. Res. **2**, 145–184 (1973)
20. Bernard, R.H., Killworth, P.D.: Scientists and crew: a case study in communication at sea. Marit. Stud. Manag. **2**, 112–125 (1974)
21. Killworth, P.D., Bernard, R.H.: Informant accuracy in social network data. Hum. Organ. **35**, 269–296 (1976)
22. Freeman, L.C., Freeman, S.C., Michaelson, A.G.: On human social intelligence. J. Soc. Biol. Struct. **11**, 415–425 (1988)
23. Freeman, L.C., Freeman, S.C., Michaelson, A.G.: How humans see social groups: a test of the Sailer-Gaulin models. J. Quant. Anthropol. **1**, 229–238 (1989)
24. Mische, A.: Relational sociology, culture, and agency. In: Scott, J., Carrington, P. (eds.) The Sage Handbook of Social Network Analysis. Sage, London (2011)
25. White, H.C.: Identity and Control. Princeton University Press, Princeton (2008 [1992])
26. Hollstein, B.: Mixed methods social network research: an introduction. In: Domínguez, S., Hollstein, B. (eds.) Mixed-Methods Social Networks Research. Design and Applications. Cambridge University Press, Cambridge (2014)
27. Tashakkori, A., Teddlie, C.: Mixed Methodology: Combining Qualitative and Quantitative Approaches. Sage, London (1998)
28. Feld, S.: The focused organization of social ties. Am. J. Sociol. **865**, 1015–1035 (1981)
29. Emirbayer, M., Mische, A.: What is agency? Am. J. Sociol. **103**, 962–1023 (1998)
30. Crossley, N.: Toward Relational Sociology. Routledge, London (2011)
31. Schutz, A.: The Phenomenology of the Social World. Northwestern University Press, Evanston (1967)
32. Allan, G.: A Sociology of Friendship and Kinship. George Allen and Unwin, London (1979)

33. Fischer, C.S.: What do we mean by "friend"? An inductive study. Soc. Netw. **3**, 287–306 (1982)
34. Spencer, L., Pahl, R.: Rethinking Friendship: Hidden solidarities today. Princeton University Press, Princeton (2006)
35. Bellotti, E.: What are friends for? Elective communities of single people. Soc. Netw. **30**, 318–329 (2008)
36. Crossley, N.: Networks of Sound, Style and Subversion: The Punk and Post-Punk Worlds of Manchester, London, Liverpool and Sheffield, 1975–80. Manchester University Press, Manchester (2015)
37. Becker, H.S.: Art Worlds. University of California Press, Berkley (1982)
38. Bourdieu, P.: Distinction: A Social Critique of the Judgement of Taste. Routledge, London (1984)
39. Wellman, B.: An egocentric network tale: comment on Bien et al. "An account of the origin and design of the East York studies". Soc. Netw. **15**, 423–436 (1993)
40. Crossley, N., Bellotti, E., Edwards, G., Koskinen, J., Everett, M., Tranmer, M.: Social Network Analysis for Ego-Nets. Sage, London (2015)
41. Simmel, G.: Conflict and the Web of Group Affiliations. Free Press, Glencoe (1955 [1922]).
42. Lin, N., Dumin, M.: Access to occupations through social ties. Soc. Netw. **8**, 365–385 (1986)
43. Van der Gaag, M., Snijders, T.A.B.: Social Capital Quantification with Concrete Items. University of Groningen Press, Groningen (2003)
44. Bernard, R.H., Killworth, P.D., Kronenfeld, D., Sailer, L.: The problem of informant accuracy: the validity of retrospective data. Ann. Rev. Anthropol. **13**, 495–517 (1984)
45. Freeman, L.C., Romney, A.K.: Words, deeds and social structure: a preliminary study of the reliability of informants. Hum. Organ. **46**, 330–334 (1987)
46. Bidart, C., Lavenu, D.: Evolution of personal networks and life events. Soc. Netw. **27**, 359–376 (2005)
47. Bernardi, L., Keim, S., Klärner, A.: Social networks, social influence, and fertility in Germany: challenges and benefits of applying a parallel mixed methods design. In: Domínguez, S., Hollstein, B. (eds.) Mixed-Methods Social Networks Research Design and applications, pp. 121–152. Cambridge University Press, Cambridge (2014)
48. Kahn, R.L., Antonucci, T.C.: Convoys over the life course. Attachment, roles, and social support. In: Baltes, P.B., Brim, O.G. (eds.) Life-Span Development and Behavior. Academic Press, New York (1980)
49. Abell, P.: History, case studies, statistics, and causal inference. Eur. Sociol. Rev. **25**, 561–567 (2008)
50. Bellotti, E., Guadalupi, L., Conaldi, G.: Comparing fields of sciences: the network of collaborations to research projects in Italian academia. In: Lazega, E., Snijders, T. (eds.) Multilevel and Network Analyses, pp. 213–244. Springer, Cham (2015)
51. Carrington, P.J.: Social network research. In: Domínguez, S., Hollstein, B. (eds.) Mixed-Methods Social Networks Research. Design and Applications. Cambridge University Press, Cambridge (2014)
52. Hollstein, B., Wagemann, C.: Fuzzy-set analysis of network data as mixed methods: personal networks and the transition from school to work. In: Domínguez, S., Hollstein, B. (eds.) Mixed-Methods Social Networks Research. Design and Applications. Cambridge University Press, Cambridge (2014)
53. Snijders, T.A.B., Bosker, R.J.: Multilevel Analysis: An Introduction to Basic and Advanced Multilevel Modelling. Sage, London (2012)
54. White, D.R., Burton, M.L., Dow, M.M.: Sexual division of labor in African agriculture: a network autocorrelation analysis. Am. Anthropologist **83**, 824–849 (1981)
55. Dow, M.M., White, D.R., Burton, M.L.: Multivariate modeling with interdependent network data. Behavior Sci. Res. **17**, 216–245 (1983)

56. Dow, M.M., Burton, M.L., White, D.R., Reitz, K.P.: Galtons problem as network autocorrelation. Am. Ethnologist **11**, 754–770 (1984)
57. Doreian, P.D.: Models of network effects on social actors. In: Freeman, L.C., White, D.R., Kimball Romney, A.K. (eds.) Research Methods in Social Network Analysis. Transaction Pub, New Brunswick (1992)
58. Robins, G., Pattison, P., Kalish, Y., Lusher, D.: An introduction to exponential random graph p* models for social networks. Soc. Netw. **29**, 173–191 (2007)
59. Lusher, D., Koskinen, J., Robins, G.: Exponential Random Graph Models for Social Networks. Cambridge University Press, Cambridge (2012)
60. Snijders, T.A.B., Steglich, C.E.G., Van de Bunt, G.G.: Introduction to actor-based models for network dynamics. Soc. Netw. **32**, 44–60 (2010)
61. Lubbers, M.J., Molina, J.L., Lerner, J., Brandes, U., Ávila, J., McCarty, C.: Longitudinal analysis of personal networks. The case of Argentinean migrants in Spain. Soc. Netw. **32**(1), 91–104 (2010)
62. Snijders, T.A.B.: The statistical evaluation of social network dynamics. Sociol. Methodol. **40**, 361–395 (2001)
63. Frey, B.B.: Logistic regression. In: Frey, B.B. (ed.) The SAGE Encyclopedia of Educational Research, Measurement, and Evaluation. Sage, Thousand Oaks (2018)
64. Montgomery, V.J., Lubell, A.M., Snijders, T.A.B., Pickup, M.: Stochastic Actor Oriented Models for Network Dynamics. Oxford Handbooks Online (2017). Accessed 12 Apr 2020. https://www.oxfordhandbooks.com/view/10.1093/oxfordhb/9780190228217.001.0001/oxfordhb-9780190228217-e-10.
65. Ragin, C.C.: The Comparative Method: Moving beyond Qualitative and Quantitative Strategies. University of California Press, Berkeley (1987)
66. Rihoux, B., Ragin, C.C. (eds.): Configurational Comparative Methods: Qualitative Comparative Analysis QCA and Related Techniques. Sage, Thousand Oaks (2009)
67. Mahoney, J., Kimball, E., Koivu, K.L.: The logic of historical explanation in the social sciences. Comp. Polit. Stud. **421**, 114–146 (2009)
68. Mackie, J.L.: The Cement of the Universe. Clarendon, Oxford (1974)
69. Berg-Schlosser, D., De Meur, G., Rihoux, B., Ragin, C.C.: Qualitative comparative analysis QCA as an approach. In: Rihoux, B., Ragin, C.C. (eds.) Configurational Comparative Methods. Qualitative Comparative Analysis QCA and Related Techniques. Sage, Thousand Oaks (2009)
70. Small, M.L.: Someone to Talk to. Oxford University Press, New York (2017)
71. McPherson, M., Smith-Lovin, M., Brashears, M.E.: Models and marginals: using survey evidence to study social networks. Am. Sociol. Rev. **713**, 353–375 (2009)
72. Small, M.L.: Unanticipated Gains: Origins of Network Inequality in Everyday Life. Oxford University Press, New York (2009)
73. Laumann, E.O.: Bond of Pluralism: Forms and Substance of Urban Social Networks. Wiley, New York (1973)
74. Fischer, C.S.: To Dwell Among Friends Personal Networks in Town and City. University of Chicago Press, Chicago (1982)
75. Wellman, B.: The community question: the intimate networks of east Yorkers. Am. J. Sociol. **84**, 1201–1231 (1979)
76. Burt, R.S.: Network items and the general social survey. Soc. Netw. **6**, 293–339 (1984)
77. Burt, R.S.: General social survey network items. Connections **81**, 19–23 (1985)
78. Small, M.L.: Weak ties and the core discussion network: why people regularly discuss important matters with unimportant alters. Soc. Netw. **353**, 470–483 (2013)
79. Small, M.L., Sukhu, C.: Because they were there: access, deliberation, and the mobilization of networks for support. Soc. Netw. **47**, 73–84 (2016)
80. Wolpert, L.: The Unnatural Nature of Science. Faber, London (1992)

Gender Salience in Women's Career-Related Networking: Interviews with Russian Women

Tatiana Kazantseva(✉), Larisa Mararitsa, and Svetlana Gurieva

St. Petersburg State University, St. Petersburg, Russia
t.kazanceva@spbu.ru, larisamararitsa@mail.ru, gurievasv@gmail.com

Abstract. Networking is closely related to personal social capital and, in turn, career outcomes. Stable reproduction of gender inequality in organizations is largely connected to career-related networking opportunities and constraints both for men and women, and this area remains under-researched. In this study we tested the hypothesis that women perceive female gender as a significant constraint in their career development. Also, we tried to find specific situations in work relationships which make networking difficult for women. The main method was qualitative analysis of 51 semi-structured interviews with working women (mean age 33 years old). The pilot study revealed that women were not willing to support "feministic talks" and could hardly recall any difficult gender-related work relationships. We found a facilitating technique through visualizing their career paths, which resulted in spontaneous flashbacks. 57% of the respondents had experienced reminders about gender from co-workers. The reality of the "gender factor" in their professional networking was encountered by women as an unpleasant discovery accompanied with feelings of powerlessness and shame. Negative or humorous gendered interactions (used mostly by men) serve for psychological distance regulation and instrumental aims, setting gender subordination and power. Situations which make networking difficult include explicit or implicit notifications from colleagues that feminine gender is an obstacle for productive work and implies certain restrictions and requirements, motherhood (expected or actual), and sexual interest from male colleagues. We concluded that women assess gender as a visible and at the same time invisible variable affecting their career. There are a number of situations which make career networking constrained for women.

Keywords: Gendered networking · Gender inequality · Gender reminders

1 Introduction

This research was built upon the intersection of three lines of contemporary organizational studies, e.g., gender inequality, career success, and networking. While the link between career and networking is well established [1–3], the correlation of social relationships and gender inequality at work is not so evident.

Gender inequality stays a relevant issue all around the world. Despite the progress in narrowing the gender gap in education and healthcare, economic inequality between

A. Antonyuk and N. Basov (Eds.): NetGloW 2020, LNNS 181, pp. 34–48, 2021.
https://doi.org/10.1007/978-3-030-64877-0_3

women and men remains high. The ratio of women's income to men's income is just over 0.6 and has not changed worldwide since the 1960s [4]. Even though the policy of equal opportunities for men and women has existed for a long time, it still does not mean that there are equal results. That is why it is so important to identify mechanisms and remove barriers which prevent men and women from taking up equal positions.

The question of how gender inequality is reproduced in organizations is one of the pressing problems in contemporary gender and organizational studies. Increasingly, research is based on the assumption that it is not a person who plays his or her gender role in a gender-neutral organization, but the organization itself, with its personnel and administrative politics which make employees' gender emphasized and salient [5]. But work relationships, unlike, for example, romantic ones, are an irrelevant context for the visibility of gender identity. The new research paradigm called "Gendered organization" focuses precisely on the phenomena of communication and interaction – those social practices that trigger discrimination and facilitate its reproduction. As D.M. Britton reflects, "we should see organizations not as gender-neutral organisms infected by the germs of workers' gender (and sexuality and race and class) identities, but as sites in which these attributes are presumed and reproduced" [6: 418].

In Russia, research in line with this approach is still very scarce, and this work is one of the attempts to apply it.

Based on a literature analysis, we also came to the conclusion that the key mechanisms for reproducing gender inequality would be mechanisms associated with the formation of social capital, but at the moment they remain poorly understood. Social capital is defined as "resources embedded in one's social networks, resources that can be accessed or mobilized through ties in the networks" [7: 4]. Social capital comprises instrumental (wealth, knowledge) or expressive resources (social support), and organizational structure and culture, formal and informal norms may help one to the detriment of others and come out in different results for men and women [8].

Networks are important in three work domains: for accomplishing tasks, gaining upward mobility, and for progress in personal and professional development [2]. Career planning and optimism are also stimulated with the support of social networks and are associated with the development of networking skills [3].

Overall, career capital consists of three ways of knowing: Knowing-how (expertise, work-related knowledge, soft skills); Knowing-why (sense of purpose, motivation, confidence) and Knowing-whom (professional and social relations) [1]. We found such a resource-orientated approach to gender inequality in the career domain very relevant and heuristic, as far as invisible barriers that limit access to various resources for women were registered in society and organizations, for example, the "glass ceiling", "glass walls" and other "glass" phenomena [9].

So, building social capital at work depends on various conditions: structural, organizational factors, and also on personality factors, which include traits [10] and skills [1].

2 Gendered Networks

It has been found that female gender impedes the formation of social capital necessary for career advancement [11, 12]. In this regard, the limited opportunities for women to build social capital could be considered one of the most powerful socio-psychological mechanisms for maintaining gender inequality in organizations.

Studies indicate differences in the structure and effectiveness of social networks of men and women [8, 13, 14]. Gender is almost always associated with homophily in personal social networks. Homophily in this case refers to a tendency to surround oneself with people similar in attitudes and behavior [15], and male networks, in contrast to female ones, are more homogeneous and closed. Homophily also characterizes the social networks of those organizations in which gender segregation is observed [16]. In the networks of managers in the USA, men's networks are characterized by more high-status individuals than those of women with similar levels of education and experience [17].

We believe gender constraints in the accumulation of social capital can be analyzed on three levels. The first level is the level of individual networking behavior, that is, real actions that people can (or cannot) perform due to their gender identity to enter social networks, develop them, maintain, change, or leave them. Women, for example, faced with intra-organizational social constraints, are more likely to establish external relations that go beyond the primary network. Thus, they can provide themselves with an alternative path to access the necessary career resources [13].

Individual psychological barriers to effective networking in women were also identified, one of which is the embarrassment of looking at social connections from an instrumental point of view [18]. One group of networking barriers appeared in a qualitative study was no motivation to socialize. Negative connotations to networking itself included waste of time, no commercial effect, uncomfortable feelings toward the process and people, and all these factors led to social apathy [19].

At the normative level, restrictions to the social capital development are determined by formal or informal rules governing trusting communication between employees of different genders and status. It is communication which is unstructured and friendly that is crucial for accelerating career advancement (in comparison with participation, for example, in official communities) [20]. But events organized for networking, as a rule, take into account only male interests: table football, racing, and beer [18].

According to in-depth interviews of men and women, specific female-only networking barriers include the following barriers to participating in networking activities: family commitment, tiredness and discouraging organizational culture, excluding people from networking activities (often junior employees) or pressing them into doing it [19]. Another common barrier to effective networking is tension in women's work-family system. The presence of a family and children creates for women a shortage of time for networking and also limits their mobility [21].

At the structural level, the opportunities of building effective relationships for representatives of different genders may differ due to the particular organizational structure (for example, homophily) and the positions that men and women occupy (these are less influential in the case of women). Another factor at this level implies structural defects that reduce the likelihood of gaining access to "elite" social resources by a woman.

One of the ways to overcome structural defects for women is the institution of mentoring, which acts as a protector, promoter, and "sponsor" of social capital [22]. An assistant in promotion should take a high-status position; the best option for a woman is a male mentor, older in age. Other corporate barriers female employees may also experience include recruitment and selection processes, and difficulties associated with working in male-dominated areas [23]. Thus, in order to overcome the gender gap in the organization, management requires special efforts to ensure the accessibility of social capital for women at every level.

3 Research Questions

The purpose of this study was to determine how gender becomes salient for a woman herself and her social surroundings in work situations, and how gender hierarchy penetrates work relationships. Will women spontaneously, without leading questions, speak about their gender as an important characteristic of difficult work situations and career growth? If they perceive the female gender as an important element of their professional careers, has it always been like that, or have some events pushed them to have such ideas? What kind of barriers do women experience in career-related networking? All these questions formed the basis for constructing an interview script for working women.

4 Method

Our study was a qualitative study by its design. Qualitative approach to networks, as E. Belotti notes, generally allows to "locate and understand formal networks in their social and cultural context" [24]. Qualitative data is mostly used for in-depth exploration of networks and network practices; for understanding network dynamics and effects. The "interpretive paradigm" of qualitative methods relies on the main properties of social reality: it is constructed, it refers to a context (meaning), it is tied to social location (or point of view), and it is dynamic [25]. Qualitative in-depth interviewing is probably the best way to fix non-evident and explicitly denied phenomena, such as gendered networking and gender inequality. We have focused on organizational network practices and their effects from women's perspective. As a result, we could formulate some suggestions for organizational management on how to change the structure of organizational and individual employee network as well as communication practices to promote gender equality.

Data collection in our study was carried out using a semi-structured interview. The content of the interview was approved by the ethics committee. Respondents signed informed consent and were convinced that there would be no personifications in texts describing the results.

There were no direct questions about gendered networking barriers for informants. The first of our respondents in a pilot study hesitated to answer such questions, stigmatizing them as "feministic talks" because they could not even remember such issues in their experience.

We changed the interview script in order to get some in-depth information. The opening question asked about life challenges our informants had encountered as women. This question excluded direct associations with gender inequality at work and left them the possibility to speak about the most relevant or sensitive topics. Then we used a facilitating technique where informants were asked to draw a graph which would visualize their career path from the moment of decision making until now, with all the turning points and comments. This technique, on one hand, had facilitated access to memories, and on the other hand increased trust toward the interviewer.

Further, the interview asked about work difficulties and advantages related to female gender, reminders of gender from those around respondents at work and at home, and then another projective technique was used asking how things would go if the informant were a man.

In particular, the interviewees were asked: "Has it happened that you felt that your gender was limiting or hindering your success, if so, in what kind of situation?", "Have people around you reminded you about your gender (at work, in the family, seriously, or in a joking manner)?" Open-ended questions were used to stimulate the description of the experience by the respondents in their own words. In addition, the women assessed the frequency of such situations on a 7-point scale, where 7 – "I am constantly reminded that I am a woman" and 1 – "never reminded".

On average, each interview lasted for 30 min and was recorded on a voice recorder. The audio recordings were then transcribed and analyzed. Interviews were collected by 7 specialists with education in psychology. All interviewers were women. We took into account only narratives that included stories about personal experiences, or personally observed episodes, as well as stories about experiences with their friends and families. We thus excluded abstract discussions on the issue.

The methodological basis of the research was the thematic analysis [26]. Themes were drawn from the coded data by a combination of inductive and deductive techniques. We assumed that women's career challenges would be defined by the structural defects in their professional networks. By defects we mean the absence of certain role positions in the network which are necessary for career development. We have applied to our data the two models of effective careers that focus on high-quality networking [27, 28]. In addition to pre-defined themes, emergent themes such as "the problem of trust between genders" were identified.

Other categories of analysis (themes) were the following: spontaneous mention of professional problems as difficulties associated with female gender, "discovering" that some career difficulties are related to being a female, types of work difficulties linked to gender, and experience of direct and indirect gender reminders from colleagues and professional communities.

5 Sample

The sample included 51 working women, mean age 33.3 ± 9.1 years old. Inclusion criteria were: age 20–60 years old, working experience not less than one year, each informant should be working a year prior to the interview (which was taken in summer 2019). All respondents were well-educated residents of the city (St. Petersburg, Russia). Sample selection procedure was snowball sampling. We aimed to make the most heterogeneous sample to get a panoramic vision of the difficulties experienced by working women in modern Russia. Participants came from a wide range of professions including those male-dominated and female-dominated (construction, finance, services, science, personnel management, law, medicine, etc.), they had different amounts of work experience, and had a variety of professional statuses (ranging from ordinary employees to directors). Each respondent signed the form of informed consent.

6 Results

6.1 Gender Salience at Work

Among the various life challenges that respondents faced as women, the majority of them (68.6%) spontaneously mentioned difficulties associated with their professional careers. These difficulties concerned mainly professional reputation and respect from colleagues, preferences for men in promotion, salary inequality, overload with work and household duties, etc.

Obstacles to career growth of this nature make women's gender "visible" to themselves. Most often, the meaning of gender in a career was experienced as an unpleasant discovery.

> I always thought that everything related to having children and marriage was just stereotypes that had sunk into the past, and we were all modern people, but it turned out that these stereotypes continue to be used for one's personal purposes! (No. 17)

This awareness mostly causes a strong "indignation": "And what is it all [gender] about, when we talk about professionalism, and the results that a person can offer to the company?" (No. 13).

This disenchantment was accompanied with feelings of powerlessness and shame: "The feeling that I was some kind of second-rate person" (No. 22); "I felt altogether wrong" (No. 23); "You can't change anything" [your gender and rules of the game] and "you get tired because you have to constantly keep defending yourself" (No. 41); "[…] the stressful state of my worthlessness" (No. 07).

So, women step on a professional path, metaphorically speaking, being "gender-blind", and certain events confront them with the need to consider the "gender factor" when building their careers. However, some women could see some advantages of being a female for work, but those advantages were very situational and strategically weak:

a. Venting feelings of guilt and shame (for mistakes or ignorance);
b. Preservation of self-image;
c. Raising the mood or self-esteem of partners: "they immediately feel like men" [caring, strong] (No. 31);
d. Reducing tension in conflict situations ("men respond positively to jokes about women" (No. 44));
e. Achievement of specific career goals: "I really love being a woman" (No. 3), "gender gives privileges" (No. 26).

The overwhelming majority of women (96%) responded positively to the question about gender reminders from other members of their networks. The average frequency of such situations, according to the subjective assessment of the respondents, was 3.39 out of 7 points. Most often these were episodes in the workplace (56.86%). In second place were comments from their husbands or partners (35.29%), then in the parental family (31.37%). A quarter of such reminders were made in the form of a joke (25.5%), but most often they were perceived by women as acts of micro-aggression.

Negative or humorous gendered interactions (used mostly by male colleagues) served for psychological distance regulation, instrumental aims, and setting gender subordination and power. Here are some examples: "In order to gratify their ego in the company, boys use you like that" (No. 29); "When he cannot adequately respond to you, and it's easier for him to say you're a woman, that's all" (No. 51).

So, as we can see, when feminine gender becomes salient, traditional gender hierarchy often replaces equal partnership at work and serves as a foundation for different networking barriers.

6.2 Networking Barriers

Emergent Themes. Though there was no direct question in the interview about networking barriers, as for example, in this very useful research [19], it seemed very significant for us that literally every woman, when talking about career obstacles, mentioned the effect of interpersonal relationships. Most of them stressed the difficulties of working relationships between men and women. So, we can conclude that these kinds of barriers are easily identified and well-pronounced by women.

Three major themes of gendered interaction emerged in the process of data analysis: the problem of trust between genders in the workplace, the problem of friendships between men and women, and the problem of getting out of professional networks due to the transition to motherhood.

The first problem is the problem of *trust between genders* in a professional context, while trust is the basic premise for interaction of any type.

Women are not trusted as professionals, and the most common form of gender reminder was: "You are a woman so you can't understand/are not able to […]" (58.82% of all cases).

This phrase questions the competence and professionalism of women (in science, business, politics, logic, technology, driving a car, and other areas that are traditionally considered being "masculine"). Women noted that their colleagues make it possible to support their position on working issues with gender arguments: "[…] that I, as a woman, do not understand something, and since he is a man, he knows better" (No. 25).

The respondents hear remarks from colleagues that "women think poorly" or "what can you expect from women, they can't do anything normally"; they feel that "expectations are lowered," and "in general, such an attitude is dismissive."

There is a more condescending attitude toward female surgeons, entrepreneurs, and lawyers. For example, according to one of the respondents, the phrase: "'Have you read the civil code?' is often slipped into court orders for female lawyers" (No. 5). The other informant noted: "It was very unpleasant when I graduated from university and wanted to apply for an internship at the Ministry […], they told me "We don't take girls" (No. 35).

The second theme is the problem of *friend-like contact* between men and women. Men are perceived as unable to be just friends as colleagues, just like the kind of friends that they can be with other men. This becomes especially evident in male-dominated groups: "Interaction with male colleagues [...] – it is always a game of the sexes [...] men clearly see that this is a woman" (No. 3).

On the one hand, women can use such interest for instrumental purposes; on the other hand, this fact arouses rejection: "Some incorrect praise, sexist remarks, yes – it was not an obstacle in terms of my career growth, but it was morally uncomfortable" (No. 46). Another respondent also shared: "I still do not understand why, well, we don't live in the Middle Ages, that it was impossible for him to say that I did an excellent job [instead of complimenting my appearance]" (No. 24).

It is not rare when a female worker becomes an object of sexual interest from male colleagues: "[when men] actively show their interest [...] you feel unsafe at this moment" (No. 27); "I was probably so shocked by what was happening that I could not even influence [the situation]" (No. 11); "it's just very unpleasant", "disgusting", "could you just treat me like a person. Well, no, you can treat me like a woman, but why like some kind of prostitute?" (No. 15).

The impossibility to be friends with men impedes entering male-only clubs, which are usually powerful or elite networks:

> And so it turned out that you understand that they are friends with each other, they have some kind of friendship with each other, but you want to have friends here, and it won't ever happen because why should they make friends with you... and they start to compete instead of accepting you as a team member. (No. 35)

Our informant also shares her vision on the situation in science: "unfortunately, in science, in political science, there is a certain hierarchy of objects of study, and, let's say, here is a female object of study, that is, the position of women in society" (No. 35).

The third theme is the *rupture of all professional network ties* during maternity leave, the obligation to choose between family and work. In Russia, paid maternity leave lasts for 18 months, and it is possible for this to be extended for up to another 18 months, guaranteeing the working position after this period ends. Only 2% of men in Russia really take "maternity leave" so it is still an exceptional case [29]. It is very hard for women to reintegrate into professional communities after such a long period, so many of them prefer to change their professional area or to stay as a housewife. According to one of the interviewees, "I see a lot of mothers around me who are beautiful and, most importantly, smart – whose career has remained within the walls of their house" (No. 1).

Women are forced to rebuild their business ties and to re-establish their informal network status.

Children require a lot of time, as well as cognitive and emotional resources: "The family factor, of course, hinders my career growth. Somehow you need to think about so many things. It would be much easier if you were completely devoted to only your business" (No. 49). Another opinion is quite similar: "I would go there, grow, and so on and so forth, and... I would have to sacrifice my family" (No. 47).

So, situations that make networking difficult for women included explicit or implicit reminders from colleagues that feminine gender is an obstacle for productive work and implies certain restrictions and requirements, sexual interest from male colleagues, and motherhood (expected or actual).

Models of Career-Related Networking. After the analysis of emerging themes, we tried to match the data with some existing models of a successful career that view success as an effect of networking.

The first taxonomy of networking barriers is based upon the model of career resources developed by one of the authors [27]. This model includes four types of social-psychological resources related to four needs in career development: socio-emotional support, explicit and implicit knowledge, protection, and role-modeling. Problems with all of them were found in the experiences of our informants, especially in companies with a larger number of men.

Socio-emotional support is often mentioned in career studies [30]. The main function of it is to assist in overcoming the career crisis and professional challenges. It is a kind of "secure base", which is necessary for advancement. A secure base is provided by people who support one's goal setting and express affection and interest.

> I don't think that it would have happened if there had been women around me, if another woman had been present in the room, perhaps I would have had some courage to voice my opinions, but they were all men, and the jokes they made – I'm embarrassed even thinking about it. [...] Realizing that you have no one to look back at, no one to rely on, and if there is anyone, this person is also in a very fragile state, such a situation. This has a very serious impact on how you feel. (No. 27)

It is not a rare situation when a husband does not offer support and even demands that his wife should leave her job or "such jobs": "her husband gave her an ultimatum, he told her that you, as my wife, should be near me, and all these business trips... it's no good for a woman and affects her family negatively. Well, she was forced to leave this job and stay at home" (No. 12).

Knowledge is provided by mentors – people who are eager to share professional experience and pass it on to their colleagues. They help to acquire the necessary competence and transfer solutions and advice. There is knowledge ("implicit knowledge") that can only be learned from face-to-face contact with another person.

From the beginning of their career path, when studying and during the selection of an academic supervisor, women are faced with the fact that mentors prefer boys: "[...] when we were assigned to supervisors... to write a thesis. Of course, more preference – of course – was given to the male gender" (No. 45).

Protection is provided by a "gatekeeper" who "protects" the entrance to the elite professional community, who can promote, suggest participating in the project, give financial and other support, etc. And considering that higher-status professionals are mostly men, it is difficult for females to build partnerships with them.

> And again, he is more often called to meetings. [...] all managers are men, and it is easier for them to communicate with him on the same wavelength. [...] The fact is that I most likely would have to be imposing in order to break into these meetings, and he is just called there. Well, it's easier for him to establish a relationship. (No. 33)

And fourth, career development is not possible without a professional who acts as a role model of possible professional development. Such people are the embodiment of the career perspectives; they push, help set new professional goals, and affect the level of claims. Their presence in one's professional network is motivating to master through the building of relationships with them so that it seems possible to approach the Ideal Self.

> There are few women who have really achieved some kind of success that could help another woman. And in general, as a rule, it seems to me that they don't really communicate with each other. [...] I had some kind of personal experience and I can't say that they are ready to help one another in any way. (No. 43)

So, some women report about the absence of a role model (successful woman) in their friendship networks, and this may impede their career growth.

Another model of career success elaborated by Randall Collins refers to the networks of eminent philosophers:

> The most notable philosophers are not organizational isolates [...]. The most notable philosophers are likely to be students of other highly notable philosophers. In addition to this vertical organization of social networks across generations, creative intellectuals tend to belong to groups of intellectual peers, both circles of allies and sometimes also of rivals and debaters. [28: 65]

So, in the network of a successful scientist (and this may be a universal model), the following roles should be present: Mentor (or teacher), Followers (or students), Secure base (support group and allies), and Oppositionists (rivals). If we integrate both models, we can add two positions of necessary roles to the list: Followers and Opposition. The analysis of interviews revealed the deficit of these roles in women's networks.

There are no people in women's networks to whom they can delegate work or to leave the post or the business – there is a shortage of followers, not only teachers and colleagues; this indicates the gaps in the network of contacts.

> [...] if you want your brainchild to exist, develop – let it go, even under another director. Perhaps right here and right now there simply isn't the person who I'm ready to pass it to completely, so that everything will be safe in the future. And this is a female thing, yes. (No. 4)

All these excerpts from respondents' answers demonstrate that the importance of career networking is well understood by women, and they really suffer from the absence of different important positions in their networks. And the reality of the gender factor is the most likely explanation of these disadvantages for them.

The problem of professional trust and friendship between genders that are reflected in women's subjective experience may be also related to the homophily of male networks. Homophily refers as well to the network closeness and gender segregation [15, 16]. That is why homophily may be a serious obstacle for women's career, especially in male-dominated environment, where men occupy major network nodes and perform most important roles.

7 Discussion

7.1 Possible Reasons for Gender Inequality Blindness

Women in our study were not willing to support "feministic talks" and could hardly remember any difficult gender-related work relationships when asked directly about them. The Russian women interviewed, as a rule, denied the very existence of the problem of gender inequality. Just recalling the details of their professional path, they could recall uncomfortable situations and evaluate them as discriminatory. It is only by delving into the interview process that women acknowledge that they have to confront patriarchy.

We hypothesized that this blindness was inherited by women after the Soviet ideology, where the myth of gender equality was widely supported [31]. Moreover, that could be a kind of cognitive bias: information that women's social status is getting better inspires women with an overly optimistic illusion of getting rid of gender prejudice and inequality [32]. Besides, we should take into account that in the modern society manifestations of sexism take subtle, ambivalent and "benevolent" forms [33].

Perhaps the first step toward eliminating the gender hierarchy in the organization should be an open recognition of the sustainable reproduction of gender discrimination and the specific practices of intra-organizational interaction that contribute to this. At the present time, we are observing the opposite. Environmental factors of reproduction of gender inequality (an organizational culture that supports traditional gender norms which penetrate the interaction of the organization's employees) force even egalitarian women to play by the rules of the gender game in order to gain career advantages.

7.2 The Effects of Gender Salience

As we have seen, the most direct reminders to a woman of her gender are, unfortunately, a common practice in organizations. It can take many forms, but there is only one result – reproduction of the gender hierarchy, and maintaining gender discrimination. Thus, an approach appealing to a "discriminatory organization" or "gendered organization" finds its empirical evidence. Employees' gender reminders may be considered one of those practices that create a context irrelevant to the working atmosphere.

The most destructive effect of gender salience is the occurrence of not only external, but also internal underestimation of the results of a woman's work compared to men. Self-stigmatization due to belonging to the "non-dominant" gender leads to serious motivational costs, known as fear of success, attribution of failure as a lack of abilities, and a decrease in self-efficacy [34]. Well-known stereotype threat effect – poorer performance out of fear of fulfilling a negative stereotype – was demonstrated for cognitive tasks [35] and lately – for complex motor tasks in women – football players sample [36].

Therefore, at the structural and normative level, the new intervention programs should target discriminatory practices and policies in organizations. And on an individual level, the woman's skills to identify these discriminatory practices and constructively overcome them.

7.3 The Cost and Return of Maintaining Social Relationships

In this section, we are trying to explain the difficulties that women experience after maternity leave and the main reason for it is not losing qualifications, but losing social ties necessary for career and professional growth.

Maintaining social relationships, according to R. Dunbar, is very costly, so the number of friends (or contacts) is limited by cognitive capacity and the time one can afford to devote to them:

> If someone is contacted less often than the defining rate (once a week for the 5-layer [of closeness], once a month for the 15-layer, once a year for the 150-layer) for more than a few months, emotional closeness to that individual will inexorably decline to a level appropriate for the new contact rate. [37: 37-38]

So, the more time one invests in certain relationships, the less time is left to keep in contact with other ties, especially distant ones. The advantage of so called "weak ties" is much underestimated, because it conceals many opportunities for professional growth. It is paradoxical, but people from all around the globe get their jobs mostly from acquaintances than friends [38].

That is why the difference in potential resources between a person with many acquaintances and a person with fewer acquaintances may be dramatic, and research shows that men and women have a huge difference in the sizes of their organizations of memberships [39].

7.4 Implications for Organizational Practice

Qualitative interviewing allowed us to get access to women's perceptions of their social (working) context that are determined by their social position. Thus, we could identify gender salience in organization and some features of inter-gender relationships in the working context (low trust, barriers to friendships and to entering men-only clubs, different resources to maintain professional ties, discriminative communication practices). We assumed that specific practices of networking produce certain network outcomes (unequal career results for men and women).

As we have seen, there are a number of situations that make career networking constrained for women. Organizations may offer training to increase women's awareness of the gendered networking differences, their own networking behaviors, and subsequent benefits for their professional development and risks [40]. We would make this training mixed by gender, as it is also important to raise awareness of all employees about communicative practices used to make gender hierarchy salient and thus reproduce unequal results.

The social environment should provide all the possibilities to communicate and build relationships in accordance with the allocated functions necessary for harmonious professional development. The following list of career-related network roles was discussed: Gatekeeper, Mentor, Role Model, Secure Base, and Followers. This proposed classification of network roles may serve as a base for assessment of the quality of the professional and social environment of women and the formation of a stimulating professional network.

Corporate culture should be revised to eliminate the practices that impede inter-gender trust and therefore reproduce homophily of high-status networks. For example, it should support a woman's right to speak out without interruptions, and not to hear arguments about gender in work discussions.

8 Conclusion

Women assess gender as a visible and at the same time invisible variable affecting their career. When feminine gender becomes salient, traditional gender hierarchy replaces equal partnership at work and serves as a foundation for different networking barriers.

The first step toward eliminating the gender hierarchy in the organization should be an open recognition of the persisting reproduction of gender discrimination, and the specific practices of intra-organizational interaction that contribute to this. The most destructive effect of gender salience is self-stigmatization, which leads to poorer performance – both personal and organizational.

Gender salience increases the risk of problems in organizational networking: low professional trust between genders at the workplace, the problem of partnerships between men and women, and the problem of women getting out of professional networks due to the transition to motherhood.

The knowledge about the deficits and advantages of women's career networking may serve as a base for managerial decisions in organizations which encourage diversity and creativity. Also, it can be fruitful to complement qualitative data with data from standardized surveys.

Acknowledgments. The authors would appreciate feedback and comments on the paper from its readers. We are very grateful to our colleagues who helped us with data collection and processing: Katrina Aleksandrova, Olga Gundelach, Elizaveta Osmak, and Uliyana Udavikhina. The authors are also very thankful to all participants in this project for their time and interest.

Funding. The reported study was funded by the Russian Foundation for Basic Research (RFBR), project number 19-013-00686 A.

References

1. Jokinen, T., Brewster, C., Suutari, V.: Career capital during international work experiences: Contrasting self-initiated expatriate experiences and assigned expatriation. Int. J. Hum. Resource Manag. **19**(6), 979–998 (2008)
2. Bartol, K.M., Zhang, X.: Networks and leadership development: Building linkages for capacity acquisition and capital accrual. Hum. Resource Manag. Rev. **17**(4), 388–401 (2007)
3. Spurk, D., Kauffeld, S., Barthauer, L., Heinemann, N.S.R.: Fostering networking behavior, career planning and optimism, and subjective career success: An intervention study. J. Vocat. Behav. **87**, 134–144 (2015)
4. Dorius, S.F., Firebaugh, G.: Trends in global gender inequality. Soc. Forces **88**(5), 1941–1968 (2010)
5. Acker, J.: Inequality regimes: gender, class, and race in organizations. Gender Soc. **20**(4), 441–464 (2006)

6. Britton, D.M.: The epistemology of the gendered organization. Gender Soc. **14**(3), 418–434 (2000)
7. Lin, N.: A network theory of social capital. Handbook Soc. Capital **50**(1), 69 (2008)
8. Forret, M.L., Dougherty, T.W.: Networking behaviors and career outcomes: differences for men and women? J. Organ. Behav. **25**(3), 419–437 (2004)
9. Gurieva, S.D., Mararitsa, L.V., Belova, O.E. Sotsialno-psikhologicheskii fenomen "stekliannogo potolka" v professionalnoi karere zhenshchin [Socio-psychological phenomenon of the "glass ceiling" in the professional career of women]. In: Gurieva, S. D., Shaboltas A. V. (eds.) Psikhologiia XXI veka [Psychology of the XXI century], vol. 1, pp. 178–185. VVM, Saint Petersburg (2018). (In Russian)
10. Tulin, M., Lancee, B., Volker, B.: Personality and social capital. Soc. Psychol. Q. **81**(4), 295–318 (2018)
11. Coleman, M.: Women at the top: challenges, choices and change. Palgrave Macmillan, London (2011)
12. Kumra, S., Vinnicombe, S.: Impressing for success: a gendered analysis of a key social capital accumulation strategy. Gender Work Organ. **17**(5), 521–546 (2010)
13. Ibarra, H.: Paving an alternative route: gender differences in managerial networks. Soc. Psychol. Q. **60**(1), 91–102 (1997)
14. Benschop, Y.: The micro-politics of gendering in networking. Gender Work Organ. **16**(2), 217–237 (2009)
15. Kilduff, M., Tsai, W.: Social Networks and Organizations. Sage, London (2003)
16. McPherson, M., Smith-Lovin, L., Cook, J.M.: Birds of a feather: homophily in social networks. Ann. Rev. Sociol. **27**, 415–444 (2001)
17. Ibarra, H.: Homophily and differential returns: sex differences in network structure and access in an advertising firm. Admin. Sci. Q. **37**(3), 422–447 (1992)
18. Greguletz, E., Diehl, M.R., Kreutzer, K.: Why women build less effective networks than men: the role of structural exclusion and personal hesitation. Hum. Relat. **72**(7), 1234–1261 (2019)
19. Broadbridge, A., Tonge, J.: Barriers to networking for women in a UK professional service. Gender in Management: An International Journal, **23**(7) (2008)
20. Shipilov, A., Labianca, G., Kalynsh, V., Kalynsh, Y.: Career-related network building behaviors, range social capital, and career outcomes. Academy of Management Proceedings, pp. 1–6
21. Duberley, J., Cohen, L.: Gendering career capital: An investigation of scientific careers. J. Vocat. Behav. **76**(2), 187–197 (2010)
22. Tharenou, P.: Does mentor support increase women's career advancement more than men's? The differential effects of career and psychosocial support. Australian J. Manag. **30**(1), 77–109 (2005)
23. Linehan, M., Scullion, H., Walsh, J.S.: Barriers to women's participation in international management. European Business Review (2001)
24. Belotti, E.: Qualitative networks: Mixed Methods in Sociological Research. Routledge, Oxon, England/New York, NY (2014)
25. Hollstein, B.: Qualitative approaches. The Sage handbook of social network analysis, pp. 404–416 (2011)
26. Braun, V., Clarke, V.: Using thematic analysis in psychology. Qualit. Res. Psychol. **3**(2), 77–101 (2006)
27. Mararitsa, L.V., Potiavina, V.V.: Socialnyi kapital kak faktor razvitiya kar'ery [Social capital as a factor in career development]. In: Social Psychology in the Educational Space. Proceeding of the 1st International scientific and practical conference (October 16–17, 2013), pp. 278–279. Moscow (2013). (In Russian)
28. Collins, R.: The Sociology of Philosophies: A Global Theory of Intellectual Change. Harvard University Press, Cambridge (1998)

29. SuperJob. Research Center. Russian men don't take maternity leave. https://www.superjob.ru/research/articles/111910/rossijskie-muzhchiny-v-dekret-ne-hodyat. Accessed 29 May 2020. (In Russian)
30. Bozionelos, N.: Intra-organizational network resources: how they relate to career success and organizational commitment. Person. Rev. **37**(3), 249–263 (2008)
31. Khasbulatova, O.A.: Tekhnologii sozdaniia mifa o ravnopravii polov: sovetskie praktiki [Technologies for creating the myth of gender equality: Soviet practices]. Woman in Russian Soc. **4**, 49–59 (2018). (In Russian)
32. Spoor, J.R., Schmitt, M.T.: "Things are getting better" isn't always better: Considering women's progress affects perceptions of and reactions to contemporary gender inequality. Basic Appl. Soc. Psychol. **33**(1), 24–36 (2011)
33. Glick, P., Fiske, S.T.: Ambivalent sexism revisited. Psychol. Women Q. **35**(3), 530–535 (2011)
34. Mednick, M., Thomas, V.: Women and achievement. Psychology of women: A handbook of issues and theories, pp. 625–651 (2008)
35. Steele, C.M., Aronson, J.: Stereotype threat and the intellectual test performance of African Americans. J. Pers. Soc. Psychol. **69**, 797–811 (1995)
36. Grabow, H., Kühl, M.: You don't bend it like Beckham if you're female and reminded of it: Stereotype threat among female football players. Front. Psychol. **10**, 1963 (2019)
37. Dunbar, R.I.: The anatomy of friendship. Trends Cogn. Sci. **22**(1), 32–51 (2018)
38. Gee, L.K., Jones, J.J., Fariss, C.J., Burke, M., Fowler, J.H.: The paradox of weak ties in 55 Countries. J. Econ. Behav. Organ. **133**, 362–372 (2017)
39. McPherson, J.M., Smith-Lovin, L.: Women and weak ties: differences by sex in the size of voluntary organizations. Am. J. Sociol. **87**(4), 883–904 (1982)
40. Wang, J.: Networking in the workplace: implications for women's career development. New Directions Adult Continuing Educ. **122**, 33–42 (2009)

Frames and Networks: Framework Narratives in 19th-century European Literature

Larisa Poluboyarinova(✉) and Olga Kulishkina

St. Petersburg State University, St. Petersburg, Russia
{l.poluboyarinova,o.kulishkina}@spbu.ru

Abstract. In this article we examine the phenomenon of the frame narrative in 19th-century continental Europe based on five national literary histories (of France, Russia, Germany, Austria and Switzerland) using social network analysis (SNA) as well as elements of the literary field and the cultural transfer analysis. Having selected three authors out of each national literary history as representatives of the respective generation of frame narrators (corresponding to the first, second and last third of the century), the relationships between fifteen chosen writers were examined on the basis of their letter collections, diaries and biographies. The analysis identified strong and weak relationships (ties) between authors (nodes), with density determined for each node of the network. It was concluded that the second generation of frame narrators, whose network activity was stimulated and maintained by the strong tie between I. Turgenev and P. Heyse, had significantly contributed to spreading of the frame technique and retaining the prestige thereof. The third generation of European frame narrators had much smaller and weaker networks compared to the second one, yet it drew on the work of their predecessors to produce framework narratives of high aesthetic value.

Keywords: European literature of the 19th century · Frame narrative · Social network analysis · Literary field · Cultural transfer · Literary networks · Ivan Turgenev · Paul Heyse

1 Introduction

1.1 Network Research in Literary Studies

Almost ten years have passed since Franko Moretti's groundbreaking network-oriented study of the constellation of characters in Shakespeare's "Hamlet", "Macbeth" and "King Lear", Charles Dickens' "Our Mutual Friend" and Cao Xueqin's "Dream of the Red Chamber" appeared [1]. Even earlier, at the beginning of the 21st century, the practical application of social network analysis began in the research of extended narratives in Marvel comics [2]. These approaches to literature dating back to Boris Yarkho [3] and Markus Solomon [4], who preferred precise methods in literary studies, are now being studied in the context of *Digital Humanities* dealing with *Big Data.* See for example

A. Antonyuk and N. Basov (Eds.): NetGloW 2020, LNNS 181, pp. 49–65, 2021.
https://doi.org/10.1007/978-3-030-64877-0_4

large international projects [5, 6] and comprehensive studies that represent actors of epic and dramatic works as well as of entire narrative cycles as networks [7–9].

This undoubtedly productive and, despite leaning towards graphs and schemes, qualitatively oriented approach develops the poetics of the works to be researched. It was apostrophized by Peer Trilcke as "literary-scientific network analysis" (*"literaturwissenschaftliche Netzwerk-Analyse" – liNA* [10]). In recent years a different point of view has been formed on literary networks that focuses less on the personality constellation system or on the interactions within the literary work (that are treated, studied and graphically represented in this case as social networks), rather than on the respective epoch-specific literary enterprise or (in the words of Pierre Bourdieu) on the *literary field* (*le champ littéraire*). The practical approbation of such an understanding of literary networks, namely as "connections between the emergence, publication and reception of literature" [11: 20] was carried out, for instance, within some projects related to the European literary system of the 18th century [11–13].

1.2 Problem Statement

Following the example of these research groups, it seems interesting to us to present and investigate the network structure that ensured and maintained the persistence of the frame narration in 19th-century Europe. Unlike the drama [5], there are no Europe-wide text corpora of this kind yet, therefore, this study is limited to a few examples from national literature of five selected countries, codified and canonically established in literary history. The authors who came to the fore and formed the actual network were in personal contact with each other (sometimes maintaining a lively exchange with each other) or were 'theoretically' aware of the literary procedure (frame narrative) of the others through (in most cases provable and verifiable) reading. Thus, it leads us to hypothesize the existence of a cross-border professional network of European frame narrators.

2 Theory

2.1 Sociology of Literature Approaches

We are interested in the problem of network-like connectivity of literary figures, the structure that has traditionally been touched, but has not been fully highlighted in the sociology of literature. Wendy Griswold in her article dating back to 1993 has summed up the development of the discipline and has radically formulated the question of the theoretical and methodological paths that could bring to light the structures of literary production, i.e. historically conditioned systems of authorship. She emphatically called for shifting the focus from the process of reception, from the "reader as hero" [14: 457] to the literary production conditions of symbolic values: "< … > there is no reason why authors, with their intentions, experiences, sociological characteristics, and 'horizons' of understanding, cannot be treated in parallel fashion to readers: as agents who interact with texts, working to encode meanings" [14: 465].

2.2 Pierre Bourdieu's *Literary Field*

This approach was realized a few years later, when Pierre Bourdieu's theory of the literary field sharpened the researcher's view on systems of literary production. On the one hand, the investigation of relationships between individual actors in the "field of literary production" came to the fore: "< ... > the analysis of the interactions in this field has to be improved and further evidence should be obtained" [15: 337]. On the other hand, the field dependence of certain literary genres or styles and techniques was pointed out: "Thus, for contemporary and historical publishers, agents of material production, CLs [= *champ litteraire*] may play a role in deciding whether or not to start a new book genre or expand the function of an existing one < ... >" [15: 341]. Furthermore, the question of transnational fields and the parameters of their interrelation was formulated, supported by the understanding that "results obtained in the research on one nation-state cannot easily be transposed to another" [15: 339].

According to Bourdieu, renowned writers who have sometimes acted as editors, translators and mediators and who, thanks to their solid *symbolic capital* (embodied in the *habitus* as self-representation style), have been able to give authority and a broad impact to literary techniques they have used. These writers as "strong" field-actors occupy the central segment of the literary field for themselves. In this way, an optimal position is gained for exercising one's own 'influence', because this motivates the young aspirants who occupy a weak position or no position at all within the field to turn to the literary genres and techniques practiced by the strong actors [16]. When we 'translate' this constellation into SNA terms, we will be able to say that influence in the literary network can be measured by node degree or other node-level measures such as different types of centrality. Accordingly, Bourdieu's theory helps us to understand the internal logic of interaction and influence within the field or network (both concepts, the literary field and the network, are brought together by Bourdieu's student and successor Gisèle Sapiro [17]) by outlining and accentuating the field of power distribution as well as the dynamics of individual positions [16].

2.3 Comparative Literature and Social Network Analysis

The phenomenon under study – the turning of a certain number of European prose writers to one and the same narrative technique over the course of several decades – traditionally belongs to the field of comparative literature. However, the genuinely literary-scientific approach is more likely to be able to cover this phenomenon 'from within', with the help of its traditional epistemes and terms such as genre, narrative technique, reception, literary relationships, influence, etc. At most, only a few authors – from two to five – can be studied in the course of a typological or genetic comparison within the framework of one and the same paradigm (see the analogy between Turgenev's frame technique and that of Th. Storm [18: 123–126] or the comparison of Turgenev's, Storm's and F. von Saar's frame narrative techniques [19: 198]). Such analysis rarely takes place outside of a single national literary historical context and is likely to aim at a hermeneutically oriented poetological analysis and interpretation (see critique [15: 335]).

First of all, the SNA does not look inside literary phenomena, but helps to comprehend their external social life as a structure. "Relations are not the properties of the individual agents, but of systems of agents" [20: 2]. As is well known, the SNA considers itself not so much as a theory, rather as a certain way of seeing and representing these social structures: "Social network analysis is neither a theory nor a methodology. Rather, it is a perspective or a paradigm. It takes as its starting point the premise that social life is created primarily and most importantly by relations and the patterns they form" [21: 22].

In this way, the SNA approach may provide a structured overview of the network of authors developing the same narrative technique across national borders. In addition to the overall comprehensive view that the SNA approach provides (at the level of the nodes), this method also focuses on ties, i.e. concrete transport channels ("pipelines" [21: 12]) and the content to be transported and its possible social dependencies:

> Flows are relations based on exchanges or transfers between nodes. These may include relations in which resources, information or influence flow through networks. Like interactions, flow-based relations often occur within other social relations and researchers frequently assume or study their co-existence. [21: 12]

Consequently, another reason why the SNA in many respects is so important for our problem is connected with the possibility of associating the intensity of the relationships between individual actors with their preference for frame narrative techniques. Since a period of several decades comes into view, the SNA analysis should also reveal how the frame narrative is transformed in its aesthetic parameters.

2.4 Michel Espagne's *Cultural Transfer* Research

When moving from the level of the entire field (network) to its individual nodes and ties, the research method of cultural transfer seems to be useful. This method focuses on the specifics of individual acts of transfer, in our case, on the mediating the technique of frame narration from one actor to another. Indeed, one of the main tasks of transfer research is precisely to "make visible" acts and samples of cultural mediation as well as "separate stages of their circulation" [22: 186], while emphasizing, first of all, the life context of intermediary, i.e. "a person with an intercultural life story", who "stand at the interface of cultures (contexts)" [23: 117]. The latter is particularly important in our case, given the expected mediation of the framework narrative process across the boundaries of the respective literary field thanks to central mediator figures.

3 Methodology

The study to be presented is the first approach to the phenomenon of the European framework narrative in the 19th century in terms of literary network analysis. Thus, it aims to outline the problem area, identifying the most important actors as nodes and denoting lines of power and distribution of ties. Nevertheless, it refrains from the all-encompassing fixation of the overall picture, including the lack of important data from the literary field of England, Spain, Italy, Scandinavia, Poland, etc., which still has to be obtained.

In this respect, in our study, we necessarily follow the 'soft' trend of the SNA, i.e. we focus "on attempting to use the concepts and theorems of network theory as analytical tools for understanding < ... > societies" [24: 3]. Nevertheless, we will operate with the basic concepts of the SNA such as social network, actor, nod and tie.

Accordingly, the nodes are writers from the European literary field who used the technique of frame narration in the 19th century. In our study, relationships between these nodes are understood as ties. We distinguish strong ties and two types of weak ties. Strong ties are created when the relevant actors are in professional contact with each other in joint projects, such as publications, translations and literary criticism, and/or have exchanged views on aesthetic and poetological issues. Weak ties are either one-sided (the first type) or based only on knowledge of a colleague's work without personal exchange with him or her (the second type).

In our research, the strong ties include the presence of a documented communication and exchange channel between two authors. In most cases, this channel was used to communicate the frame narratives as works or the technique of frame narration as such (see examples in Sect. 6.2). Of course, the strong ties also include the mutual reading of the works in question (see further weak ties). In most cases, this exchange was also reflected in the (often mutual) creative fertilization at various levels of reception, as has been proven by relevant comparative case studies. We considered the relevant collections of letters as sources of evidence and identification of a strong tie. In addition, the (mostly classical, belonging to the common property of literary studies) primary texts themselves were also consulted, which show various similarities in terms of narrative technique. Other important sources were text-historical commentaries on possible framework narratives in historical-critical (academic) editions of works, which record the respective 'contacts' or 'influences'.

The weak ties, on the other hand, may be suggested above all when the real exchange either had place only sporadically, mostly through the intermediaries (the first type of weak ties) or failed completely (the second type of weak ties).

The one-sided tie (the first type of weak ties) arose in cases where only one author from the two had an interest in common projects or in a closer exchange with the other. This can be documented in the offer of exchange by letter, which was not accepted or not continued. A one-sided relationship is also involved when an interest in the work of the counterpart is attested in an exchange with another colleague (mediator), who in turn is in personal or written contact with the object of interest (see Sect. 6.3).

One may assume that the author was acquainted with the work of the respective frame narrator through reading (the second type of weak ties). As evidence of this acquaintance, we consider both the traces of reception that can be found in the respective texts themselves, as well as mentions of the work titles in letters and diaries of the respective authors, as well as the references contained in the relevant special literature or work commentaries. In the case of the so-called precedent texts, which were widely known, quickly translated into other languages and received intensive international attention, such as E. Th. A. Hoffmann's novels, one can usually assume that the younger generation of writers are familiar with or read the text. For the verification of such reception phenomena that are valid beyond national borders, the cultural transfer procedure is very helpful. Similarly, one can assume that the later generations of writers are familiar with the classical or precedent texts of their own national literature, as in the case of Balzac's for the French, Pushkin's for the Russians, Grillparzer's for the Austrians, and Gotthelf's for the Swiss.

We assume that the pairs of writers can only be connected by one type of relationship, especially since the 'theoretical' acquaintance with the text of the respective writer colleague through reading is assumed for both the strong and weak ties. The strong ties, however, are based on reciprocity, while the weak ties of both types remain one-sided in most cases. Hereby they remain mostly directed ties: an author was acquainted with another author's works, but not vice versa.

Focusing the research view on the ties, whether they are strong or weak (both types), is in accordance with the traditional ideas of comparative literature and research on the transfer of the international dissemination mechanisms of literary materials, motifs and techniques (the so-called genetic or contact relationships). The current study aims to illustrate and analyze this transfer process with the help of network instruments.

4 Data

In determining which authors should be considered as representative frame narrators, we focused on the corresponding national literary histories (for example, from the Metzler series) of Germany, Switzerland, France, and Russia [25–28], as well as on the table presented in the overview study by Andreas Jäggi [29], which lists the key authors and works of frame narrative. Accordingly, for each of the five national literary fields we selected a representative figure from the first, second and last third of the century which correspond approximately to the first, second and third generation of frame storytellers in the respective national literary field. This produced the following list: H. de Balzac, P. Mérimée, G. de Maupassant for France; E. Th. A. Hoffmann, P. Heyse, Th. Storm for Germany; A. Pushkin, I. Turgenev, N. Leskov for Russia; F. Grillparzer, L. von Sacher-Masoch, M. von Ebner-Eschenbach for Austria and J. Gotthelf, G. Keller, C. F. Meyer for Switzerland.

We assume that one of the authors has produced at least one frame story (in fact, there were several). Since individual works do not function as nodes in our case (although such a view is possible), we will further turn to some works as examples or proofs without listing them herein. As mentioned in 4, the nature of the relations between the authors (whether strong or weak ties) was verified and proved mainly thanks to the letter [30–35] and diary corpora [37], as well as to data in biographies of the authors [38, 39]. To demonstrate the spread of the narrative technique, we sometimes turn to other authors outside the list of selected authors.

5 Frame Narrative of the 19th Century: Overview

5.1 Frame Narrative as a Literary Tradition

The object of research in our study is European framework narratives of the 19th century, especially as they appear in the literary histories of Germany, Russia, France, Austria and Switzerland. A frame story is defined in the Metzler Literary Encyclopaedia as "a narrative form which presents a fictional narrative situation in an enclosing epic unit (frame), which becomes the occasion for one or more internal narratives embedded in the frame" [40: 373].

The historical development of the frame narrative is traced back to the large-scale frame cycles such as "1001 Nights" (around 250), "Book of the Seven Wise Masters" (around 9th century), Geoffrey Chaucer's "The Canterbury Tales" (the end of the 14th century) and Boccaccio's "Decamerone" (1352–1354) and resulted in single framed narratives, dominated in the 19th century. This tendency, already noted in a remarkable study by Hans Bracher in 1909 [41], has also been confirmed by recent research.

5.2 Frame Narrative of the 19th Century

The end of the 18th - beginning or the first third of the 19th century are characterized by a number of still impressive framed cycles such as J. W. Goethe's "Unterhaltungen deutscher Ausgewanderter" (1795), Ch. M. Wieland's "Das Hexameron von Rosenhain" (1803–1804), L. Tieck's "Phantasus" (1812–1817), E. Th. A. Hoffmann's "Serapions-brüder" (1819–1821) in Germany, the large-scale project of the "Comédie Humaine" (1829–1850) by O. de Balzac in France, A. Pushkin's "Povesti Belkina" (1830), N. Gogol's "Vechera na khutore bliz Dikanki" (1831–1832) and others.

In the second half of the 19th century there was a growing interest throughout Europe in the single framed narrative, as its outstanding aesthetic potential was increasingly recognized by the end of the century. A single, manageable aesthetic structure could provide much more scope for a modern narrative design as well as for experiments in the configuration of the frame and its references to the internal plot:

> The expansion of narrator's role appearing in the frame narrative through reflective insertions, readings and manuscript fictions up to the thematization of the narrative situation itself, characterizes the development of the frame narrative as a highly artificial, experimental art form. [40: 373–374]

The tendency to convert a single frame narrative into an experimental site for modern narrative ideas went hand in hand with the general orientation of the culture of the second half of the 19th century which was heading towards aesthetic modernity, with an increasingly differentiated way of representing the states of individuals' inner world. In essence this tendency was nourished and steered by the pioneering model works of this type, such as Ivan Turgenev's, Theodor Storm's, Guy de Maupassant's, C. F. Meyer's frame stories and by personal and postal exchanges of these authors, many of whom were intertwined. However, of great importance was also an institutional support by the dissemination among the ever-growing bourgeois reading public of literary journals suitable for family reading, such as "Revue des deux Mondes", "Gartenlaube", "Deutsche Rundschau", "Grenzboten", "Biblioteka dlya chteniya", "Otechestvennye zapiski" and others. The voluminous project "Novellenschatz" by Paul Heyse and Hermann Kurz [42], realized between 1871 and 1884, that will be discussed below, subsequently also expressly favored the form of the frame story throughout Europe.

6 The Network of Frame Narrative Transmission in the 19th Century

6.1 Research Situation

Despite the above-mentioned importance of the framework narrative in terms of literary history and literary typology across Europe, this narrative genre has so far only been taken seriously in German literary studies. The corresponding studies, which began relatively early [41, 43], including the most recent ones, unfortunately remain focused on the German-speaking area [29, 44–47] without integrating and paying due attention to the international context. The fact that the framing technique was a Europe-wide phenomenon in the 19th century, that was sometimes also used in the literature of the USA (see E. A. Poe's "The Oval Portrait" (1842)) is neither reflected nor perceived as an impulse for research. In this sense, network analysis can be useful in identifying relevant ties that tended to extend beyond borders in the 19th century.

6.2 National and International Transmission-Lines

In order to identify these internationally important relevant nodes, links and clusters of the European network of framework narratives, we will deal with national lines of development of this genre on the basis of five national literary histories chosen for being studied.

Even a glance at the national literary field has revealed a number of regularities that are inherent in every national literature to be analyzed and can therefore be described as a transnational, Europe-wide tendency. One of these peculiarities is the presence in every selected national literature of authentic 'classical' examples of frame narratives, which, as a narrative cycle or as a single frame narrative, dated back to the first third (or, less common, to the first half) of the 19th century, that are also regarded by their successors in the second and last third of the century as a beacon.

In addition to these, in the second and last third of the 19th century 'successors' were influenced by more and more active foreign impulses due to lively and ever-more intensive publishing, translation and print media business, as well as to intensified correspondence between authors. For example, when Paul Heyse invited one or the other author of frame stories into his collection of several volumes of novellas (*Novellenschatz*), he also discussed the poetics and technique of these works with this fellow writer by letter [31, 32, 34]. This was also the case with Mérimée, who translated Turgenev's frame stories into French and discussed the details of these works with him [35]. Or in the case of the letter partnerships Turgenev and Storm, Keller and Storm, Storm and Meyer [33–35], who in their letters discussed their own frame-novels with each other. The reading of the published frame novella of the guild colleagues could either precede or follow these brief meeting acts. The reception of foreign and the production of one's own frame narratives was not necessarily a linearly, mechanical process that can be reconstructed. Nevertheless, in many representative cases, clear lines of influence can be drawn. One of the significant examples is Turgenev's inspiration by Storm's early frame stories for his own frame narratives, which in turn influenced Storm's late frame novellas [18, 19].

Finally, in the last third of the 19th century, in all five cases there is a deliberate and reflected complication and refinement of the frame narrative technique, so that "the development of the frame narrative as highly artificial, experimental art form, which can lead to associatively conditioned chain forms or in perspective to change operating forms" [40: 374] can be registered among representatives of the frame narrative in each selected national literature. Further, we will use concrete examples to introduce the 'tripartite division' of the frame narrative tradition for each national literature. Our task is to identify the type of relations between authors within the national literary tradition and on the international level. In the visualization of the network below, authors are represented as nodes. Node shape indicates generation and node color indicates country. Dashed lines are passive weak ties, thin solid lines are active weak ties, and thick solid lines are strong ties (Fig. 1).

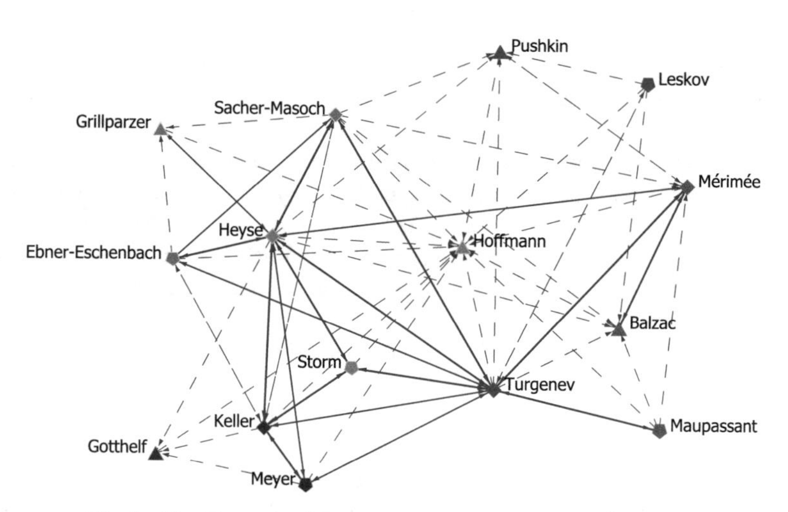

Fig. 1. The Network of European frame narrators in 19th century

Germany. The earlier narrative cycles of the German literary tradition (by J. W. Goethe, L. Tieck, E. Th. A. Hoffmann, Ch. M. Wieland) listed in paragraph 5.2 were within the first generation and then promptly replaced by single framework narratives such as W. Hauff's frame stories, e.g. "Das Wirtshaus im Spessart" (1829), "Die Karawane" (1825) and "Der Scheich von Alessandria" (1826), which "contain the first attempt to make the fiction of the frame aesthetically effective" [41: 11].

The orientation towards the type of a single frame narrative is also characteristic of the second generation leading representative, Munich 'literary pope' Paul Heyse (1830–1914), who used this method in the course of his long literary career in his novellas committed to Romanesque patterns and materials from the 1850s to the 1880s, such as "Die Einsamen" (1857), "Helene Morten" (1858), "Der Kreisrichter" (1858), "Die Witwe von Pisa" (1865). Rather epigonally inclined, Heyse hardly manages "to avoid certain trivialities and bad taste" [41: 11], which brings with it such an exquisite and subtle form as a frame novella. Thus, his own achievement as an author of frame narratives remains not so much qualitative as quantitative.

Noticeably higher in aesthetic value are frame stories by Theodor Storm (1817–1888). In his early frame novel "Immensee" (1849) the narrative frame seems to be interwoven with the fine lyrical state of the first-person narrator, who recalls his childhood love. Similarly, in his later texts, such as "Späte Rosen" (1860), "In St. Jürgen" (1867), "Psyche" (1875), "Aquis submersus" (1876), "Renate" (1878), "Zur Chronik des Grieshuus" (1884) and many others, an increasing differentiation and refinement of the organization of the narrative framework and its integration into the inner plot can be observed.

In this process, which is completed and crowned by the late masterpiece – the 'double-framed' narrative "Der Schimmelreiter" (1888) – the constant exchange with domestic and foreign prominent frame stories writers such as P. Heyse, I. Turgenev and G. Keller has undoubtedly played the role that can hardly be overestimated [48].

If the highest aesthetic relevance in the organization of the narrative frame goes to Storm, then Heyse clearly exceeds it in the intensity of national and international relations, which served to a significant extent to propagate and disseminate the frame narrative type. A studied Italian-Romanist, Heyse was a profound expert and outspoken advocate of Boccaccio's frame storytelling technique [42]. He propagated this technique not only through his own work, but also in the foreword to the internationally renowned collection "Novellenschatz", of which he was the publisher. In this way, the centrally positioned and densely networked author increasingly helped to gain authority and the broad impact of the frame narrative technique.

As the visualization shows, P. Heyse has strong ties with five out of fourteen other selected frame narrators and has weak ones – with seven of them.

Conversely, Th. Storm compared with Heyse clearly has insufficient contacts. He is connected with other authors only by three strong ties and has one weak tie.

France. In the French tradition, the masterful organization of the narrative framework in the novella "Gobseck" (1830) by Honoré de Balzac (1799–1850) set the tone and marked the first generation. The differentiation and refinement (sometimes complication) of the narrative framework and the intensification of its relation to the internal plot simultaneously take place in later decades in literary works by Prosper Mérimée (1803–1870), especially in his novellas "La Partie de trictrac" (1830), "Carmen" (1845) and "Lokis" (1869) (the second generation). In the second half of his life, Mérimée was connected with Ivan Turgenev, the Russian master of frame storytelling, through several joint literary projects and personal friendship [35, 39]. Guy de Maupassant (1850–1893) representing the third generation of the French frame narrative tradition, regarded Turgenev as his literary mentor [49]. Through Maupassant, who in the period between 1870 and 1891 published several collections of novellas, and as every second of them was a frame story, this genre really flourished in France. His collections of short stories such as "Mademoiselle Fifi" (1882) and "Claire de Lune" (1883) are worth mentioning as they are experimental in terms of inter-plot connections, as well as in the point of establishing connections between the narrative frame and the internal plot.

However, the distribution of contacts with the other frame storytellers remains similar to that of the German authors. In particular, Mérimée maintains a total of five relationships, two of which are strong. Maupassant, on the other hand, has only four relationships, including one strong tie.

Russia. The tradition of Russian narrative cycles mentioned in paragraph 5.2 that was introduced by Pushkin and Gogol, was adopted by Ivan Turgenev (1818–1883) and successfully continued in his "Zapiski okhotnika" (1852). Turgenev as a frame narrator was related to the idol of his youth Alexander Bestuzhev-Marlinski (1797–1837), whose romantic narrative prose often resorted to the frame form of narration, e.g.: "Vecher na bivuake" (1823), "Vtoroi vecher na bivuake" (1823), "Krov' za krov'" (1825), "Zamok Venden" (1821), "Vecher na Kavkazskikh vodakh v 1824 godu" (1830) and others.

The tendency towards complex frame narratives typical for Turgenev in the second third of the century, for example, in "Dnevnik lishnego cheloveka" (1850), "Pervaya lyubov'" (1860), "Stepnoy korol' Lir" (1870), "Veshnie vody" (1872) etc., was shown in the last third of the 19th century by representatives of the third generation such as Lev Tolstoy (1828–1910) ("Kretserova sonata", 1890) and Anton Chekhov (1860–1904) ("O lyubvi", 1898 and "Skuchnaya istoriya", 1899). This form acquires particularly differentiated and sophisticated expression in literary works by Nikolai Leskov (1831–1895) [50]. "Zapechatlennyy angel" (1872), "Ocharovanny strannik" (1873) "Tupeiny khudozhnik" (1883) and his other works can be taken as an example.

Despite the structural complexity that Leskov's frame stories are imbued with, this author has tenuous connections with other narrators, so only weak ties are observed. In contrast, Turgenev, who in terms of complexity and subtlety of frame narrative technique hardly reached the level of Leskov, was an active networker, with five strong and seven weak connections to be observed.

Austria. In the Austrian tradition of narrative prose, that is known to begin relatively late, actually only in the 19th century, there were some brilliant examples of frame stories dating back to the first half of the century, such as "Das Kloster von Sendomir" (1829) and "Der arme Spielmann" (1847) by Franz Grillparzer. Grillparzer represents the first generation of Austrian frame narrators. The most important Austrian prose writer of the second third of the 19th century, Leopold von Sacher-Masoch (1836–1895) – the second generation – knew his work, but was knowledgeable also of the variations of frame stories practiced abroad especially by Paul Heyse and Ivan Turgenev. In both cases, the research of frame narrative forms has been remarkably fruitful. Sacher-Masoch sought to bring together his frame stories such as "Don Juan von Kolomea" (1868), "Venus im Pelz" (1870), "Die Liebe des Plato" (1870) and others within the framework of voluminous and overarching unity – in the narrative cycle "Das Vermächtnis Kains" (1870–1878).

Really innovative and experimental variations of narrative frame forms are to be found in M. von Ebner-Eschenbach's stories (the third generation), including "Krambambuli" (1883), "Die Resel" (1883), "Er lasst die Hand küssen" (1886), "Oversberg" (1892), "Die Reisegefährten" (1901) and many others, whose originality and modern potential were praised in a special study shortly after the author's death [43].

In this case it is also clear that the representative of the second generation maintains significantly more active relations with each other even beyond the borders of national literature (see Sacher-Masoch with two strong and six weak ties) than the representative of the third generation (Ebner-Eschenbach), for whom, except for her contacts with Heyse, weak connections are more evident (four ties).

Switzerland. As in the other cases, Switzerland also has a native forerunner in the genre of frame stories, in the person of Jeremias Gotthelf (1797–1854) who wrote "Berner Erzählungen" popular in the 1830s – early 40s. He was replaced by Gottfried Keller (1819–1890), as a representative of the second generation of frame storytellers in Switzerland, whose "Die Leute von Seldwyla" is considered an outstanding frame narrative cycle. Also, in Switzerland in the last third of the 19th century there was an author who was able to give his frame stories a special creative touch, particularly through subtle structure and varied design of the narrative frame which remained connected to the internal plot. It was Conrad Ferdinand Meyer (1825–1898), whose frame novellas "Der Heilige" (1880), "Plautus im Nonnenkloster" (1882), "Die Hochzeit des Mönchs" (1884) stand out in this respect [34].

The second and third generations had 'external' contacts and influences in the form of correspondence with Paul Heyse, as Keller and Meyer did. Nevertheless, as Keller corresponded also with Th. Storm, his contacts appeared to be somewhat more active than those of Meyer, who was characterized by a predominance of weak ties (only one strong tie by four weak ties).

6.3 Weak Ties: One-Sided Type

The weak relations of the second type need an extra explication. It is exciting that most of them were relations across national borders. They arose in a few cases and were either business-like or signaled an aspirant interest in the work of a recognized master.

For example, P. Heyse, as a prospective director of "Novellenschatz" project, had a business-related interest in contacts with F. Grillparzer, P. Mérimée and C. F. Meyer, which he initiated directly or through intermediaries such as G. Keller and I. Turgenev. These contacts, even when they were by letter (as with Grillparzer and Meyer [34, 38]), were short-lived, not accompanied by poetological exchanges, and ended when the goal (in this case, the inclusion of a story in the collection) was achieved.

Of a different kind was M. von Ebner-Eschenbach's interest in contacts with established authors of the second generation of writers. As a budding author and writing woman in the male-dominated field, unsettled, she sought support and advice from them. Of the established writers she was interested in, only P. Heyse was responsive to a lasting contact, so that the correspondence between the two became a strong tie [51]. Turgenev reacted to her enthusiastic letter politely but reservedly [52], what did not presuppose a continuation of the contact. Her interest in the works of G. Keller and L. von Sacher-Masoch and in contacts with these authors, as evidenced in her correspondence with Ferdinand von Saar (1833–1906) [36] and in her diaries [37], had no continuation.

6.4 Central Nodes: Heyse and Turgenev

We can measure the importance of the node in our transnational European network of frame stories authors by the originality of their frame design and reflexivity of the aesthetic claim. In this case from them all it is the later authors (the third generation) such as Storm, Maupassant, Leskov, Ebner-Eschenbach and C. F. Meyer who would make themselves noticed. However, these very authors in the context of the entire network had predominantly weak connections with other writers, because as individuals they were less active in communication than Mérimée, Keller and Sacher-Masoch, and, certainly, they were not practitioners of poetological information popularization, unlike Heyse and Turgenev. Our visualization highlights the networking of the letter two writers (five strong relationships each), who became undoubtedly one of the strongest impulses in the European literary field in the second third of the 19th century, particularly for the promotion and dissemination of the framework narrative.

Paul Heyse. P. Heyse, the most renowned figure in 19th-century German-language literature, was the first German author to be awarded the Nobel Prize in 1910. Even if his own frame stories, like all his other literary works, are considered trivial nowadays [53], he remains, in terms of literary history, one of the most powerful figures in the literary field of Europe in the late 19th century, in accordance with his *habitus,* symbolic capital, and network density.

In between 1870 and 1880, Heyse realized together with Hermann Kurz and Ludwig Laistner a project of the "Novellenschatz" ("Treasure of Novellas"). It was a collection of initially only German-language ("Deutscher Novellenschatz", 1871–1876, 24 volumes; continued as "Neuer Deutscher Novellenschatz", 1884–1887, 24 volumes), but then also foreign novellas ("Novellenschatz des Auslandes", 1877–1884, 14 volumes [54]), the most solid part of which (the exact percentage has yet to be determined) consisted of Heyse's favorite framed stories. Thus, among selected writers, the large number of eminent authors with their frame novellas (Turgenev, Keller, Meyer, Storm, Ebner-Eschenbach, Mérimée, Sacher-Masoch, Balzac, Gotthelf, Pushkin) were represented in these collections, that was tantamount to the canonization of both the work and the author.

Ivan Turgenev. Ivan Turgenev's unique function as a mediator has been repeatedly confirmed and documented in the extensive research literature at various levels [18, 19]. In the context of the network of authors of European framework narratives, Turgenev's role along with that of Heyse has proved to be a leading one. His personal contacts, as well as thorough correspondence with Heyse and Storm, strengthened during his several years' stay in Germany, have led to the formation of a significant cluster. Within this cluster, the hub 'Heyse–Turgenev' was particularly significant. The further branching contacts of this hub to and between the nodes of German-language literature (via Heyse to C. F. Meyer, via Storm to Keller, via Turgenev himself to Sacher-Masoch and Ebner-Eschenbach, both of whom were also directly connected to Heyse) revealed relationships with the French literary network of Mérimée and Maupassant.

7 Conclusion

The first network-specific approach to the structure of the European framework narratives has led to three conclusions that are considerably important for further research of this problem.

Firstly, the network provided as an example has been composed of a total of fifteen selected frame narrators (nodes) from continental Europe and the relationships between them (ties). Assessing their density, weakness or strength to get an insight into the structure of the literary network, we made conclusions about the specific developmental dynamics of this narrative genre.

Secondly, the network analysis of this structure uniting three generations helped to realize the importance of the second-generation tie 'Heyse–Turgenev' (the second third of the 19th century) as the main impetus for spreading and popularization of the framework narrative in Europe. They were both not only authors of influential frame stories, but also maintained active publishing and translating activities accompanied by intensive correspondence.

Thirdly, our analysis has shown that the flourishing of sophisticated frame narrative in the third generation of writers (the last third of the 19th century) was accompanied with much less network activity. The lack of interest in horizontal relations among third-generation representatives can be partly explained by the fact that they were able to take advantage of the already established authority of the frame narrative in Europe (embodied particularly in Heyse's major *Novellenschatz* project), and to which their predecessors (the second generation) made significant contributions as more active networkers.

This phenomenon can be interpreted with the help of Bourdieu's field theory, when it is projected onto international – in this case supra-European – literary processes by means of the cultural transfer method. In the French literary field, Bourdieu excludes an opposition between two poles, "the pole of pure production, where the producers tend to have as clients only other producers (who are also rivals) < ... > and the pole of large-scale production, subordinated to the expectations of a wide audience" [16: 121]. The representatives of the third generation of frame storytellers are clearly tending towards the first pole throughout Europe, as they concentrated primarily on the production process. In doing so, they disregarded or ignored offers of cooperation and thus the participation in audience-effective (also financially successful) projects. On the contrary, the increased communication activity of their older guild colleagues, especially Heyse and Turgenev, led to the formation of the (after Bourdieu) "second pole", where the broad effect could be achieved at the expense of aesthetic sophistication.

The perspective of the analysis carried out can be seen in the future, firstly, in its spread to other national literatures. Thus, the inclusion of the literature of England and the USA, especially via Henry James (1843–1916), seems almost appropriate. This involvement can be all the more productive because this Anglo-American author of frame narratives communicated with Turgenev in Paris in 1876–77 and was under Turgenev's influence.

A further application of the method presented in this study would be possible secondly if a different genre type is chosen as subject matter, e.g. a martyr drama in the 17th century, the Bildungsroman at the end of the 18th – in the 19th century or a verse type, e.g. vers libre in the 20th century.

Acknowledgment. This article resulted from the research within the framework of the scientific institute partnership between the Universities of St. Petersburg and Freiburg "Writers' Networks of the 19th Century" was supported by the Alexander von Humboldt Foundation.

References

1. Moretti, F.: Network Theory, Plot Analysis. New Left Rev. **68**, 80–102 (2011)
2. Alberich, R., Miro-Julia, J., Rossello, F.: Marvel universe looks almost like a real social network. (2002). https://arxiv.org/abs/cond-mat/0202174. Accessed 31 May 2020
3. Yarkho, B.: Metodologiya tochnogo literaturovedeniya. Izbrannye trudy po teorii literatury. [Methodology of precise literary studies. Selected works on the theory of literature]. Yasyki slavyanskikh kultur, Moskva (2006)
4. Solomon, M.: Mathematische Poetik [Mathematical Poetics]. Athenäum, Frankfurt/M (1973)
5. Fischer, F., Orlova, T., Skorinkin, D., Palchikov, G., Tyshkevich, N.: Introducing RusDraCor, A TEI-Encoded Russian Drama Corpus For The Digital Literary Studies. In: Proceedings of the International Conference «Corpus Linguistics–2017», pp. 28–31. St. Petersburg (2017)
6. Fischer, F., et al.: Programmable Corpora: Introducing DraCor, an Infrastructure for the Research on European Drama. Presented at the Digital Humanities 2019: "Complexities" (DH2019), Zenodo, Utrecht (2019). https://doi.org/10.5281/zenodo.4284002
7. Algee-Hewitt, M.: Distributed Character: Quantitative Models of the English Stage, 1550–1900. New Literary History **48**(4), 751–782 (2017). https://doi.org/10.1353/nlh.2017.0038
8. Falk, M.: Making connections: network analysis, the bildungsroman and the world of the absentee. J. Lang. Lit. Cult. **63**, 107–122 (2016). https://doi.org/10.1080/20512856.2016.124
9. Miranda, P.J., Murilo, S.B., Ely de Souza, S.: The Odyssey's mythological network. PLOS ONE, **13**(7) (2018). https://doi.org/10.1371/journal.pone.0200703. Accessed 29 May 2020
10. Trilcke, P.: Social Network Analysis (SNA) als Methode einer textempirischen Literaturwissenschaft [Social Network Analysis (SNA) as a method of text empirical literature studies]. In: Ajouri, Ph., Mellmann, K., Rauen, A.Ch. (ed.): Empirie in der Literaturwissenschaft [Empiricism in Literary Studies], Mentis, Münster, pp. 201–247 (2013)
11. Knapp, L.: Literarische Netzwerke im 18. Jahrhundert: theoretisch, empirisch, metaphorisch. Zur Einleitung [Literary networks in the 18th century: theoretical, empirical, metaphorical. To the introduction]. In: Knapp, L. (ed.): Literarische Netzwerke im 18. Jahrhundert [Literary Networks in the 18th Century]. Jahrhundert. Aisthesis Verlag, Bielefeld, pp. 7–34 (2019)
12. Thomalla, E., Spoerhase, C., Martus S.: Werke in Relationen. Netzwerktheoretische Ansätze in der Literaturwissenschaft: Vorwort. [Works in Relation. Network Theoretical Approaches in Literary Studies: Preface] Zeitschrift für Germanistik. **29**, 7–23 (2019). https://doi.org/10.3726/92164_7
13. Edmondson, Ch., Edelstein, D. (eds.): Networks of Enlightenment: Digital Approaches to the Republic of Letters. Oxford University Studies in the Enlightenment, Liverpool University Press (2019)
14. Griswold, W.: Recent Moves in the Sociology of Literature. Ann. Rev. Sociol. **19**, 455–467 (1993)
15. van Rees, K., Dorleijnb, G.J.: The eighteenth-century literary field in Western Europe: The interdependence of material and symbolic production and consumption. Poetics **28**, 331–348 (2001)
16. Bourdieu, P.: The rules of art. Stanford University Press, Stanford California, Genesis and Structure of the Literary Field (1995)
17. Sapiro, G.: Réseaux, instituition(s) et champs [Networks, institution(s) and fields]. In: Marneffe, D. De, Benoît, D. (ed.): Réseaux littéraires [Literary networks], Le Cri/CIEL, Brüssel, pp. 44–59 (2006)
18. Thieme, G.: Ivan Turgenev und die deutsche Literatur. Sein Verhältnis zu Goethe und seine Gemeinsamkeiten mit Berthold Auerbach, Theodor Fontane und Theodor Storm [Ivan Turgenev and German literature. His relationship to Goethe and his similarities with Berthold Auerbach, Theodor Fontane and Theodor Storm]. Peter Lang, Frankfurt/M (2000)

19. Gerigk, H.-J.: Turgenjew. Eine Einführung für den Leser von heute [Turgenev. An introduction for the reader of today]. Winter, Heidelberg (2015)
20. Scott, J.: Social Network Analysis, 2nd edn. SAGE Publications, London, Thousand Oaks, New Delhi (2000)
21. Scott, J., Carrington, P.J. (eds.): The SAGE Handbook of Social Network Analysis. SAGE, London, Thousand Oaks, New Delhi (2011)
22. Espagne, M.: Quelques aspects actuels de la recherche sur les transferts culturels [Some current aspects of research on cultural transfers]. In: Arnoux-Farnoux, L., Hermetet, A.-R. (eds.): Questions de réception [Questions of reception]. Nimes, Champ social Editions, pp. 163–190 (2009)
23. Keller, Th.: Kulturtransferforschung: Grenzgänge zwischen den Kulturen [Cultural transfer research: Crossing borders between cultures]. In: Moebius, S., Quardflieg, D. (eds.): Kultur. Theorien der Gegenwart [Culture. Contemporary theories]. 2nd ed., Wiesbaden, Verlag für Sozialwissenschaften, pp. 106–117 (2006)
24. Rollinger, Ch.: Prolegomena: problems and perspectives of historical network research and ancient history. J. Hist. Network Res. **4**, 1–35 (2020). https://doi.org/10.25517/jhnr.v4i0.72
25. Beutin, W., Beilein, M., et al.: Deutsche Literaturgeschichte: von den Anfängen bis zur Gegenwart [German literary history: from the beginnings to the present]. 9th ed. J.B.Metzler, Berlin (2019)
26. Brinker-von der Heyde, C., Rusterholz, P. (eds.): Schweizer Literaturgeschichte [History of Swiss Literature]. Stuttgart, Weimar, J.B.Metzler (2007)
27. Grimm, J., Hartwig, S. (eds.): Französische Literaturgeschichte [History of French Literature], 6th edn. Weimar, Stuttgart (2014)
28. Städtke, K., Engel, Ch.: Russische Literaturgeschichte [History of Russian Literature], 2nd edn. J.B.Metzler, Weimar, Stuttgart (2011)
29. Jäggi, A.: Die Rahmenerzählung im 19. Jahrhundert. Untersuchungen zur Technik und Funktion einer Sonderform der fingierten Wirklichkeitsaussage [The frame narrative in the 19th century. Investigations into the technique and function of a special form of fictitious reality narrative]. Peter Lang, Bern; Berlin; Frankfurt/M (1994)
30. von Ebner-Eschenbach, M.: Briefwechsel 1851–1908: Kritische und kommentierte Ausgabe [Correspondence 1851–1908: Critical and commented edition]. De Gruyter, Berlin, Boston (2016)
31. Heyse, P., Storm, Th.: Briefwechsel [Correspondence]. J.F. Lehmann, München (1917)
32. Keller, G., Heyse, P.: "Du hast alles, was mir fehlt". Gottfried Keller im Briefwechsel mit Paul Heyse ["You have everything I need". Gottfried Keller in correspondence with Paul Heyse]. Th. Gut & Co., Stäfa (Zürich) (1990)
33. Laage, K.E.: Theodor Storm und Iwan Turgenjew: persönliche und literarische Beziehungen, Einflüsse, Briefe, Bilder [Theodor Storm and Ivan Turgenev: Personal and literary relations, influences, letters, pictures]. Boyens, Heide (1967)
34. Meyer, C.F.: Briefwechsel [Correspondence]. In 4 Vol. Wallstein, Göttingen (2014–2020)
35. Turgenev, I.: Polnoe sobranie sočinenij i pisem [Complete collection of works and letters]. In 30 Vol. Pis´ma [Letters]. In 18 Vol. 2nd ed. Nauka, Moskau (1982–2020)
36. von Saar, F., von Ebner-Eschenbach, M.: Briefwechsel [Correspondence]. Wiener bibliophilen Gesellschaft, Wien (1957)
37. von Ebner-Eschenbach, M.: Tagebücher. 1862–1916 [Diaries. 1862–1916]. In 6 Vol. Niemeyer, Tübingen (1989–1997)
38. Krausnick, M.: Paul Heyse und der Münchner Dichterkreis [Paul Heyse and the Munich Circle of Poets]. Bouvier, Bonn (1974)
39. Zajcev, B.: Žizn' Turgeneva [Turgenev's life], YMCA-Press, Paris (1949)
40. Hübner, K.: Rahmenerzählung [Frame story], pp. 373–374. Metzler Literaturlexikon. Metzler, Stuttgart (1990)

41. Bracher, H.: Rahmenerzählung und Verwandtes bei G. Keller, C.F. Meyer und Th. Storm: ein Beitrag zur Technik der Novelle [Frame story and related material by G. Keller, C. F. Meyer and Th. Storm. A contribution to the technique of the novella]. Haesel Verlag, Leipzig (1909)
42. Heyse, P., Kurz, H.: Einleitung [Introduction]. Deutscher Novellenschatz. Vol. 1. Oldenberg München, V–XXIV (1971)
43. Eiltzer, A.: Rahmenerzählung und Ähnliches bei Marie von Ebner-Eschenbach (Beitrag zur Kennzeichnung ihrer Erzählungskunst und zugleich zur Technik der Novelle) [Frame narration and the like in Marie von Ebner-Eschenbach (contribution to marking her narrative art and at the same time to the technique of the novella)]. Zeitschrift für den deutschen Unterricht. **33**(12), 539–549 (1919)
44. Stratmann, G.: Rahmenerzählungen der Moderne: Situation und Gestaltung einer Erzählform zwischen 1883 und 1928 [Modernist Frame Narrative: Situation and Design of a Narrative Form between 1883 and 1928]. Tectum-Verl, Marburg (2000)
45. Mielke, Ch.: Zyklisch-serielle Narration: erzähltes Erzählen von 1001 Nacht bis zur TV-Serie [Cyclical-serial narration: Narrative from 1001 nights to the TV series]. de Gruyter, Berlin (2006)
46. Beck, A.: Geselliges Erzählen in Rahmenzyklen: Goethe – Tieck – E.T.A.Hoffmann [Social storytelling in frame cycles: Goethe – Tieck – E.T.A.Hoffman]. Winter, Heidelberg (2008)
47. Kleinschmidt, Ch., Japp, U. (eds.): Der Rahmenzyklus in den europäischen Literaturen: Von Boccaccio bis Goethe, von Chaucer bis Gernhardt [The frame cycle in European literature: From Boccaccio to Goethe, from Chaucer to Gernhardt]. Universitätsverlag Winter, Heidelberg (2018)
48. Jackson, D.A.: "Sie können Ihren eigenen Augen doch nicht mißtrauen": Noch einmal zum zweiten Rahmenerzähler in Theodor Storms "Der Schimmelreiter" ["You can't mistrust your own eyes after all": Once again to the second frame narrator in Theodor Storm's "Der Schimmelreiter"]. Schriften der Theodor-Storm-Gesellschaft. **64**, 53–73 (2015)
49. Halpérine-Kaminsky, E. (ed.): Ivan Tourguéneff, d'après sa correspondance avec ses amis français: m-me Viardot, Gustave Flaubert, m-me Commanville, George Sand, Emile Zola, Guy de Maupassant, Taine, Renan, Ch. Edmond, Théophile Gautier, Sainte-Beuve, Ambroise Thomas, Jules Claretie, André Theuriet, etc. [Ivan Tourguéneff, according to his correspondence with his French friends] E. Fasquelle, Paris (1901)
50. Benjamin, W.: The storyteller. In: Benjamin, W. (ed.): Illuminations: Essays and Reflections (Trans. Zorn, H.), pp. 83–107, Harvard University Press and Hancourt, New York (1968)
51. Alkemade M.: Die Lebens- und Weltanschauung der Freifrau M. von Ebner-Eschenbach, mit Anlage von ihrem Briefwechsel mit Paul Heyse [The life and world view of Baroness M. von Ebner-Eschenbach, with annexes from her correspondence with Paul Heyse]. Stiasny, Graz (1935)
52. Geserick, I.: Marie von Ebner-Eschenbach und Ivan Turgenev [Marie von Ebner-Eschenbach and Ivan Turgenev]. Zeitschrift für Slawistik. **3**(1), 43–64 (1958)
53. Hillenbrand, R.: Heyses Novellen: Ein literarischer Führer [Heyse's Novellas: A literary guide]. Lang, Frankfurt/M (1998)
54. Heyse, P., Kurz, H. (eds.): Novellenschatz des Auslandes [Novella treasure of foreign countries]. Vol. 1–14. Oldenberg München (1877–1884)

Stefan George's Poetry Cycle 'The Star of the Covenant' as a Product and Instrument of Networking

Elizaveta Burmistrova-Jennert(✉)

St. Petersburg State University of Culture, St. Petersburg, Russia
burmel11@gmail.com

Abstract. The poetic cycle of the German symbolist poet Stefan George (1868–1933) 'The Star of the Covenant' (*Der Stern des Bundes,* 1914) is his eighth poetic book, which was originally addressed to the poet's closest entourage – the so-called *George-Kreis*, the 'George Circle'. In this article, Stefan George's poetic cycle 'The Star of the Covenant' will be considered, on the one hand, as a product of social (personal) networking, on the other hand, as an instrument of the latter, actively and successfully used by the 'leader' or 'Master' George to strengthen and expand the public relations of the association. Using elements of social network analysis combined with historical-literary method and close reading, we cover such aspects as Strong Ties and the integration of the paradigms represented by key members of the association into the poetic cycle, personalities and myths of the Circle, and their reflection in the thematic-figurative content of the cycle. In particular, we show how symbolic schemes of interaction between participants of the George Circle are 'ciphered' in the poetic book and in the contemporary comments on it written by Ernst Morwitz and Friedrich Gundolf.

Keywords: Networking · Stefan George · Literary group

1 Introduction and Problem Statement

The poetic cycle of the German symbolist poet Stefan George (1868–1933) 'The Star of the Covenant' (*Der Stern des Bundes,* 1914) is a vivid example of how a work may grow out of the interaction of members of a literary association, while also becoming something like a code of laws for this association, and an instrument of influence both on the members of the union and on potential new 'adherents'.

'The Star of the Covenant' is Stefan George's eighth poetic book. Published seven months before the outbreak of World War I, it, as the author writes in his foreword, was originally addressed to the poet's closest entourage [1] – the so-called *George-Kreis*, the 'George Circle', which included intellectual youths, who saw in the poet their leader and enthusiastically accepted his mystical and political projects for the cultural transformation of Germany and the creation of a 'new spiritual elite' from adherents of the George Circle.

A. Antonyuk and N. Basov (Eds.): NetGloW 2020, LNNS 181, pp. 66–77, 2021.
https://doi.org/10.1007/978-3-030-64877-0_5

Speaking about the nature of the organization of the George Circle itself, it should be noted that researchers have not yet fully determined the mechanisms for its constitution. This association, as Mayatsky writes, 'was scattered across a number of cities (…) represented a series of circles or pair relations: teacher/student, guide/the guided, but often without a strict distribution of roles' [2]. Initially inspired by the example of medieval monastic and knightly orders, as well as other types of 'male unions' (*Männerbünde*) as public institutions, the association gradually begins to focus on Plato's Academy as an ideal structural model. Claude David notes the special aesthetic practices of the Circle when, by means of joint reading of poems (primarily George's works) and of conversations about the literary preferences of young men, the latter were 'introduced (…) into the mindset of novices, into their morality and law' [3].

The publication of 'The Star of the Covenant' provoked conflicting responses, from enthusiastic to sharply critical [4]; at the same time, many of them were united by the idea expressed about the programmatic nature of the book for the members of the 'George Circle'. Ulrich Rauscher, for instance, writes in a review that appeared in the Frankfurter Zeitung: 'What Stefan George published under the heading "*Der Stern des Bundes*" is not a collection of verses: this is the newest covenant. This is a dialogue with his friends and followers, the charter of the order, testament' [5].

In this article, Stefan George's poetic cycle 'The Star of the Covenant' will be considered, on the one hand, as a product of social (personal) networking, on the other hand, as an instrument of the latter, actively and successfully used by the 'leader' or 'Master' George to strengthen and expand the public relations of the association, in order to legitimize the aesthetic and social practices of the Circle. This problem statement seems to be especially relevant because there have never been any attempts to analyze both the structure and activity of the *George-Kreis*, and the present poetic cycle in the aspect of networking.

2 Theoretical Background

There are a number of scientific works, predominantly of German origin, that are devoted to both the phenomenon of the George Circle and the literary analysis of the 'The Star of the Covenant'.

The foundation of the sociological understanding of the phenomenon of the George Circle was laid by Max Weber, who stood at the origins of German sociology as a science. He was personally familiar with George, and after having entered the Circle, he generalized the concept of a sect and extended it to associations outside the religious field, including the George Circle, which offer adherents both an almost divine object of worship and a set of rules of conduct. He also argues that the type of a sect to which, in his view, the George Circle belongs is closer to the community (*Gemeinschaft*) than to society (*Gesellschaft*) due to more intimate, 'family' ties [6]. Also relevant to Weber was the concept of 'charisma' in relation to the sociological type of power that existed within the Circle.

Subsequently, Herman Schmalenbach added a third concept to the binary opposition of community and society: the union (*Bund*), which, according to him, is based not on kinship or contractual relationships, but on enthusiasm for the ideal, the readiness for self-sacrifice and self-denial [7].

These concepts have largely determined the nature of discussions around the George Circle in later research literature, most of which belongs to the field of literary studies and which operates with historical-literary methods of analysis (Braungart, Breuer, Groppe, Karlauf, Kolk).

Among the purely sociological works, two works should further be mentioned. George's contemporary, Joachim Wach, relying on Weber's theses, emphasizes the idea of salvation/deliverance as the leading one in the George Circle, as well as the importance of the role of relationships between the 'master' and the 'disciple' [8, 9]. Furthermore, in the 1970s, the Norwegian sociologist Arvid Brodersen again summarized Weber's ideas and defended their importance [10].

In the 1990s, Günter Baumann emphasized the vectors of influence of the George Circle, which, in his opinion, went beyond the literary association and extended into the field of scientific, political and new religious phenomena. For him the term 'Circle' is less adequate than the terms 'poetic school' (with an emphasis on continuity of the latter) or 'Secret Germany' (due to the multidirectional nature of ties within the Circle and due to its political aspirations) [11]. Following Michael Winkler, he postulates the idea that there was not only one George Circle, but many 'Circles', not one closed entity, but, on the contrary, many entities expanding concentrically both in space and time. This aspect is considered in the chapter dedicated to the George Circle in Baumann's monograph 'Poetry as a way of life: Wolfgang Frommel between George Circle and Castrum Peregrini' (1995) [12].

Recently, as already mentioned, the George Circle has not only become an object of purely sociological research. In literary research works, however, the aspects of the sociological approach are very rarely used. At the same time, it is social network analysis that seems to be extremely productive in relation to the material under consideration, since it simultaneously clarifies the specifics of the structure of the association and contributes to a better understanding of Stefan George's poetic cycles. Also, having analyzed the main works of literary criticism in George's time, devoted to the analysis of 'The Star of the Covenant', we found that, for example, Friedrich Gundolf uses concepts comparable to the modern terminology of network science to describe structural phenomena. Therefore, in our research, we apply the terminology and approaches of social network analysis in combination with traditional literary research methods. Developing the idea of Wouter de Nooy about 'stories as social action and models for social action' [13], we refocused the vector of influence outlined by the researcher and considered Stefan George's poetic cycle 'The Star of the Covenant' in two of its functions – not only as a networking tool/model, but also as a product/result of relationships within the association.

3 Methodology

When analyzing the material, various methods were useful to explore the topic in the deepest and most versatile way. These are, first of all, elements of social network analysis, used in reliance on the works of Peer Trilcke [14], Bruno Latour [15], Ingo Schulz-Schaeffer [16] and Mark Granovetter [17]. To review the stages of creation of 'The Star of the Covenant' and to study the relationships between members of the George

Circle, we used the historical-literary method of analysis. In the process of analyzing the poetic texts of 'The Star of the Covenant', a poetological analysis with elements of close reading was used, which made it possible to single out certain figurative and structural elements of the text, correlating with network theory. There was also carried out an analysis of the reception of 'The Star of the Covenant' by George's contemporary philologists who were members of the George Circle.

4 Analysis

4.1 'The Star of the Covenant' as a Product of Networking

Let us first consider the structural features and elements of the content of 'The Star of the Covenant' that may allow us to call the latter a 'product' of networking.

The Title of the Cycle as a Reflection of the Philosophy of the George Circle
The very name of the cycle arouses research interest – 'The Star of the Covenant'. It can be assumed that in the title there is already implicitly displayed the function intended by the author. Initially, the work was planned to be entitled '*Lieder an die heilige Schar*' ('Songs to the Holy Order'). Mikhail Mayatsky, in his work on the influence of Plato on the George Circle, notes that this self-definition, which became one of many internal names of the Circle, dates back to *hieros lokhos*, a military detachment formed in Thebes in the 6th century BC of 150 couples, and destroyed in the Battle of Chersonesos [2]. In the future George would reject such a name (possibly due to a more or less obvious homoerotic background, perhaps due to insufficient 'solidity') and give the cycle the title '*Der Stern des Bundes*'.

The image of a star did not cause any difficulties in the interpretation for the adherents of the Circle. This image was to be understood as an allegorical reference to the novice poet Maximilian Kronberger, also a member of the Circle, deceased at the age of 16, who posthumously was, in fact, deified by the members of the union and became the object of their worship. At the same time, the symbolic image of the star can also be perceived as an autological metaphor for the book 'The Star of the Covenant' itself, which performs a 'guiding' function for all members of the George Circle. And by 'union' (*Bund*), of course, one should understand the George Circle. It was the figure of Maximin that became the 'product' of the association (the union literally 'gave birth' to a 'star' from their own ranks).

It can be noted that already in the title of the cycle implicitly appear, in fact, two key concepts of the modern theory of networking: the group and its connections ('union') and the central actor [14] ('star'). From this point of view, it is not without interest to analyze other names of the association created by George, which were used by his followers: '*Reich*' (empire, kingdom), '*Staat*' (state), '*Kreis*' (circle). The listed names are clearly divided into two categories according to the type of their internal connections: associations with vertical connections, hierarchically organized (kingdom, state) and associations of equal members with horizontal relationships (circle, union). Perhaps – if we adhere to the thesis about the autological nature of the name – it was the name from the second category that was more appropriate for the poetic cycle, where the constituent parts (poems) are in equal relations to each other.

'The Star of the Covenant' is divided by the author into three parts ('books') of 30 poems each, an Introduction (*Eingang*) consisting of nine poems, and the so-called 'Final Chorus' (by the way, this formal arrangement mirrors the structure of Dante's 'Divine Comedy'). Whereas a reader from outside would most likely perceive 'The Star of the Covenant' as an enigmatic, 'dark' book, adherents of the Circle were easily able to decipher the allegories and allusions in the cycle, sometimes to concrete persons and situations.

Strong Ties and the Integration of the Paradigms Represented by Key Members of the Association into the Poetic Cycle

It should be noted here that, although not explicitly postulated by the 'Master', there still existed a certain hierarchy of adherents in the Circle on such features as achievements in activities for the benefit of the association or successes in art and/or academic activities. Members of the Circle who enjoyed George's special favor also directly or indirectly participated in the planning of strategies for the further development of the association. By the time of the creation of 'The Star of the Covenant' there were three such key figures: the German literary critic and professor of the University of Heidelberg, Friedrich Gundolf (1880–1931), the lawyer and writer Ernst Morwitz (1887–1971), and the historian, poet and translator Friedrich Wolters (1876–1930), nicknamed the 'Apostle Paul' of the George Circle [18] for actively promoting the poet's work wherever possible, for instance, in circles that were organized by himself in Berlin, Marburg and Kiel. In relation to these three figures, it would seem appropriate to use the term 'strong ties' as proposed by Mark Granovetter [17].

Kai Kauffmann in his biography of Stefan George writes quite bluntly and for good reason that 'George probably sought to integrate into the concept of "The Star of the Covenant" as many points of view as possible at what is most essential in its spiritual mission to increase the circle's density' [19]. According to Kauffmann, the poet was able to combine with each other a concept of the divine eros, represented by Gundolf, a concept of pedagogical eros, represented by Morwitz, and direction of cultural and political struggle, propagated by Wolters, presenting them as different stages of one big spiritual movement [19].

Personalities and Myths of the Circle and their Reflection in the Thematic-Figurative Content of the Cycle

But even ordinary adepts of the Circle, reading 'The Star of the Covenant', were happy to recognize the familiar 'myths' of the association (Maximin's cult), and sometimes also situations and conversations that they themselves experienced (although the names and corresponding realities were deliberately absent in the text). In his biography of George Thomas Karlauf ironically remarks: 'For many of them, reading the Star has become a kind of déjà vu. Some even thought that they would recognize the event or the evening hour to which the corresponding poem referred' [20]. A key source for the researchers here is Morwitz, who in 1960 published a 'Commentary to the work of Stefan George', which thoroughly commented poetic cycles of the 'Master', including 'The Star of the Covenant' [21].

The Introduction of the cycle is almost entirely devoted to the short earthly life of Maximin, his early departure and rebirth as the deity of the George Circle. Already in the first poem of the Introduction, the deity is characterized as a 'sprout from our own trunk' (*spross aus unsrem eignen stamm*) [1], that means it was generated by the George Circle itself. In the next poem, Maximin (by the way, in the cycle he is never mentioned by name) with the help of allegories is compared with ancient deities and biblical characters – Jacob, Narcissus and Endymion. There is clearly present a homoerotic subtext – Narcissus symbolizes an attraction to his likeness, and in connection with Endymion it is not his beloved Selena that is mentioned, but Zeus, who gave him eternal sleep. At the end of this poem the members of the George Circle 'kneel in front of the body/in which the birth of the god was performed' (*vorm leibe knieten/in dem geburt des gottes sich vollzog*) [1]. We see how an unremarkable life of a Munich schoolboy elevated to the myth that for a long time would carry the functions of identification and integration in the George Circle.

The seventh poem of the Introduction is noteworthy. Here the lyrical subject encounters the riddle: 'how come he is my child, and I am the child of my child' (*wie er mein kind ich meines kindes kind*) [1]. In addition to the obvious allusion to the Oedipus myth, we see the ambivalence of relations between George and Maximin – the young man is the poet's 'child', and George, in turn, is the 'child of his child'. That means that George is at the same time adept and creator of his own god (or God the Father in relation to the Son of God, Jesus Christ). This point is crucial for clarifying the networking scheme that was carried out within the Circle. We have already seen that in the title itself there is an image of the center/middle and union/connections. Now it turns out that in the Circle there are basically two centers – George himself and his deity Maximin. The connection between these two central actors is not vertical, but rather horizontal. The vertical connections of the other members of the Circle are directed both to Maximin and to the 'Master' Stefan George.

The last poem of the Introduction ends with the following lines: 'He is a son, conceived by the stars/born by the new middle from the spirit' (*Der sohn aus sternenzeugung stellt ihn dar/Den neue mitte aus dem geist gebar*) [1]. Here, the George Circle is designated as the 'new middle', which, being 'inseminated' by the spirit, gives birth to a new god. This god, Maximin, in turn, in the very first poem of the Introduction, is characterized as follows: 'You are always the beginning and the end and the middle' (*Du stets noch anfang uns und end und mitte*) [1]. In this perspective, just like Maximin functions as the center ('middle') for the Circle, the community of 'ordinary' members of the Circle constitutes the center for the entire German society, which means, the structure is projected one step lower.

In his work 'George', Friedrich Gundolf lists a number of functions with which the deified Maximin (instrumentalized for integration purposes) was endowed and states, as one of them, the following: 'collects ghostly forces around the living core' [22]. This statement is fully correlated with the above considerations.

At the end of the first part (book), in its last nine poems, the association is also mentioned, which once played an important role in the constitution of the George Circle – the circle of 'Munich cosmists' (*Kosmiker*), whose key figures were Alfred Schuler (1865–1923), Ludwig Klages (1872–1956) and Karl Wolfskehl (1869–1948). George was ideologically close to the cosmists, however, by the change of the years 1903 and 1904, for a number of reasons, there was a mutual estrangement, and later even a break in relations. In 'The Star of the Covenant', the poet, without mentioning specific names, in a number of poems indicates the reasons why this connection broke apart, which at one time was so useful for the association. Among these reasons, there is a peculiar escapism (*Schwärmer aus Zwang*), a lack of active principles, as well as Klages's anti-Semitic views, to which George polemically replied in the 21st poem by representing Germans and Jews as two 'poles' equally distant from the 'center' of civilization (that is the Mediterranean region, the cradle of ancient Roman civilization) [21]. Here again, there is the concept of the 'center', 'middle', which is very important for the structure of the George Circle and for building links within it.

Here is an interesting detail: George, in one of the poems addressed to the cosmists (again without mentioning the name of the person), criticizes another of their representatives, the writer Ludwig Derleth. For his excessive commitment to Catholicism, the poet compares him, among other things, with the theologian Bernard Clairvaux. Later, George confesses to Morwitz that he mentioned this historical character mainly in connection with the young Bernard Uxkull, whose first poetic experiments at that time caught the 'Master's' eye [21]. The ambivalence of this reference as an element of 'The Star of the Covenant' is evident – on the one hand, it is a 'product' of the Circle's social life (George reflects his interest in Uxkull's poetry by covertly mentioning his name in the book), on the other hand, it is an instrument for influencing potential adherents (one may suppose that George was counting on Bernard's positive reaction if he could unravel this 'cipher' left by the poet).

In the second book of the cycle, the focus shifts from the social sphere of the Circle's existence to the level of private, personal relations within the association – of course, these are the relationships between the Master and some of his followers. Gundolf writes that in the second book of the cycle George is 'embodied as living among his like-minded, he loves and suffers along with people who are not devoid of weaknesses, clear, simple and free'; and further he states: 'we are present during the education of the beloved by the lover himself' [22]. Many poems of the second part are written in a dialogical form; when being read by the 'unprepared' recipient it is sometimes impossible to determine which phrases belong to the 'Master', and which to his admirer.

In addition to the obviously growing tendency to formalize relations within the Circle in the manner of the Plato's Academy, one more circumstance is interesting for our study. It is known that in the George Circle it was actively discussed who was depicted by the Master in his poems as his beloved. Ernst Morwitz in his 'Commentary' gives very definite answers to these questions. According to him, the first ten poems are based on conversations between George and himself, the second and third group of poems are devoted to relationships with Ludwig Thormaehlen and Percy Gothein. Morwitz points out that the author himself can no longer say for sure which adherent he had in mind in each specific poem. However, he writes that in some cases this question can be answered fairly accurately, for both followers of George were of different ages (Thormaehlen then turned 21, and Gothein was fifteen) and they had a different style of speech that George sometimes sought to convey.

We can again conclude that the second part of 'The Star of the Covenant' is a 'product' of events that actually took place in the Circle, that arose, intensified and weakened the links between the center ('Master') and the periphery of the association. At the same time, the insufficiently clear 'personalization' of these poems even for adepts of the

Circle and their complete lack of clarity for the 'uninitiated', in our opinion, makes these texts a productive networking instrument. There is evidence that members of the union argued, wanting to prove whom exactly the Master had in mind; beginners were given an exemplary of the relationship between teacher and student, which was acceptable for and encouraged in the Circle. Kauffmann calls these poems 'role poems' [19]. These texts emphasize yet another important element of the 'teacher–student' relationship – the simultaneous dissolution and acquisition of one partner in another (images of pliable clay in the hands of the Creator, etc.). In Robert Norton's words,

> he thought himself to be entering into a 'spiritual marriage' with each of his newly acquired friends. He implanted himself in them, shaped them by his own consent, so that the 'rebirth' they experienced would inevitably change their attitude towards him, turning them, one might say, from lovers to sons (…). All these verses describe scenes of a radical alteration of one's own self, but always in the image of another. [23]

Co-writing Poems

Another strategy that is specifically applied in 'The Star of the Covenant' is, in fact, the writing of individual poems of the cycle by Stefan George himself together with one of his followers. Morwitz admits in his 'Commentary' that the 28th poem in the third book partially belongs to him and treats this decision as a kind of sacrifice, as a proof of his willingness to devote his life to serving the 'Master' [21]. Thus, this text, when co-written, ultimately strengthens the connection between George and his very follower.

4.2 'The Star of the Covenant' as an Instrument of Networking

We now turn to the analysis of 'The Star of the Covenant' as a kind of networking 'instrument' in the George Circle. We were able to identify several ways of how this work has influenced members of the union and their interaction.

The Role of Systematic Disclosure of the Concept of 'The Star of the Covenant' in the George Circle

Firstly, many researchers emphasize the role of oral reproduction and reception by ear (recitation, listening) of Stefan George's texts in the Circle. In his standard work, 'Literary group formation' Rainer Kolk mentions the 'oral tradition' (*Mündlichkeit*) in the union and writes about it as follows:

> The practice of reading out loud in the George Circle seems to be a revival of pre-Modern forms of reading; a phenomenon that indicates the specific role of group-like associations, which not only reproduce social processes, but claim to have their own logic of their internal dynamics. [24]

Kauffmann, while describing the creation process of 'The Star of the Covenant', mentions that from 1911 to 1913, George was organizing 'solemn readings' of single poems, which subsequently would be included in the poetic cycle [4]. The poet was also gradually revealing the idea of the work: firstly, Ernst Morwitz was invited to proofread the manuscript, then, by the end of November 1913, when ten signal copies had been printed, he sent them to those members of the circle closest to him [4]. In visual terms, it would be appropriate to depict this process in the form of expanding concentric circles. One can imagine the joy of recognition and the feeling of 'uniqueness' experienced by adherents as they were reading a cycle which was just out of print and suddenly came across texts that were already familiar to them. Thus, the network of connections created in the George Circle became at the same time a kind of 'fishing net', by which the 'Master' intentionally 'caught' his potential new followers.

The Function of the Foreword to the Cycle

Obviously, a brief introduction serves the task to create for the George Circle an aura of inaccessibility and simultaneous attraction. It contains the following lines:

> The Star of the Covenant was first intended for a close circle of friends, and only the consideration that the concealment of the once told is hardly possible in our days, made me prefer publicity as the most reliable protection. Then the world events that broke out immediately after the publication of the cycle made the souls of people belonging to other classes as well susceptible to this book, which could have remained a secret book for years. [1]

This example shows the ambivalence of George's aspirations in the aspect of networking – apparently, he wanted the common reader to get acquainted with the book, but at the same time, such an introduction also attracted the youth of the 'noble' class who wanted to become chosen, deciphering the secret meanings of the book.

Friedrich Gundolf also writes about the role of mystery as an instrument for the production of new ties in the union and for the spread of its influence, and, surprisingly, already in 1920, he is actually using for this purpose the key concepts of present-day network theory:

> Mystery is not something done, mystery is the law. Since mystery is the center of everything fulfilled, undivided, of an indivisible being, then the field of its influence is the circle. Not a separate atom that revolves around itself, but a living force, which, being bound in itself and closed, is radiating to the outside, forms and includes close or distant beings permeable by rays, gives them independence and absorbs them, but since this force is simple and holistic, everything that it achieves – even in the distance – becomes, having connected with it, simple and holistic as well; a circle can grow, but completeness and connection cannot. [22]

Elements of the Cycle as a 'Cipher' for Initiates

The text of the poetic cycle is also permeated with enigmatic, 'dark' elements requiring interpretation, with metaphors and symbols, the decoding of which is possible (and not even fully) only for George's followers. This process of decoding gave readers a sense of involvement in the secrets of the Circle. For example, in the cycle 'The Star of the Covenant' there is a complex of euphemistic concepts that indicate the process of conception, bearing and birth of a 'new man' and at the same time of a new god. The point here, of course, is the forming of a new, perfect personality. Norton notes: 'In the experience of spiritual creation, he, the poet, assumes the role of both sexes, becoming a mother and father to himself and to his divine offspring at the same time' [23].

Ernst Morwitz and Friedrich Gundolf as Commentators of 'The Star of the Covenant'

Speaking about the function of 'The Star of the Covenant' as a networking instrument, one cannot fail to mention the 'intermediate links' of this process. These are the followers of the inner circle, those same 'Strong Ties', some of which acted as intermediaries between George, uttering 'dark prophecies', and the broad masses of his potential followers. In the case of 'The Star of the Covenant', these are, first of all, Ernst Morwitz and Friedrich Gundolf. If Morwitz, as was said above, assiduously, methodically, and somewhat naively, in a descriptive manner, comments on each poem, Gundolf on the contrary, being a university professor, goes beyond the scope of a simple description and explanation. He gives a visual diagram of the poetic cycle, which can be used to study this collection of texts in the aspect of networking:

> If we imagine the 'Star of the Covenant', like all the works of George, to be in the form of a ball (…) then the Introduction will be its center, the first book its surface, which limits the sphere not only from the outside, but also from the inside, the second book will be the radiation of the center reaching the surface, and the third book will be the total content of the sphere, its center, radiation and surface. [22]

Continuing to discuss the structure and impact of 'The Star of the Covenant' on its readers, Gundolf builds another scheme using concepts such as 'bonds' (i.e. connections) and 'center':

> Sensual renewal is united with a spiritual change that (…) dissolves any scab and scale, cancels the inherited ties and obligations, cancels the family, the people, the profession (…) former unions are collapsing because their center has died (…). Each new center causes the new need and a new freedom (…). The same pulse beats in what moves an individual person and in what he himself moves. Obsolete bonds and freedoms in the new alliance are not valid (…). [22]

Thus, Gundolf manifests network connections within the George Circle, in fact, as the only 'living' and relevant ones in modern society.

Symbolic Schemes of Interaction Between Participants of the George Circle

As for the relations between the members of the circle, 'The Star of the Covenant' gives symbolic schemes designed to encourage community members to interact. So, in the 20th poem of the third book, the image of the George Circle (*runde, liebesring, kreis*) appears, in which the process of birth of new creatures takes place: 'From this love ring where nothing gets lost/Every Templar derives new strength/And shoots his own – the greater one – into everyone/And flows backwards into the circle' (*Aus diesem liebesring dem nichts entfalle/Holt kraft sich jeder neue Tempeleis/Und seine eigne – grössre – schiesst in alle/Und flutet wieder rückwärts in den kreis*) [1].

In addition to imagery, which can conjure up a picture of a collective orgy, the concept of a circle of like-minded men, which also has a female, 'generative' function, is also interesting. Each member draws his power from it and simultaneously 'injects' his own energy in all the other members and then 'pours' back into the circle. Of course, we are not talking here about vertical, hierarchical connections in their pure form, but about the spiritual interaction of 'everyone with everyone' (horizontal) and at the same time with the center (vertical, first of all, to the deified Maximin).

As one of the most powerful texts of 'The Star of the Covenant', which had a strong impact on the followers of the George Circle, we can name the final poem of the cycle (the so-called 'Final Chorus'). If we try to visually display the content and structure of the poem, then in the end we will come to the same scheme consisting of a 'center/central actor' (*Gott,* 'God') and of 'connections with other actors' (*wir,* 'we'), e.g.: 'God's peace is in our hearts/God's power is in our chests' (*Gottes ruh in unsren herzen/Gottes kraft in unsrer brust*) [1]. Moreover, in this text several vectors are sent away from the central actor who is at the same time influencing each member of the union (their different body parts as well, symbolizing the different areas of the physical, intellectual and spiritual activity). The structure of the poem recalls a prayer or a spell, and it is fully legitimate that this particular text was recited in a chorus by members of the George Circle at the poet's funeral.

5 Conclusion

Having considered the poetic cycle 'The Star of the Covenant', we come to the conclusion that this literary work is a product of networking (in this case the interaction between the members of the literary association) and is simultaneously its instrument (a means of establishing new and varied relationships both between the members within the association as well as outside of it). This article opens up a wide field of research; in particular, in our opinion, it may be productive to consider this issue in a diachronic aspect, since after the death of Stefan George and the departure of the most active participants from the Circle, the interaction of the descendants of the 'Georgians' continued, attracting representatives of the young generation into the union, and these texts by George were being continuously recited and rethought.

References

1. George, S.: Der Stern des Bundes [The Star of the Covenant]. Werke:Ausgabe in zweiBänden. Band 1 [Works: Edition in Two Volumes. Volume 1], pp. 345–394. Deutscher Taschenbuch Verlag, München (2000)
2. Mayatsky, M.: Spor o Platone: Krug Stefana George i nemetsky universitet [Controversy about Plato: Stefan George and the German University]. NIU WSE, Moskwa (2011)
3. David, C.: Stefan George: Sein dichterisches Werk [Stefan George: His poetic work]. Carl Hanser Verlag, München (1967)
4. Kauffmann, K.: Der Stern des Bundes [The Star of the Covenant]. In: Aurnhammer, A., Braungart, W., Breuer, S., Oelmann, U. (eds.) Stefan George und sein Kreis: ein Handbuch, vol. 1 [Stefan George and His Circle: A Handbook, vol. 1], pp. 191–203. Walter de Gruyter, Berlin/Boston (2012)

5. Rauscher, U.: Neue Lyrik [New Lyrics]. Frankfurter Zeitung und Handelsblatt 185 (1910)
6. Weber, M.: Geschäftsbericht auf dem ersten Deutschen Soziologentage in Frankfurt 1910 [Annual report at the First German Sociological Conference in Frankfurt in 1910]. In: Weber, M. (ed.) Gesammelte Aufsätze zur Soziologie und Sozialpolitik [Collected Essays on Sociology and Social Policy], pp. 431–449. Mohr Siebeck, Tübingen (1988)
7. Schmalenbach, H.: Die soziologische Kategorie des Bundes [The sociological category of the association]. Die Dioskuren **1**, 35–105 (1922)
8. Wach, J.: Der Erlösungsgedanke und seine Deutung [The thought of salvation and its interpretation]. Hinrichs, Leipzig (1922)
9. Wach, J.: Meister und Jünger: Zwei religionssoziologische Betrachtungen [Master and Disciple: Two religious-sociological considerations]. Pfeiffer, Leipzig (1924)
10. Brodersen, A.: Stefan George und sein Kreis: Eine Deutung aus der Sicht Max Webers [Stefan George and his Circle: An interpretation from Max Weber's point of view]. Castrum Peregrini **19**, 5–24 (1970)
11. Baumann, G.: Der George-Kreis [The George Circle]. In: Faber, R., Holste, C. (eds.) Kreise, Gruppen, Bünde: zur Soziologie moderner Intellektuellenassoziation [Circles, Groups, Alliances: On the Sociology of Modern Intellectual Association], pp. 65–84. Königshausen & Neumann, Würzburg (2000)
12. Baumann, G.: Dichtung als Lebensform: Wolfgang Frommel zwischen George-Kreis und Castrum Peregrini [Poetry as a way of life: Wolfgang Frommel between George Circle and Castrum Peregrini]. Königshausen und Neumann, Würzburg (1995)
13. de Nooy, W.: Stories and social structure. A structural perspective on literature in society. In: Schram, D. and Steen, G. (eds). Psychology and Sociology of Literature, pp. 359–377. Benjamin, Amsterdam (2001). https://doi.org/10.1075/upal.35.20noo
14. Trilcke, P.: Social Network Analysis (SNA) als Methode einer textempirischen Literaturwissenschaft [Social Network Analysis (SNA) as a method of text-empirical literary studies]. In: Ajouri, P., Mellmann, K., Rauen, C. (eds.) Empirie in der Literaturwissenschaft [Empiricism in Literary Studies], pp. 201–247. Mentis, Münster (2013)
15. Latour, B.: On actor-network theory: a few clarifications. Soziale Welt **47**, 369–381 (1996)
16. Schulz-Schaeffer, I.: Zugeschriebene Handlungen: Ein Beitrag zur Theorie sozialen Handelns [Attributed actions: A contribution to the theory of social action]. Velbrück, Weilerswist (2007)
17. Granovetter, M.: The strength of weak ties. Am. J. Sociol. **78**, 1360–1380 (1973). https://doi.org/10.1086/225469
18. Salin, E.: Um Stefan George: Erinnerung und Zeugnis [About Stefan George: memory and testimony]. H. Küpper, München/ Düsseldorf (1954)
19. Kauffmann, K.: Stefan George: Eine Biographie [Stefan George: A Biography]. Wallstein Verlag, Göttingen (2014)
20. Karlauf, T.: Stefan George: Die Entdeckung des Charisma [Stefan George: The discovery of the charisma]. Pantheon, München (2008)
21. Morwitz, E.: Kommentar zu dem Werk Stefan Georges [Commentary on the work of Stefan George]. Helmut Küpper vormals Georg Bondi, Düsseldorf und München (1969)
22. Gundolf, F.: George. Georg Bondi, Berlin (1920)
23. Norton, R.: Taynaya Germaniya: Stefan George i ego krug [Secret Germany: Stefan George and his Circle]. Nauka, Sankt-Peterburg (2016)
24. Kolk, R.: Literarische Gruppenbildung: Am Beispiel des George-Kreises: 1890–1945 [Literary group formation: Using the example of the George Circle: 1890–1945]. Max Niemeyer Verlag, Tübingen (1998)

Computational Network Analysis

Topological Data Analysis Approach for Weighted Networks Embedding

Irina Knyazeva[1,2(✉)] and Olga Talalaeva[1]

[1] St. Petersburg State University, St. Petersburg, Russia
iknyazeva@gmail.com, o.talalaeva@yandex.ru
[2] Institute of Information and Computational Technologies, Almaty, Kazakhstan

Abstract. Efficient node representation for weighted networks is an important problem for many domains in real-world network analysis. Network exploration usually comes down to description of some structural features which give information about network properties in general, but not about specific nodes. Whereas information about node profile is very important in any network with attributed nodes. Recently, the network embedding approach has emerged, which could be formalized as mapping each node in the undirected and weighted graph into a d-dimensional vector that captures its structural properties. The output representation can be used as the input for a variety of data analysis tasks as well as for individual node analysis. We suggested using an additional approach for node description based on the topological structure of the network. Some interesting topological features of such data can be revealed with topological data analysis. Each node in this case may be described in terms of participation in topological cavities of different sizes and their persistence. Such representation was used as an alternative node description and demonstrated their efficiency in the community detection task. In order to test the approach, we used a weighted stochastic block model with different parameters as a network generative process.

Keywords: Node embedding · Weighted networks · WSBM · Topological data analysis

1 Introduction

A network can be denoted as a graph-based model $G = (V, E)$ [1], where V stands for the set of vertices, and E corresponds to the set of existing edges. In social and biological networks vertices play the main role in generating the network's large-scale structure. Vertices with a similar structural role form group-level connectivity patterns. These groups form communities in a broad sense, not in an assortative-only definition. Detection of such communities, consisting in grouping vertices according to their structural roles, is an important area in social and biological networks. There are plenty of different algorithms for community

A. Antonyuk and N. Basov (Eds.): NetGloW 2020, LNNS 181, pp. 81–100, 2021.
https://doi.org/10.1007/978-3-030-64877-0_6

detection, both data-driven [2] and model-based [3]. Recently, a network embedding approach has emerged, which can be formalized as mapping each network node into a d-dimensional vector that captures its structural properties [4]. If there is vector representation for each node, the community detection task is narrowed down to the clustering task in a well-studied vector space.

In this paper, we describe how weighted networks can be analysed and explored with the topological data analysis approach. Representation of a network as a graph serves as an optimal way to encode and describe dyadic relationships between particular objects. Nonetheless, it fails to capture complex and nonlinear polyadic relations, which most frequently underlie network structure and formation process [5]. To overcome these graph nature-based restrictions, it is essential to indicate the approach which expands the number of possible forms of elementary interactions in a system. This can be accomplished by proceeding from pairwise interaction to structures, which involve larger sets of nodes. Algebraic topology techniques make that shift possible using the concepts of a simplexes and simplicial complex. Features received with these techniques can be further used for node representation.

Efficient node representation for weighted networks is an important problem for many domains in real-world network analysis. Network exploration usually comes down to the description of structural properties which gives information about networks in general, rather than a distinct node. In its turn information about node profiles is very important in every network with actors, for example, in social or neuroscience networks. Node embedding approach [6,7] can be formalized as mapping each node into a d-dimensional vector that captures its structural properties. The output representation can be used as the input for a variety of data analysis tasks, as well as for individual node analysis. We suggest using a complementary approach for node description based on the topological structure of the network. In this framework, a weighted network is pictured as a collection of separated points in a finite metric space, where every point corresponds to an original network node. Weighted connections in such a space are presented by the notion of point distance or similarity. Such data structure can be pictured as a 'point cloud'. Persistent homology features, being the main topological features of a 'point cloud', can be revealed with topological data analysis [8]. In this framework data described with topological invariants, including connected components, cavities, voids, so each node can be described in terms of participation in topological invariants and their persistence.

In summary, the contributions of this paper are as follows.

- We propose topological data analysis (TDA) framework for the description of weighted networks in general as well as for individual nodes.
- Possible qualitative analysis along with node embedding schema based on node participation in topological cavities of size 0 and 1 is presented.
- We demonstrated the efficiency of TDA representation in community detection tasks and compared the results with node2vec embedding on model networks.

The rest of the paper is structured as follows. We first describe weighted stochastic block model (WSBM) in Sect. 2, then discuss the topological analysis framework and its application to networks generated with WSBM in Sect. 3, then provide network embedding methods description in Sect. 4. The results of embedding efficiency in node clustering tasks are presented in Sect. 4.3, and conclusions of the paper are listed in Sect. 5.

2 WSBM Model

Real-world networks usually have weights. Weights often come from attributed networks, where nodes possess not only labels, but also some additional profiling information about the connections between them, like in social [9] or brain networks [10]. The presence of node attributes expands and deepens studying of network organizational properties: it may reveal a certain modular structure, where nodes with similar structural roles are placed closely to each other. To create a proper synthetic dataset, which reflects these real network properties, in this particular study we decided to pick a weighted stochastic block model (WSBM) [3]. WSBM is frequently applied as a generative template for learning network community structures. The original stochastic block model [11] is a popular generative model of pairwise interactions among n vertices. It can produce a wide range of networks including large heterogeneous systems, where smaller and more homogeneous patterns interact with each other. Classic SBM model assumes that each vertex belongs to one of the K latent blocks denoted by z_i and each edge A_{ij} is sampled from Bernoulli distribution. Its probability parameter $\theta_{z_i z_j}$ depends only on the group membership of the connected vertices. However, SBM models work strictly with binary networks, where edges can be either present or absent. Information about weights, if they exist in the analysed network, is typically discarded by thresholding.

The generalization of the SBM model, which takes edge weights into account, is called weighted stochastic block model [3]. Specifically, in this model each edge A_{ij} is sampled not from Bernoulli distribution with binary outcomes but from parametric exponential distribution; thus, the parameters depend on the group membership and block structure. One could pick Normal distribution for real-valued weights, Poisson – for positive discrete values, Exponential – for positive real values respectively. WSBM model is fully described with the next set of parameters:

1. Number of nodes in the network;
2. Vector of cluster membership z_i;
3. $\theta_{z_i z_j}$ parameter, that defines strength of connection inside a group and between groups. In this paper, we used one-parametric Poisson distribution for edge sampling. Examples of networks that were generated by using the WSBM model are shown in Fig. 1.

3 Topological Data Analysis

Topological data analysis (TDA) refers to statistical methods that find structure in data. One of the principal assumptions underlying topological data analysis is

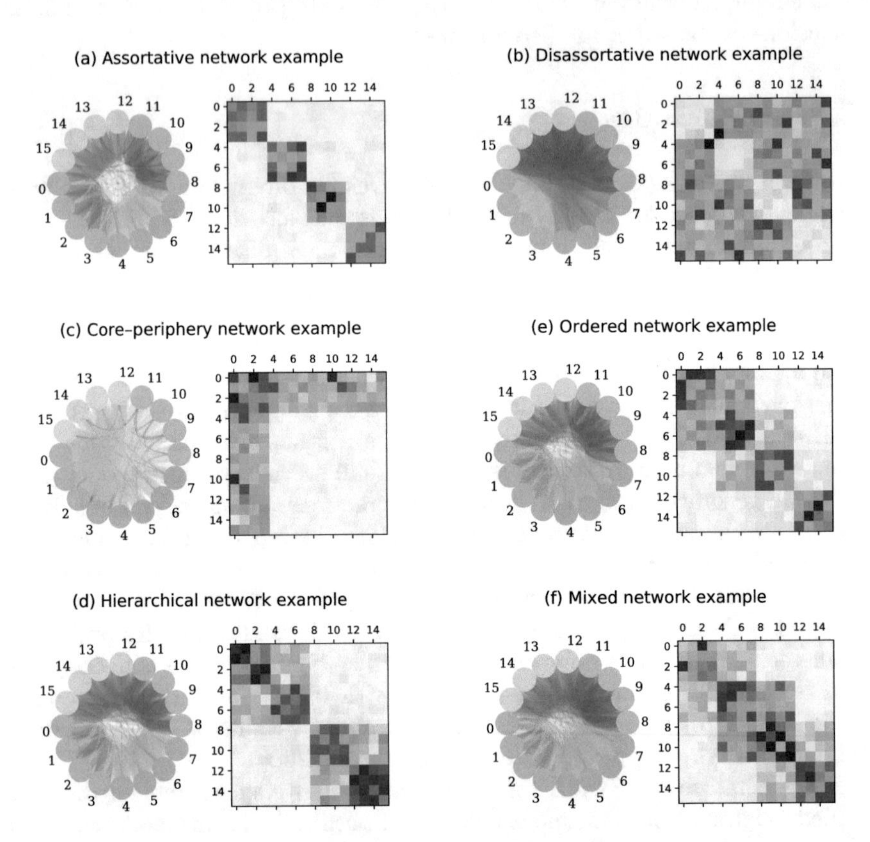

Fig. 1. Examples of WSBM-generated networks. Every picture displays a weighted network model in a circle layout (left) and a corresponding weighted adjacency matrix (right), nodes in both are sorted by community membership. Nodes belonged to one group was colored in the same color, link color was unique for each pairwise group connection, so for group 1–1, 1–2, 1–3, etc. Inside the matrix, the darkness of cell color indicates the level of connection strength. Six following types of networks are presented: a) Assortative structure: strong connections within groups; b) Disassortative structure: connections mainly exist between distinct groups rather than within them; c) Core-Periphery structure: the 'core' (green group) connects mainly within itself as well as with the 'periphery' (gray, orange and pale mauve); d) Ordered structure: besides strong in-group connections, all groups are connected sequentially. The first (green) group is linked to the second (gray) one, the second group, likewise, is connected to the third (pale mauve) one, etc.; e) Hierarchical structure: two main blocks with strong inner connections, each divided into two smaller blocks with even stronger links within them; f) Mixed model: hierarchical and ordered structure are combined within the same network.

that data can be represented as a geometric shape and studying descriptive features of that shape could clarify and explain data generation process. Here TDA is used to describe particular method from algebraic topology called persistent homology.

3.1 Algebraic Topology Methods

Simplexes and Simplicial Complices. Simplicial complex can be viewed as a generalization of network or graph model. It is a combination of k-simplexes or fully connected finite sets of $(k+1)$ vertices, where k denotes the dimension of simplex, under certain rules. For instance, a single vertice is a 0-simplex, a standard graph edge – 1-simplex, 2-simplex is a triangle, 3-simplex – a tetrahedron, etc. [12,13]. Triangle is considered as a maximum base simplex as long as all the other polyhedron simplexes could be constructed with triangles. The examples of previously listed basic simplexes are shown in Fig. 2. Simplicial complex is defined by a set of simplexes analogously to the way how network is constituted by a set of edges. Formally, simplicial complex should satisfy two following conditions: 1) any subset of a simplex which initially belongs to a simplicial complex K also belongs to K; 2) intersection of any two simplexes in K either belongs to complex or is empty [14]. The difference between a common set of simplexes and simplicial complex is particularly demonstrated on Fig. 3.

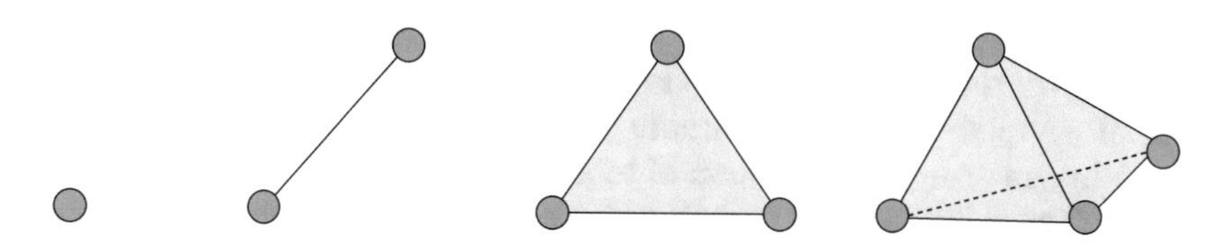

Fig. 2. Left to right: illustrations of 0-, 1-, 2-, 3-simplexes [15]. Simplicial complex is constructed by 'gluing' together base simplexes.

From Binary Graph to Simplicial Complex. Over the last decade, there have been developed multiple approaches for constructing simplicial complexes out of raw data [5]. One of the popular ways to encode raw data is called clique complex: the procedure consists of extracting cliques – all-to-all connected subgraphs – from a graph or a network adjacency matrix, and replacing it with corresponding simplexes built on the vertices participating in the clique [16]. Thus, clique complex preserves the information underlying graph, while constructing additional space for revealing organizational principles of network that cannot be distinguished by standard graph metrics. Clique complex is a widely applied tool for dealing with raw network data presented as a binary incidence matrix [5].

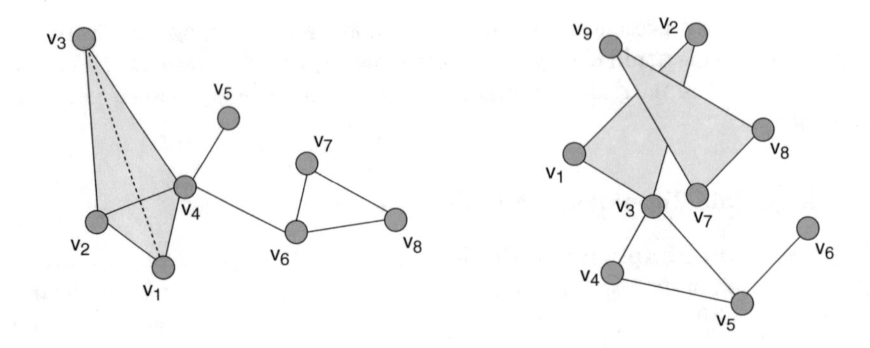

Fig. 3. Two collections of simplexes in a eucledian metric space. On the left, conditions for simplicial complex existence are met – all the simplexes are 'glued' together, so, their subsets and intersections become part of the complex; on the right [15], however, conditions are violated, as long as the intersection of v_4, v_5, v_6 triangle and $v1, v2$ edge does not form a simplex at all, and, thus, does not belong to the complex.

Simplicial Homology. There is a number of topological features which are commonly extracted from the simplicial complex for further estimation: one of them is called simplicial homology [16]. Simplicial homology represents a sequence of measurements: therefore, n-th homology of a simplicial complex is the set of n-cycles, formed out of corresponding n-simplexes. Informally, n-cycle can be defined as a group of n-simplexes arranged in a such way that they create an n-dimensional 'hole' on the topological surface. Thus, for every simplicial complex one could compute a collection of n-th homologies, whilst constructing an ordered description of topological invariants. Information about those topological invariants is contained in the vectors of homology generators or Betti numbers: n-th homology group generator or n-th Betti number in a given simplicial complex consists of a list of n-dimensional cycles. Thus, β_0 denotes the number of connected components in a complex, β_1 is a number of 1-cycle, or 'circular' holes, and β_2 counts 'voids' or 'cavities' – 2-dimensional cycles, which could represent information flow properties of a network [13]. Vectors of Betti numbers for several simplicial complexes are shown in Fig. 4.

3.2 Persistent Homology

TDA Filtration. Clique complex was already mentioned as a common method for simplicial complex construction from binary graphs. The casual way of dealing with weighted networks is thresholding: transforming original data to binary graph by applying a threshold which leaves only significant connections in a network. Unfortunately, this approach implies partial information loss of original data. To handle this complication, TDA offers a procedure called filtration: every weight value of original matrix is sequentially applied as a threshold ε, leaving only connections with values $< \varepsilon$, afterwards translating initial network to a family of unweighted graphs, which retains all of the information.

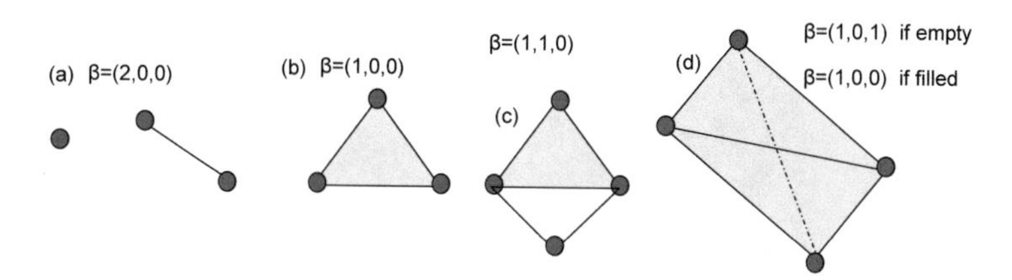

Fig. 4. Vectors of Betti numbers for different simplicial complexes. (a) Two connected components, no 1- or 2-cycles (voids), so $\beta_0 = 2$ and others 0; (b) One connected component. As long as it is a 'filled' complex, there are no observed cycles; (c) One connected component; 2 cycles, the large one includes all the nodes in a complex, and the smaller one, being empty, forms a circular 'hole'; (d) Tetrahedron; one connected component is formed. Faces of the tetrahedron are filled, so there are no 1-cycles. If the complex is 'empty' inside, there will be 'void', or 2-cycles, formed by the 1-cycle faces of complex, so $\beta_2 = 1$. If it is filled, there are no either 1- or 2-cycles, so $\beta_2 = 0$ [14].

Extension of homology concept to such filtrated sequence is called 'persistence' and was introduced by [17] (refined by [16]). Figure 5 shows an example of filtration process for a toy network built from 5 vertices.

Persistent Homology and Barcodes. The presence of filtration requires tool modification for describing topological features. Applying standard homology concepts to each complex in a filtered sequence creates dynamic 'persistence' features, and a method called 'persistent homology'. Persistence of cycles in a filtered complex of given dimension is computed by tracking changes in network cycle structure as long as the threshold value varies. Throughout the filtration, new connected components, cycles and cavities can form, merge, decompose and change their shapes. Observing a sequence, it is possible to define the 'birth' and 'death' threshold values, and time intervals between them, where homologies perform various complex interactions. There is a small uncertainty in the merging process of connected components. At the initial stage, if both components have equal birth time, any component can die, but usually the one with the lowest index absorbs the other. As a result, the component which appeared last 'dies' and the first one continues living. Persistent homology is a sequence of intervals containing the combination of birth and death time with participated simplexes for each dimension. The members of these simplexes are called cycle representatives. The longer the interval exists, the higher the persistence of homology is. Possible graphical representation for persistent homology is as a collection of horizontal line segments, called barcodes [13]. Horizontal axis corresponds to threshold values, and vertical axis displays arbitrarily ordered homology genera-

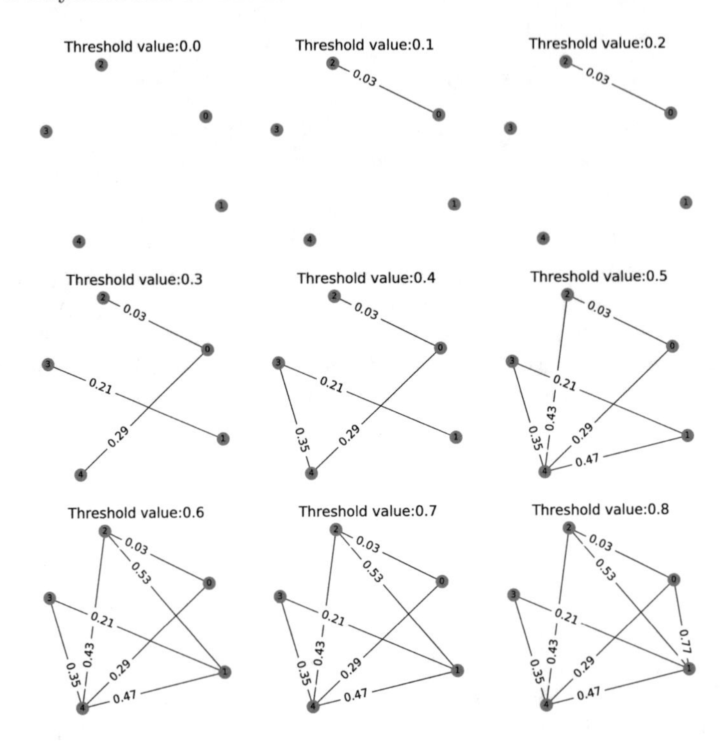

Fig. 5. Sequential steps of filtration for a toy network containing 5 nodes. Threshold value increases by 0.1 on each step of filtration.

tors in a given dimension. For better visualization we ordered connected component barcodes along vertical axes by length. Figures 6–11 show barcodes, annotated with cycle representatives, accompanied by topological description based on these barcodes for the WSBM network models, previously listed and displayed in Fig. 1. The resulted picture of barcode representation annotated with the cycle representatives can be summarized as follows:

1. Assortative network (Fig. 1(a)). The barcode plots annotated with cycle representatives are shown in Fig. 6. Left image represents a barcode plot with lifetimes of connected components. From this plot, it can be seen that, on early stages of filtration, connected components were forming and merging around the nodes from the same groups. At the first stage inner group nodes merged to each other, until only four connected components were left, which reflects the original community structure. After that outer connections started to merge the groups until the number of components decreased to 1. On the right, there is a barcode plot with cycles' lifetimes for the same network.

There are two short-lived cycles, which united nodes from the first and the second group; one short cycle appeared inside the fourth group; one long cycle including nodes from the third and the fourth group, inside of which another short cycle appeared; and on the later stages another long-lived cycle appeared, connecting nodes from the first, the second and the third group simultaneously. The last longest cycle denoted that there were strong connections inside group clusters and weak outer connection which corresponds to assortative structure.

2. Disassortative network (Fig. 1(b)). From the left image in Fig. 7 representing a barcode plot with lifetimes of connected components, it could be seen that the more the threshold value increased, the more nodes from separate blocks were linked together, in this case there were practically no observed inner connections. On the right plot, there were a lot of cycles, co-existing at the same time, and appearing inside each other.
3. Core-periphery network (Fig. 1(c)). Again, the network structure could be observed throughout the barcode picture (Fig. 8). The components merged as follows: the nodes from the first group were inferred firstly with inner and then with outer connections to other groups, while the nodes from the other groups did not merged with each other. On the right plot, there are three cycles, the longest one appears on the earlier stage, and after that two short cycles appeared inside.
4. Ordered network (Fig. 1(d)). Barcode picture (Fig. 9) reflects the structure of a given ordered network. While components merged with each other, links within and between groups appeared simultaneously, some blocks became connected faster than the others, and links between groups appeared before the groups were completely merged within, as it was in case of assortative. On the right plot, there are several short-lived cycles on earlier stages, which indicate sequential connection between particular blocks in a given order: in the same cycle, first group appeared along with the second, second group with the third, etc. One long cycle at the later filtration stages showed presence of weaker linkage between all four groups of nodes in a network.
5. Hierarchical network (Fig. 1(e)). Like in the assortative structure, connected components (see Fig. 10 on the left side) started to form simultaneously between the nodes in each group, and then the links between groups appeared, but in the hierarchical model, as can be seen from the upper bars, these connections were made strictly sequentially. On the right plot, there are several short-lived cycles, and two long-lived on the later filtration stages; cycle structure is also similar to the assortative case.

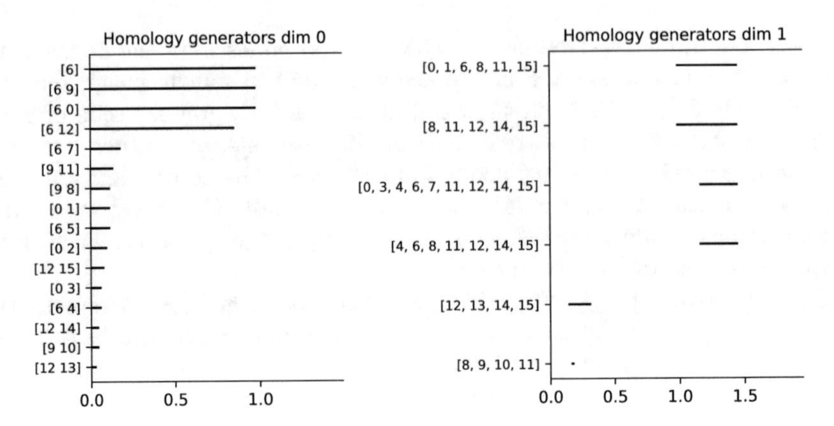

Fig. 6. Barcode plots annotated with cycle representatives for assortative network shown in Fig. 1(a). Left image represented a barcode plot with lifetimes of connected components, right – birth and death time of 1 cycles.

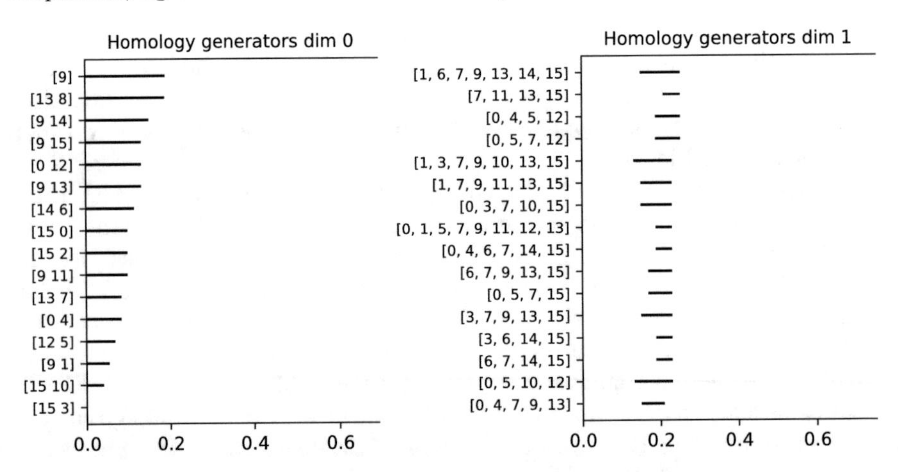

Fig. 7. Barcode plots annotated with cycle representatives for disassortative network shown in Fig. 1(b). Left image represented a barcode plot with lifetimes of connected components, right – birth and death time of 1 cycles.

6. Mixed network (Fig. 1(f)). In the given mixed network, components (see Fig. 11) were formed and merged similarly to the ordered network example. On the right plot, there are three short-lived cycles, which include nodes from the overlapping blocks, but there are no cycles between groups, that were not overlapped. So all the outside interactions were performed through the overlapping nodes.

The computation of homology groups and cycle representatives were done with the Eirene package for homological algebra [18].

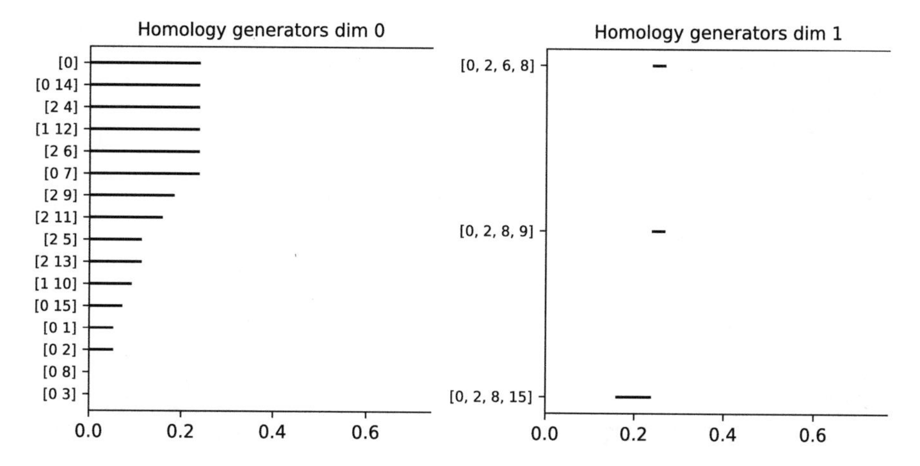

Fig. 8. Barcode plots annotated with cycle representatives for core-periphery network shown in Fig. 1(c). Left image represented a barcode plot with lifetimes of connected components, right – birth and death time of 1 cycles.

3.3 TDA Descriptors

TDA proposes various techniques for extracting and visualizing information about discovered topological invariants for shapes, in our case, point cloud network representation. Barcode plots represent topological invariants collected at each dimension, as a sequence of horizontal bars, which are placed between the birth and the death times of the feature. Persistence diagram provides another perspective, from which every feature can be displayed as a point on the diagram, while coordinate values correspond to birth and death values respectively [19]. Given such representation, the lifetime of a feature is determined by its closeness to the diagonal: the farther a point is located, the more persistent this feature is. Another widely-used way for TDA results summarizing is Betti numbers. We already discussed in Sect. 3.1 the nature and representation of Betti numbers, while being applied to standard homology concept. In persistent homology, Betti numbers gain specific importance because of their dynamic representation. Betti curve of dimension n displays the changes in number of connected components, cycles or 'voids' at each step of filtration. While statistical processing of persistent barcodes is still quite a challenging issue in the TDA field, according to their complicated geometric representation, Betti curves are more interpretable and could be analyzed with various statistical metrics. Such feasibility for numerical estimation presents an opportunity for making certain implications about the growing rate of persistent homologies, their distribution along the filtration, and other significant properties. Usually, different statistics based on statistical summarizing listed above representation are used. There is a number of successful applications of that approach for different real problems in astrophysics [20, 21].

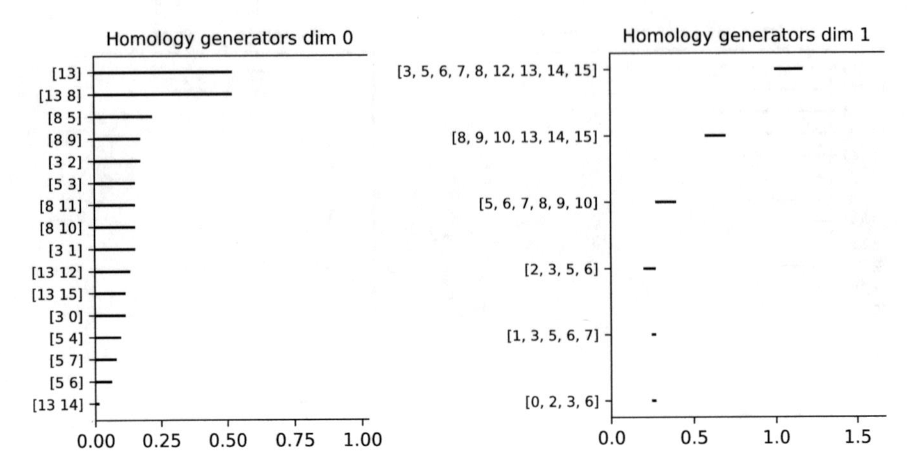

Fig. 9. Barcode plots annotated with cycle representatives for ordered network shown in Fig. 1(d). Left image represented a barcode plot with lifetimes of connected components, right – birth and death time of 1 cycles.

However, such representation does not provide enough information in case of performing node embedding task: more detailed, node-oriented description of persistent homologies is required. One way to do that is to characterize every distinct node in terms of its participation in the collected components, cycles or 'voids'. To accomplish that, in the first place, information about 'class representatives' should be gathered; 'class' p means the corresponding p-th persistent homology class in the barcode interval list, and, consequently, node that engages in p-th class is called 'p-th class representative'. In the Figs. 6–11 each barcode corresponds to their cycle representatives. We suggest to use this information for node encoding. The most intuitive way of creating a particular node description is to construct a feature vector, where each feature correspond to participation of a node in an observed cycle. Efficiency of this encoding was compared to other node embedding techniques. Detailed description of relevant node embedding methods and further topological encoding are presented below in Sect. 4.

4 Network Embedding

Network embedding approach poses a challenge for discovering an efficient low-dimensional representation of network, while capturing its structural and topological properties at the level of individual node. Such representation plays a key role in variety of data analysis tasks applied to networks, such as node classification and clustering, network visualization, link prediction, etc. Most of them are based on three different approaches: matrix factorization, random walk and deep learning.

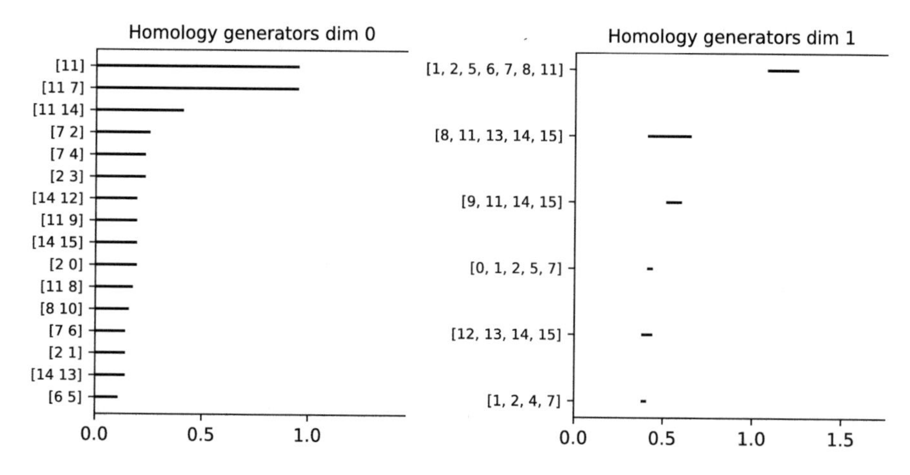

Fig. 10. Barcode plots annotated with cycle representatives for hierarchical network shown in Fig. 1(d). Left image represented a barcode plot with lifetimes of connected components, right – birth and death time of 1 cycles.

4.1 Related Work for Network Embeddings

Matrix Factorization. Matrix factorization (MF) can be considered as a dimensionality reduction technique, which explores structural network properties in the form of a network graph matrix representation. Graph Factorization model, introduced in 2013 in [22], may be considered as an example of direct adjacency matrix factorization. Another algorithm, GraRep [23], applies MF to the transition probability-based matrix, which is constructed by running random walks with specified length, thus preserving higher-order proximity in a network. Furthermore, HOPE (High-Order Proximity Embedding) model [24] also takes into account asymmetric transitivity of directed graphs, which makes it applicable for directed social or webpage networks. Both GraRep and HOPE use SVD (singular value decomposition) as an optimization technique for an objective function. SVD is often used in matrix factorization algorithms as the optimal solution for low-rank approximation task. However, there are several models based on other decomposition tools, for instance, non-negative matrix factorization (N-NMF). N-NMF model, introduced in [25], aims to preserve mesoscopic community structure of a network by combining community detection framework with non-negative matrix factorization.

Deep Learning. The principal target of network embedding lies in discovering such mapping function for network transition to low-dimensional space, that it would efficiently preserve original structural and topological network properties. Some methods, like matrix factorization, use an assumption that this mapping procedure is linear. Deep learning approach has demonstrated significant progress in various data analysis tasks and has also emerged in network

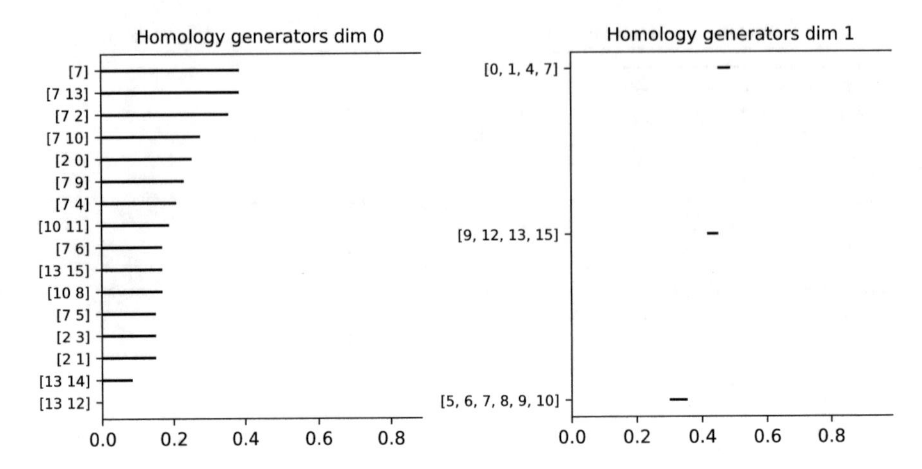

Fig. 11. Barcode plots annotated with cycle representatives for mixed network shown in Fig. 1(f). Left image represented a barcode plot with lifetimes of connected components, right – birth and death time of 1 cycles.

embedding field. It proposes tools which take into account possible nonlinearity in the network formation process. SDNE (Structural Deep Network Embedding) is the architecture based on semi-supervised deep autoencoder model, consisting of multiple non-linear mapping functions, transforming original data to highly non-linear latent space. Supervised and unsupervised components capture 1st and 2nd order proximities respectively. For learning 2nd order proximity, SDNE feeds to autoencoder high-dimensional neighborhood node structures, and then minimizes the reconstruction loss. Laplacian eigenmaps are used as a loss function for supervised component [26]. Another model, deep neural networks for graph representations (DNGR), applies deep learning method to the positive pointwise mutual information (PPMI) matrix. More accurately, the whole embedding process consists of the following steps: running a random surfing procedure on network, further calculation of PPMI matrix, and feature reduction using stacked denoising autoencoders (SDAE) [27]. Besides autoencoders, generative-adversarial networks (GAN) are frequently used in embedding model construction. Thus, adversarial framework ANE (adversarial network embedding), introduced in [28], consists of two main parts: structure preserving component and adversarial learning component. For structure preserving, an inductive DeepWalk model with parameterized function for embedding vectors generation is exploited. Then, adversarial learning works as a regularization process for learning robust feature extraction. [29] represents an example of adversarial network embedding applied for a cross-network node classification. Different types of architectures could be also combined in a single model.

Random Walks. Another way to extract information about structural relations in a network while focusing on the local neighbourhood structure of a distinct node is a random walk method. Random walk algorithms are largely based on the ideas of word embedding. According to skip-gram approach, incorporated in diverse natural language processing algorithms, the similarity of words is intuitively defined by their co-occurrence rate in a sentence. Skip-gram presumes the ability of a particular word to reconstruct its neighbourhood word vectors in a certain context window [7]. In case of network embedding, node sequences, captured throughout truncated random walks, are interpreted as sentences, and the node becomes an equivalent for a word. Hence, the node neighbourhoods could be reconstructed according to their co-occurrence rate in the random paths. The DeepWalk model, introduced in [6] in 2014, exploits an abovementioned combination of Skip-gram and truncated random walks. Another method, node2vec [30], modifies the random walk procedure by adding some flexibility in interpolating between breadth-first search (BFS) and depth-first search (DFS) strategies. Such flexibility allows to capture both local and global proximities in a network. Walklets [31] represents an extended version of DeepWalk with an opportunity for skipping some nodes in a random walk, aiming to preserve global proximity information. Besides, there is a DCA (diffusion component analysis) model [32], which consists of diffusion component analysis and random walks with restart: this algorithm was initially proposed for accurate functional annotation of objects in biological networks. Random walk and matrix factorization are often combined with or optimized by each other in particular models. There is no strict differentiation between them as for the baseline methods.

Node2vec Embedding. In the previous section various random walk methods were briefly reviewed, including node2vec model. This feature learning framework, introduced in [30], proposes several modifications to standard node neighborhood sampling strategies. More precisely, node2vec node-neighborhood construction algorithm utilizes biased random walks regularized by certain hyperparameters, thus efficiently interpolating between BFS and DFS sampling strategies. The node2vec model was successfully applied as a feature extraction framework in a variety of data analysis tasks, like item recommendation [33], gene prediction [34], community detection [35], etc. and is actively exploited in the node embedding field. Besides, this model has accessible and easy-operated implementations: source code is available at github.

4.2 TDA Embedding Algorithm

In network analysis with TDA approach, usually persistence-diagram vectorization is used as a data description tool. In this case we got embedding for networks in general. We suggested to use a cycle representative for each homology group as a node embedding technique. Below are the details of said approach.

- **Input data preprocessing**. Raw data, presented by a weighted adjacency matrix of a network, where each weight reflects the level of connection strength, was converted to a distance matrix, which defines network representation in a finite metric space in the form of a point cloud. So, all original connections were initially normalized and rescaled to interval $[0, 1]$. After that, $\log_{10}$ transform was applied and the absolute values of the results were stored in a final distance matrix. Thus, bigger distance values denoted weaker

connections, and zero distance corresponded to maximum level of connection, which equals to 1.
- **TDA procedure**. We utilized TDA for a preprocessed input distance matrix. The Eirene package was used for the analyses [18]. Persistent barcodes as well as cycle representatives for dimension 0 and 1 were collected.
- **Node embedding, dimension 0**. Feature extraction and embedding matrix construction slightly vary for TDA results collected from different dimensions. The number of persistent barcodes arising from dimension 0 corresponds to the number of nodes in a particular network. Indeed, we initiated the filtration process from separate points with zero distance to themselves, and after that started the merging process, during which we achieved the closing point, which disappeared and reappeared in the component that it merged with. So we encoded each node with a vector length n – number of nodes. We inputted the value of barcode length at the position j equal to the index of the closing point, and in case of absence of interaction we placed 0 value. We placed 1 at the diagonal elements, which corresponds to the interaction of node with itself. The barcode length was rescaled back to the range of connection strength. For example at Fig. 6 node with index 2 absorbed nodes 0,1,3, and was encoded as vector $[0.64, 0.72, 1, 0.58, 0..., 0]$
- **Node embedding, dimension 1**. For dimension 1 we used binning approach for barcode encoding. We split interval $[0, 1]$ into equally-sized elements, mapped them to the distance space with log 10 transform and counted for each node and for each bin the number of intervals where they participated. The same procedure potentially can be applied for higher dimensions.

4.3 TDA and Node2vec Embedding Comparisons

We tested TDA embedding by running an earlier described algorithm on different configurations of the WSBM model, mentioned in Sect. 2. We also constructed a node2vec embedding for each of these networks. The quality of embedding is hard to compare in the absence of specific tasks. One of the typical tasks in network analysis is community detection with a group membership of each node. As community detection task in network analysis is reduced to clustering task in vector space, we used dimension reduction techniques for visual representation of cluster structure. We decided against using clustering quality metrics, since we analysed networks with predefined properties and community structure. All networks, except assortative ones, demonstrated complex community interaction structure, which can still be examined visually. For visual representation in 2-dimensional space we used UMAP technique [36] as a dimension reduction tool. Our experimental setup for each case was as follows:

- computation of node2vec network embedding
- computation of TDA embedding
- visual inspection of embeddings with UMAP dimensional reduction

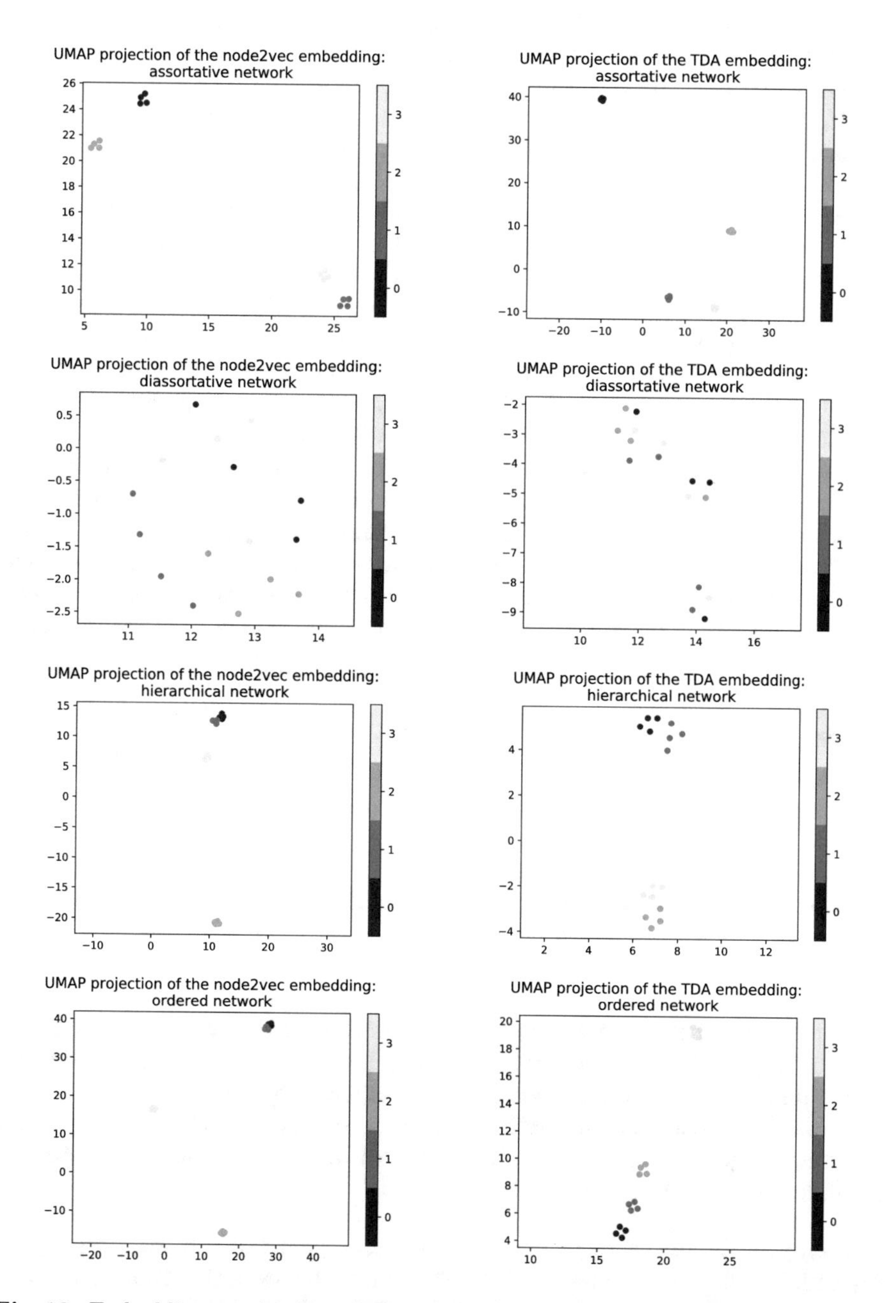

Fig. 12. Embedding results for WSBM network models. On the left side, node2vec embedding projections in UMAP are placed; on the right, there are UMAP projections for TDA embedding. Projection picture reflects network structure both for TDA and node2vec embedding.

Figure 12 displays the UMAP visualization of node embedding performed by node2vec and TDA algorithms for 4 different networks generated by WSBM models with assortative, disassortative, hierarchical and ordered structures. It can be seen that the node clusters, which have been formed after TDA embedding, highly reflect the original structure of the assortative, hierarchical and ordered models. In case of applying TDA embedding to a disassortative network, there were defined several node groups, that correspond to the image of the initial model, but it is hard to make any implications about the clustering structure of the network in whole – both for node2vec and TDA algorithms. These results show that TDA embedding can be used as an alternative approach to node description in community detection task. There is an additional meaningful interpretation of each element of vector embedding (see figures with barcode description). For example, in case of 'core-periphery' network Fig. 8 for node 0 we got connected component embedding: $[1, 0.88, 0.88, 1, 0, 0, 0, 0.57, 1, 0, 0, ..., 0.58, 0.85]$ which can be interpreted to mean strong connections within first four nodes (so we can suggest the existence of block to which node 0 belongs), and strong connection to several others nodes, which corresponds to core-periphery structure. From the dimension 1 embedding we can read participation of this node in three short-lived cycles, which can be interpreted as playing a 'bridge' role in the connection of different groups of nodes.

5 Conclusion

All things considered, it is crucial to take into account both network structure information and individual node characteristics for learning informative representation of weighted networks. As a source for individual node description we proposed using so-called 'cycle representatives' additionally to homology generator parameters. A possible node storytelling approach based on cycle representative analysis can be used for small-to-medium networks with attributed nodes. TDA embedding can be used as an alternative or addition to other network embedding methods, for example node2vec. According to the examples with the weighted stochastic block models, we showed that connected components along with 1-dimensional cycles can be used to reveal clusters in networks and even to point out the connection strength in the clusters, and to characterize network structure in general. This embedding can also be used as a feature vector for graph neural network modelling.

Acknowledgments. The reported study was funded by Russian Foundation for Basic Research (RFBR), project number 19-07-00337 and by the Institute of Information and Computational Technologies (Grant AR05134227, Kazakhstan).

References

1. Newman, M.E.J.: Analysis of weighted networks. Phys. Rev. E **70**, 056131 (2004)
2. Leskovec, J., Lang, K.J., Mahoney, M.: Empirical comparison of algorithms for network community detection. In: Proceedings of the 19th International Conference on World Wide Web, pp. 631–640 (2010)
3. Aicher, C., Jacobs, A.Z., Clauset, A.: Learning latent block structure in weighted networks. J. Complex Networks **3**(2), 221–248 (2015)
4. Cui, P., Wang, X., Pei, J., Zhu, W.: A survey on network embedding. IEEE Trans. Knowl. Data Eng. **31**(5), 833–852 (2018)
5. Giusti, C., Ghrist, R., Bassett, D.S.: Two's company, three (or more) is a simplex. J. Comput. Neurosci. **41**(1), 1–14 (2016)
6. Perozzi, B., Al-Rfou, R., Skiena, S.: Deepwalk: Online learning of social representations. In: Proceedings of the 20th ACM SIGKDD International Conference on Knowledge Discovery and Data Mining, pp. 701–710 (2014)
7. Mikolov, T., Chen, K., Corrado, G., Dean, J.: Efficient estimation of word representations in vector space. arXiv preprint arXiv:1301.3781 (2013)
8. Sizemore, A.E., Giusti, C., Kahn, A., Vettel, J.M., Betzel, R.F., Bassett, D.S.: Cliques and cavities in the human connectome. J. Comput. Neurosci. **44**(1), 115–145 (2018)
9. Wang, S., Aggarwal, C., Tang, J., Liu, H.: Attributed signed network embedding. In: Proceedings of the 2017 ACM on Conference on Information and Knowledge Management, pp. 137–146 (2017)
10. Faskowitz, J., Yan, X., Zuo, X.N., Sporns, O.: Weighted stochastic block models of the human connectome across the life span. Sci. Rep. **8**(1), 1–16 (2018)
11. Karrer, B., Newman, M.E.: Stochastic blockmodels and community structure in networks. Phys. Rev. E **83**(1), 016107 (2011)
12. Giusti, C., Pastalkova, E., Curto, C., Itskov, V.: Clique topology reveals intrinsic geometric structure in neural correlations. Proc. Natl. Acad. Sci. **112**(44), 13455–13460 (2015)
13. Ghrist, R.: Barcodes: the persistent topology of data. Bull. Am. Math. Soc. **45**(1), 61–75 (2008)
14. Knyazeva, I., Poyda, A., Orlov, V., Verkhlyutov, V., Makarenko, N., Kozlov, S., Velichkovsky, B., Ushakov, V.: Resting state dynamic functional connectivity: network topology analysis. Biol. Insp. Cognit. Archit. **23**, 43–53 (2018)
15. Rabadán, R., Blumberg, A.J.: Topological Data Analysis for Genomics and Evolution: Topology in Biology. Cambridge University Press, Cambridge (2019)
16. Zomorodian, A., Carlsson, G.: Computing persistent homology. Discr. Comput. Geom. **33**(2), 249–274 (2005)
17. Edelsbrunner, H., Letscher, D., Zomorodian, A.: Topological persistence and simplification. In: Proceedings 41st Annual Symposium on Foundations of Computer Science, pp. 454–463. IEEE (2000)
18. Henselman, G., Ghrist, R.: Matroid Filtrations and Computational Persistent Homology. ArXiv e-prints, June 2016
19. Edelsbrunner, H., Harer, J.: Persistent homology-a survey. Contemp. math. **453**, 257–282 (2008)
20. Wasserman, L.: Topological data analysis. Ann. Rev. Stat. Appl. **5**(1), 501–532 (2018)
21. Makarenko, I., Shukurov, A., Henderson, R., Rodrigues, L.F., Bushby, P., Fletcher, A.: Topological signatures of interstellar magnetic fields–i. betti numbers and persistence diagrams. Monthly Not. Roy. Astronom. Soc. **475**(2), 1843–1858 (2018)

22. Ahmed, A., Shervashidze, N., Narayanamurthy, S., Josifovski, V., Smola, A.J.: Distributed large-scale natural graph factorization. In: Proceedings of the 22nd International Conference on World Wide Web, pp. 37–48 (2013)
23. Cao, S., Lu, W., Xu, Q.: Grarep: Learning graph representations with global structural information. In: Proceedings of the 24th ACM International on Conference on Information and Knowledge Management, pp. 891–900 (2015)
24. Ou, M., Cui, P., Pei, J., Zhang, Z., Zhu, W.: Asymmetric transitivity preserving graph embedding. In: Proceedings of the 22nd ACM SIGKDD International Conference on Knowledge Discovery and Data Mining, pp. 1105–1114 (2016)
25. Wang, X., Cui, P., Wang, J., Pei, J., Zhu, W., Yang, S.: Community preserving network embedding. In: Thirty-first AAAI Conference on Artificial Intelligence (2017)
26. Wang, D., Cui, P., Zhu, W.: Structural deep network embedding. In: Proceedings of the 22nd ACM SIGKDD International Conference on Knowledge Discovery and Data Mining, pp. 1225–1234 (2016)
27. Cao, S., Lu, W., Xu, Q.: Deep neural networks for learning graph representations. In: Thirtieth AAAI Conference on Artificial Intelligence (2016)
28. Dai, Q., Li, Q., Tang, J., Wang, D.: Adversarial network embedding. In: Thirty-Second AAAI Conference on Artificial Intelligence (2018)
29. Shen, X., Dai, Q., Chung, F.l., Lu, W., Choi, K.S.: Adversarial deep network embedding for cross-network node classification. arXiv preprint arXiv:2002.07366 (2020)
30. Grover, A., Leskovec, J.: node2vec: scalable feature learning for networks. In: Proceedings of the 22nd ACM SIGKDD International Conference on Knowledge Discovery and Data Mining, pp. 855–864 (2016)
31. Perozzi, B., Kulkarni, V., Skiena, S.: Walklets: Multiscale graph embeddings for interpretable network classification. arXiv preprint arXiv:1605.02115 (2016)
32. Cho, H., Berger, B., Peng, J.: Diffusion component analysis: unraveling functional topology in biological networks. In: International Conference on Research in Computational Molecular Biology, pp. 62–64. Springer (2015)
33. Palumbo, E., Rizzo, G., Troncy, R., Baralis, E., Osella, M., Ferro, E.: Knowledge graph embeddings with node2vec for item recommendation. In: European Semantic Web Conference, pp. 117–120. Springer (2018)
34. Peng, J., Guan, J., Shang, X.: Predicting Parkinson's disease genes based on node2vec and autoencoder. Front. Genet. **10** (2019)
35. Hu, F., Liu, J., Li, L., Liang, J.: Community detection in complex networks using node2vec with spectral clustering. Physica A: Stat. Mech. Appl. 123633 (2019)
36. McInnes, L., Healy, J., Melville, J.: Umap: uniform manifold approximation and projection for dimension reduction. arXiv preprint arXiv:1802.03426 (2018)

A Technique to Infer Symbolic and Socio-symbolic Micro Patterns

Artem Antonyuk[1,2](✉), Kseniia Puzyreva[1], Darkhan Medeuov[1], and Nikita Basov[1,2]

[1] Centre for German and European Studies, St. Petersburg State University, St. Petersburg, Russia
{a.antonyuk,n.basov}@spbu.ru, kseniiapuzyreva@gmail.com, darkhan_medeuov@yahoo.com

[2] Faculty of Sociology, St. Petersburg State University, St. Petersburg, Russia

Abstract. The interplay between symbolic and social structures in groups is often analysed at the whole-network level of their semantic and socio-semantic networks, e.g. via comparison of graph distributions, multidimensional scaling, or QAP correlations. Meanwhile, the interplay between the symbolic and the social operates through the usage of signs (e.g. words) and their associations by interacting individuals. Hence, structural properties of the whole network can be explained by analysing specific instances of symbolic and socio-symbolic micro patterns – elementary configurations linking signs, and signs and individuals – occurring in practical contexts. This paper introduces a technique and a customisable pattern retriever tool (an R script) to (1) programme socio-symbolic patterns of theoretical importance, (2) use them as 'search terms' to query network data, (3) extract from the data instances of the patterns and text quotes corresponding to them, (4) store and represent these instances and quotes in a form convenient for their subsequent qualitative analysis – to uncover the contextual meanings of the patterns. We illustrate the proposed technique with an analysis of a mixed dataset on the interplay between expert and local symbolic structures in the context of social structures of two local groups engaged in flood risk management in 2019 England.

Keywords: Pattern retrieval · Socio-semantic network · Symbolic structure · Socio-symbolic structure

1 Introduction

Increasing attention is being paid to the co-evolution of symbolic and social structures [1–6]. The interplay between these structures in social groups is often approached at the whole-network level of structure and content of groups' (socio-)semantic networks, for instance, via comparison of graph distributions [7, 8], multidimensional scaling and QAP correlations [9], or hyperbolic spaces [10]. Meanwhile, the symbolic and the social interplay through the exchanges of signs (e.g. words) and their meaningful associations in utterances by particular individuals interacting in practical contexts [11, 12]. Therefore,

A. Antonyuk and N. Basov (Eds.): NetGloW 2020, LNNS 181, pp. 101–119, 2021.
https://doi.org/10.1007/978-3-030-64877-0_7

structural properties and content of symbolic and social structures should be explained through an analysis of their micro patterns – elementary configurations linking signs, and signs and socially tied individuals – as they occur in concrete verbal expressions and interactions against the backgrounds of broader cultural and social contexts [13, 14]. So far, however, there has been a lack of techniques and tools to facilitate the systematic extraction of such patterns from empirical data.

The present paper addresses this gap and introduces a technique and a customisable pattern retriever tool – an R [15] script – for semi-automatic computer-aided extraction of (socio-)symbolic micro patterns from empirical network data. The technique and the tool enable the specification of patterns of theoretical importance, accurate and fast retrieval of instances of specified patterns from semantic and socio-semantic networks, and extraction of textual contexts (phrases and sentences) of the retrieved patterns' components from the data for subsequent qualitative analysis [see 14]. As the majority of operations are carried out within the R environment, this tool minimises the possibility of unintentional errors and data loss because of conversion between different data formats. The technique and the tool can retrieve patterns from longitudinal as well as cross-sectional data.

The paper is organised as follows. First, we conceptually introduce patterns of the interplay between symbolic structures, and between symbolic and social structures. Second, we introduce the technique and the tool for extracting instances of patterns from the data. Finally, we illustrate the technique and the tool using our mixed dataset on local and expert groups engaged in flood risk management gathered in England in 2019.

2 Symbolic and Socio-symbolic Patterns

Social groups use signs to refer to objects, actors, actions, and situations. By associating signs in particular ways, groups express their specific identities and perspectives on reality [16]. For example, by associating 'river' with 'flooding', group members express their shared understanding of a river as something that might flood surrounding areas. This meaning of the river may be irrelevant to groups that occupy riverbanks but never experienced floods. Signs and associations continuously used by a group constitute a symbolic structure of this group. The signs and associations between them are the focal subject for the analysis of the interplay between symbolic structures of different groups [14].

The interplay between symbolic structures involves the mutually induced reproduction and/or change of signs and their associations used by different groups. For instance, consider symbolic structures of expert groups, e.g. scientists, politicians, or representatives of NGOs. Usually developed in a comprehensive and professional manner, symbolic structures of expert groups contain authoritative definitions of issues and solution proposals in corresponding fields of expertise [17]. These definitions, often expressed in scientific research, political programmes, and laws, are imposed through authoritative language to be adopted and enacted by local groups (e.g. ordinary citizens, indigenous people) in a field [e.g. 18].

Meanwhile, symbolic structures of local groups are developed in a more spontaneous manner in everyday practice [e.g. 19, 20]. These symbolic structures include local news and rumours, stories, jokes, etc., which are mostly relevant for a particular local group and have little relevance for others. Owing to the difference in social statuses of expert and local groups, locals are subjected to significant institutional pressures to adopt certain meanings (such as definitions of situations, issues, models, and best practices) from experts [21]. Thus, local symbolic structures can be regarded as 'dependent' in relation to 'independent' expert symbolic structures.

At the same time, local groups often resist institutional pressures [22–24] and reinterpret elements of expert symbolic structures, instantiating them locally [14, 25]. This way, locals simultaneously meet institutional expectations [26, 27] and preserve their own definitions of situations and of their place in them [22, 26]. Transformation of 'dependent' symbolic structures under the influence of 'independent' symbolic structures reveals itself in language and can be traced using *symbolic patterns* that capture addition, change, and removal of signs and meaningful associations between mutually defining signs over time.

Individuals constitute their common perspectives on reality by using and combining signs as they interact in a group and refer to its common context [12]. They may reproduce the existing signs and associations between signs, recombine signs, or introduce new ones [28, 29]. This mostly happens in the context of dyadic and triadic social ties between group members and, therefore, the group's symbolic structure relies on the structure of social ties within the group [30–34]. Hence, the effect of social structure on the symbolic structure should be controlled for when examining the interplay between symbolic structures. It can be traced through *socio-symbolic patterns* that combine social ties between individuals with signs and associations between signs they use.

Based on the existing literature, we theorised a number of symbolic [35] and socio-symbolic [36] patterns of the interplay between different groups' symbolic structures in the context of their local social structures. For illustrative purposes, the further presentation of our pattern inference technique relies on one pattern of each type.

The symbolic pattern *loose coupling* reflects how a sign from one symbolic structure is reinterpreted in another symbolic structure [37]. This pattern represents a process when actors from one group (e.g. locals) reproduce and reinterpret signs used by another group (e.g. experts), such as specific terms, in the context of their own symbolic structure, fitting them to their purposes. More specifically, it implies that a group's sign used at t_1 is reproduced by another group at t_2 in a new association with a pre-existing sign specific to the second group (see Fig. 1).

As reflected in Fig. 1, locals may appropriate from experts the idea of producing plans, start to use the sign 'plan', and adapt it to their own local context by associating it with the sign 'neighbourhood'. Simultaneously, they ignore the original experts' understanding of this idea represented by the association between the signs 'plan' and 'management'.

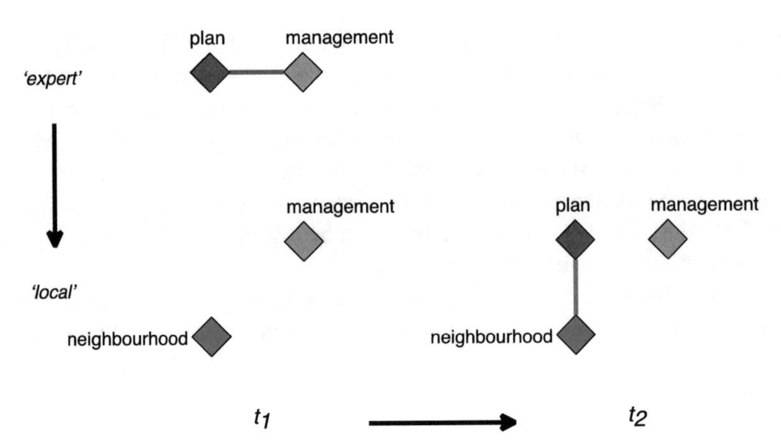

Fig. 1. Symbolic pattern 'loose coupling'. Blue diamonds = signs used only by locals; red diamonds = locally reproduced expert sign; orange diamonds = sign used by experts and locals; red line = expert-specific sign association; blue line = new local-specific sign association.

The socio-symbolic pattern *contagion* implies the reproduction of signs and/or associations between signs used by a single member of a group by other group members who were not yet using them. Such signs and/or associations become shared as a result of direct interaction [38–43] and, hence, become parts of group symbolic structure. Specifically, contagion implies that a sign and/or an association between signs used by one individual at t_1 is reproduced at t_2 by another individual socially tied to the first one (see Fig. 2).

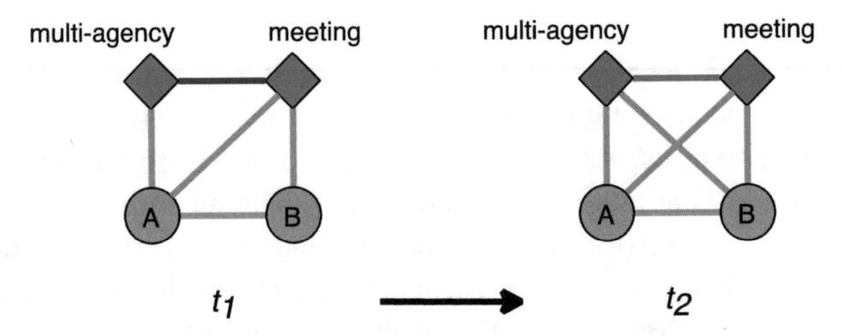

Fig. 2. Socio-symbolic pattern 'contagion'. Green circles = individuals; green lines = social tie; blue diamonds = signs; purple line = unshared association between signs; blue line = shared association between signs; grey lines = usage of signs.

Figure 2 represents a situation when a local group member, A, interacts with another group member, B, and uses the term 'multi-agency meeting' that refers to a type of meeting with officials. At the next point in time, B learns this special term and reproduces the sign 'multi-agency', associating it with the previously known sign 'meeting'.

3 Technique and Tool to Extract Instances of Symbolic and Socio-symbolic Micro Patterns

Symbolic and socio-symbolic patterns can be traced in semantic and socio-semantic networks that contain information on social ties and/or usage of signs and meaningful associations between them in different social groups[1].

The proposed technique and the pattern retriever tool to infer symbolic and socio-symbolic patterns involve the construction of semantic and socio-semantic networks using empirical data; programming of network configurations for symbolic and socio-symbolic patterns; semi-automatic search, storage, and visualisation of instances of patterns in the networks, and extraction of textual contexts in which the patterns occurred, for further qualitative analysis.

3.1 Producing Semantic and Socio-semantic Networks

To produce 'independent' and 'dependent' semantic and socio-semantic networks for two points in time, we map semantic, social, and sign usage networks from empirical data that include texts and sociometric surveys.

Semantic networks are produced from textual data based on the co-occurrence of signs (words) within a certain textual context in sentences. First, the texts representing a symbolic structure of a group at a certain point in time are combined in a corpus using the *quanteda* [44] package in R. Then, the texts in the corpus are converted into sets of all grammatical forms of words encountered in the texts, using the *UDPipe* [45] package. The words are tagged with a corresponding part of speech (POS) and then converted into their dictionary forms through lemmatisation. The lemmatised words are then combined with POS tags (e.g. 'flood_verb' and 'flood_noun') to allow distinguishing between different meanings of the same word when it is used as a noun, a verb, or an adjective. We consider these lemmatised words combined with their POS tags as signs. Additionally, punctuation is removed except for full stops to preserve sentence boundaries. Then, signs other than nouns, verbs, and adjectives, as well as signs from a customised stop list, are marked to be omitted from the networks. Finally, the co-occurrence of signs within a textual context ('window') of several signs (unless separated by a full stop, i.e. sentence boundary) is counted for each text in the corpus. Note that the size of the 'window' has to be adjusted to optimally capture meaningful associations between signs depending on the type and amount of original textual data [see 44, 45]. These co-occurrence counts are used to produce lists of nodes (signs) and links (sign associations) that are then converted into semantic networks using *igraph* package [46]. The semantic networks are then binarised using researcher-defined threshold values to retain stable meaning structures. Different binarisation threshold values can be chosen, depending on a type

[1] Note that to trace symbolic patterns in the most accurate way, the 'independent' symbolic structure has to be captured at least at a single point in time, and the 'dependent' symbolic structure has to be captured at least at two points in time. This way, it is possible to trace changes in appearance of signs and sign associations in the 'dependent' symbolic structure at t_2 compared to the same symbolic structure at t_1 and to the 'independent' symbolic structure at t_1. While longitudinal data are desirable, cross-sectional data can also be used.

and size of textual data in a corresponding corpus. For example, if texts in a corpus contain many complex associations between signs (e.g. large written texts such as documents), a higher threshold can be chosen. If texts contain fewer complex associations between signs (e.g. sources from oral speech such as transcripts of loosely structured interviews), a lower threshold can be used.

After the creation of semantic networks per each text in the corpus, the semantic networks corresponding to the corpus are combined in a threshold-based merge semantic network. This network includes all links that appear in at least n semantic networks, where n is a researcher-defined threshold. For our purposes, we create merge networks using n of 1 and 2. The merge semantic network based on the threshold of 1 includes signs and associations between signs that occur in one or more semantic networks of texts in the corpus. This merge network preserves information on exclusive and common signs and associations between signs in the corpus that is needed to trace the socio-symbolic patterns. The merge semantic network based on the threshold of 2 contains signs and associations between signs used in two and more semantic networks of texts in the corpus. It preserves information only on common signs and associations between signs and dismisses those used only in a single text as irrelevant or idiosyncratic (this applies for tracing symbolic patterns). Furthermore, isolated signs in the merge networks are deleted.

Social networks are mapped by importing sociometric matrices in R and converting them into networks using *igraph* package. Then, the resulting social networks are binarised.

Bipartite sign usage networks are produced by creating nodes representing different texts in the corpora and linking them with the signs used in those texts. Since the texts can be associated with individuals from the social networks produced earlier, the bipartite network represents individuals and all signs they used.

A socio-semantic network is produced as a union of a social network, a bipartite sign usage network, and a threshold-based merge semantic network.

Further, to find and extract symbolic and socio-symbolic patterns of the interplay between different symbolic structures in the context of social ties, we construct combined quasi-longitudinal semantic and socio-semantic networks. To enable the construction of such networks, we have developed a special coding scheme that reflects usage of signs and associations between signs by two types of actors, as well as the occurrence of social ties at different points in time.

3.2 Coding Scheme

The coding scheme is used to apply codes to the threshold-based merge semantic and socio-semantic networks and to interpret the codes in the combined quasi-longitudinal networks. For illustrative purposes, we describe the scheme that encodes social ties of local actors, the usage of signs by local actors, and occurrence of associations between signs in expert and local texts at one and two points in time.

The coding scheme uses three exclusive sets of numerical codes (see Table 1). The first one is the 'basic' set, containing three codes, '1', '2', and '5'. Each code in this set indicates the usage of signs and associations between signs by locals or experts at a single point in time captured in the locals' networks at t_1 and t_2 and the experts' network at t_1, respectively. The first two codes in the set also indicate occurrence of social ties at t_1 and t_2. The second set is 'cumulative', containing four codes, '3', '6', '7', and '8'. The codes from the 'cumulative' set capture all possible cases of the usage of signs and their associations in longitudinal data at more than one point in time by more than one type of actor, as well as the occurrence of social ties. Each code in the 'cumulative' set corresponds to a sum of two or three code values from the 'basic' set. For example, the code '3' is a sum of the values of the codes '1' and '2', which represent locals' node and link usage at the first and the second point in time, respectively. Furthermore, each code in the 'cumulative' set corresponds to only one possible combination of code values from the 'basic' set. Thus, one may unambiguously interpret the 'cumulative' code values of nodes and links in the combined quasi-longitudinal networks to find out to which point(s) in time and to which type(s) of actor(s) the sign usage links and associations between signs correspond, and at which point in time the social ties occur. In addition, the third 'node type' set contains two codes, '10' and '11', that are used to designate the types of nodes in the quasi-longitudinal socio-semantic networks.

Table 1. The coding scheme for threshold-based merge semantic and socio-semantic networks. Codes indicate social ties and local and/or expert usage of signs and associations between them at the first and/or the second point in time, as well as the type of nodes.

Code	Description
	'Basic' set
1	Locals at t_1
2	Locals at t_2
5	Experts at t_1
	'Cumulative' set
3	Locals at t_1 and t_2
6	Experts and locals at t_1
7	Experts at t_1 and locals at t_2
8	Experts at t_1 and locals at t_1 and t_2
	'Node type' set
10	Individual
11	Sign

3.3 Constructing a Combined Quasi-longitudinal Semantic Network

As the input for creating a combined quasi-longitudinal *semantic* network, three threshold-based merge *semantic* networks are used: 'independent' expert network at t_1 (based on the threshold of 1) and 'dependent' local networks at t_1 and t_2 (based on the threshold of 2). The 'independent' semantic network contains common as well as exclusive signs and associations between signs that are used relatively regularly (i.e. the frequency of their usage is above a binarisation threshold chosen at the stage of semantic mapping). The signs and associations between signs in the 'dependent' semantic networks are relatively common (used in at least two texts in the corresponding corpus) and relatively regular (as defined by a chosen binarisation threshold). In all networks, the signs are associated with at least one other sign (i.e. there are no isolated signs).

Using the *igraph* package in R, for each merge semantic network, the codes are assigned to all signs (as node attributes) and associations between signs (as link values) according to the developed coding scheme (see Table 1). Then, the networks are converted into data frame format to enable further manipulations. The data frames are combined, automatically summing the assigned code values. Then, the resulting data frame is converted back into the network format. This procedure results in the *combined quasi-longitudinal semantic* network where sign and sign association codes indicate a point in time at which they were used in the 'independent' and the 'dependent' symbolic structures (see the schematic representation of the union procedure in Fig. 3).

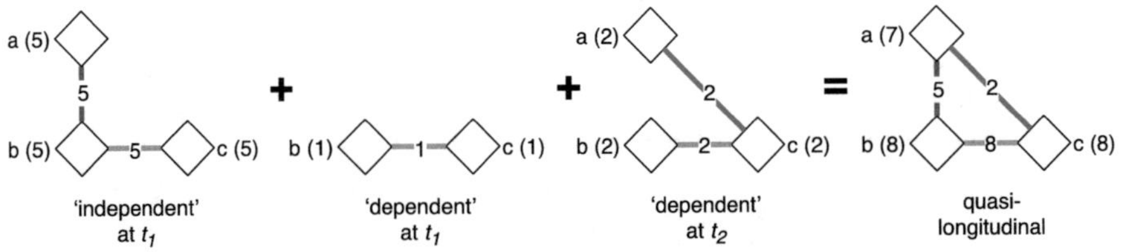

Fig. 3. Procedure for creating a combined quasi-longitudinal semantic network. Toy semantic networks, left to right: 'independent' at t_1, 'dependent' at t_1, 'dependent' at t_2, combined quasi-longitudinal. Diamonds = signs; red lines = association between signs used only in the 'independent' network; blue lines = association between signs used only in the 'dependent' networks; orange line = association between signs used in both types of networks. Letters in node labels shown for illustrative purposes; sign and sign association codes are shown as node and link labels.

3.4 Constructing a Combined Quasi-longitudinal Socio-semantic Network

As the input for constructing a combined quasi-longitudinal *socio-semantic* network, we use 'dependent' merge *semantic* networks based on the threshold of 1 (where signs and associations between signs are not necessarily common[2]), *sign usage* networks and *social* networks for t_1 and t_2.

In R, for each association between signs in each merge semantic network, the *igraph* package is used to code whether a particular association between signs occurred in a particular text from a corpus and hence was used by a particular individual or not. Specifically, for each association between signs, we add attributes with codenames of all individuals in a group, such as 'A', 'B', 'C', and fill the attributes with binary values indicating their (non-)usage by corresponding individuals. For example, for an association between signs used only by A and C, the attributes 'A', 'B', and 'C' would have the codes '1', '0', and '1', respectively.

[2] Considering unshared signs and associations between them allows us to trace their introduction into group symbolic structure presupposed by several socio-symbolic patterns.

Then, for each point in time, the threshold-based merge semantic, sign usage and social networks are combined into a *socio-semantic* network that represents ties between individuals using particular signs and associations between signs. Next, in each resulting socio-semantic network, a code is assigned to all social ties, sign usage links, and associations between signs (as link values), reflecting their usage at a specific point in time according to the coding scheme (see Table 1).

Finally, the socio-semantic networks at t_1 and t_2 are combined through a union procedure, automatically summing the code values. In the resulting network, additional codes are assigned to all nodes (as node attributes) indicating whether a node represents an individual (coded as 10) or a sign (coded as 11). This procedure results in a *combined quasi-longitudinal socio-semantic* network (see the schematic representation of the union procedure in Fig. 4).

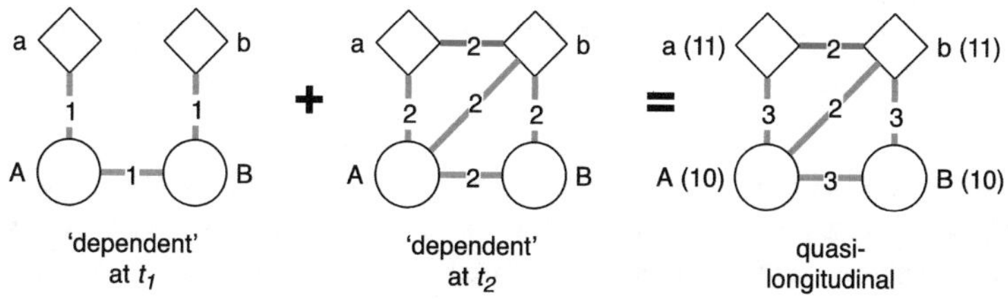

Fig. 4. Procedure for creating a combined quasi-longitudinal socio-semantic network. Toy socio-semantic networks, left to right: 'dependent' at t_1, 'dependent' at t_2, quasi-longitudinal. Diamonds = signs; circles = individuals; blue lines = association between signs; green lines = social tie between individuals; grey lines = usage of signs by individuals. Letters in node labels shown for illustrative purposes; sign association codes shown as link labels. Node codes reflect node types.

3.5 Programming Configurations for Symbolic and Socio-symbolic Patterns

To represent theoretically derived symbolic and socio-symbolic patterns in a computer-readable format, we programme network configurations for each pattern using *igraph* package. Programmed network configurations represent patterns as nodes connected by links. The nodes and the links in a pattern are assigned codes according to the coding scheme (see Table 1), based on the theoretical description of a pattern.

In programmed network configurations for symbolic patterns, nodes represent signs and links represent associations between signs. Consider the example of the configuration for the symbolic *loose coupling* pattern that concerns the interplay between expert and local symbolic structures. The pattern indicates that locals reproduce experts' sign *c* associating it with their pre-existing sign *a,* while ignoring the experts' association between *c* with *b* (see Fig. 5). The programmed configuration for this pattern consists of three nodes *a, b,* and *c* connected by two links overlapping through the node *c.* All information about the usage of signs and associations between signs characteristic of the loose coupling pattern is reflected in the codes assigned to the nodes and the links in the programmed configuration according to the coding scheme. For example, the node *c* is assigned the code '7' that indicates that experts use the sign *c* at the first point in time (coded as '5') and that locals use the same sign at the second point in time (coded as '2'). Another node, *a,* (coded as '3') corresponds to the sign *a* that locals use at both points in time (coded as '1' and '2') while experts do not use at all, and that becomes associated with experts' sign *c* at the second point in time (the sign association coded as '2'). The remaining node and link are coded according to the same logic.

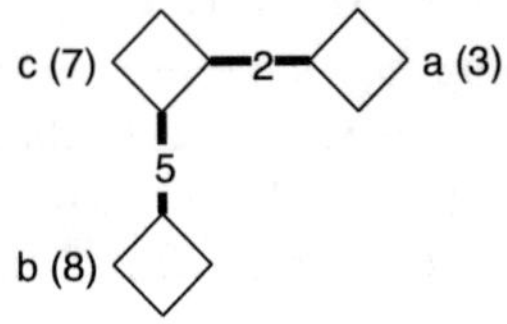

Fig. 5. Programmed network configuration for pattern 'loose coupling'. Diamonds = signs; lines = associations between signs. Sign and sign association codes shown as node and link labels. Letters in node labels are not part of the programmed configuration and are shown for illustrative purposes.

In programmed network configurations for socio-symbolic patterns, nodes represent individuals and signs, and links represent social ties, sign usage by individuals, and associations between signs. Links are assigned with codes indicating their occurrence at one or two points in time according to the coding scheme (see Table 1). In addition, nodes are assigned with codes indicating whether they represent an individual ('10') or a sign ('11').

Programmed configurations are stored in R as *igraph* objects. They will be used to find instances of patterns in empirical data.

3.6 Finding, Extracting, and Visualising Instances of Patterns

In this section, we describe the pattern retriever, an R script for semi-automatic retrieval of patterns from network data. The tool allows us to find, extract, and visualise instances of symbolic and socio-symbolic patterns of the interplay between symbolic structures in the context of social ties, as well as to find and extract textual contexts of associations between signs part of the instances of the patterns. The pattern retriever conducts the following operations.

1. **Network pre-processing.** The combined quasi-longitudinal networks are pre-processed: links are symmetrised, and multiple edges and self-loops are removed.
2. **Pattern retrieval.** Using standard *igraph* functions, the previously programmed network configurations are used as 'search terms' to find parts of the combined quasi-longitudinal semantic and socio-semantic networks that correspond to the symbolic and socio-symbolic patterns. This operation is implemented as follows. First, for each pattern, the algorithm looks up parts of the quasi-longitudinal network that correspond to a programmed configuration for a pattern and extracts a list of nodes corresponding to the pattern. The difficulty is to extract the links that correspond to the pattern, given that not all links connecting the extracted nodes in the quasi-longitudinal network correspond to this pattern. Consider an instance of the pattern

a–c–b found in a quasi-longitudinal network. While the quasi-longitudinal network contains nodes *a, b,* and *c,* it may also contain a link *a–b* that does not correspond to the pattern. To retrieve only those links that correspond to a pattern, the algorithm makes use of link codes stored in the programmed network configurations and the quasi-longitudinal networks. First, the algorithm extracts *all* links connecting the derived nodes, including those that do not correspond to a queried pattern. Then, to remove unrelated links, the algorithm filters the extracted links based on the code values of links in the programmed network configuration for the queried pattern. Finally, the algorithm constructs networks for separate extracted instances of the pattern and for all its instances, which are stored as separate *igraph* objects in R.

3. **Pattern visualisation.** To enable qualitative analysis, visualisations of the extracted instances of patterns are created. For each pattern, *all* its instances are visualised in a single network plot (see example in Fig. 6). These visualisations allow the in-depth examination of the interplay between symbolic structures as well as between social and symbolic structures. They enable an understanding of the structural organisation of patterns. For example, visualisations allow us to identify the most central signs (that appear in many instances of a pattern) that would have to be subjected to further qualitative analysis.

   In addition, every single instance of a pattern is visualised in a separate network plot (see Fig. 7). For each pattern, all these visualisations are exported for further qualitative analysis.
4. **Textual contexts extraction.** Finally, to facilitate further qualitative analysis, for each instance of a pattern, functions in the *quanteda* package are used to semi-automatically extract all textual contexts containing the associations between signs appearing in that instance of a pattern from a corresponding textual corpus.

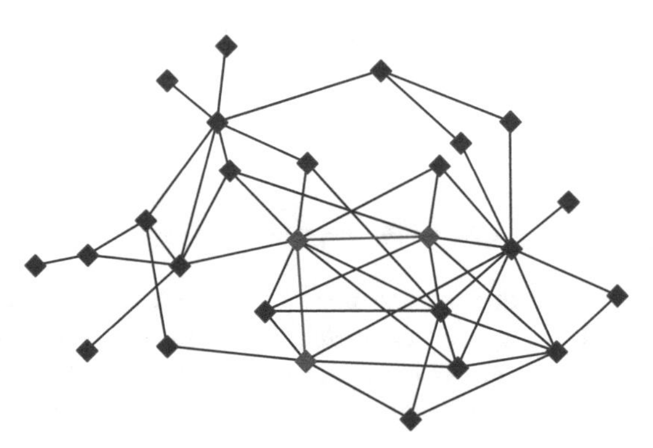

Fig. 6. Extracted toy network containing all instances of the 'loose coupling' pattern. Diamonds = signs; lines = associations between signs. Sign labels and codes are hidden. A single instance of the pattern is shown in red for illustrative purposes.

In the next section, we illustrate the application of the technique and the pattern retriever tool to analyse the interplay between expert and local symbolic structures in the context of social ties in two local groups engaged in flood risk management.

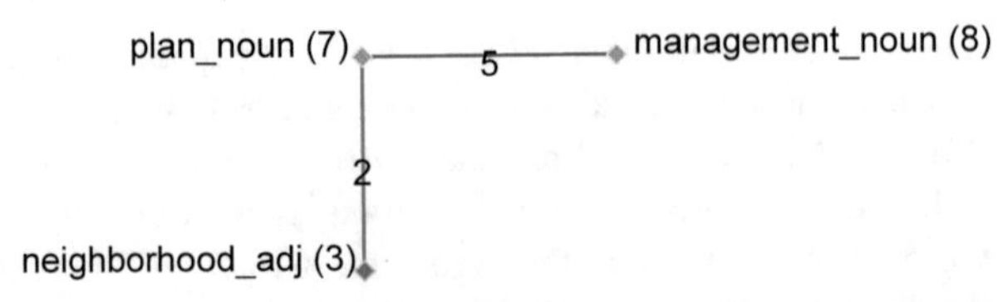

Fig. 7. A single instance of 'loose coupling' pattern. Blue diamond and line = sign and association used only in the 'dependent' network; orange diamonds = signs used in 'independent' and 'dependent' networks; red line = association between signs used only in the 'independent' network. Sign labels show lemmas, part-of-speech information, and codes; link labels show sign association codes.

4 Illustration

4.1 Data

The source of test data is an ethnographic study of two local flood management groups in England. The data were collected during six weeks of fieldwork in two villages in the County of Shropshire, England.

The dataset consists of cross-sectional textual data on expert and local symbolic structures as well as of sociometric data on relationships between local flood group members. The data representing expert symbolic structures were collected in the form of relevant documents (totalling around 316,000 words) produced by official flood risk management agencies and authorities to inform local groups and other stakeholders about flood risk management measures, activities, and strategies. Information on local symbolic structures was collected through semi-structured interviews with 15 members of the two 'local flood groups' voluntarily involved in flood risk management in the two villages (henceforth, LFG I and LFG II). The corpus of interview transcripts contains around 186,000 words. The data on social relationships, i.e. friendship and collaboration, within the local groups were collected using sociometric surveys. The data were processed as described in the previous section to produce combined cross-sectional semantic and socio-semantic networks[3].

4.2 Symbolic and Socio-symbolic Patterns in Two English Flood-Prone Groups

Applying the described technique and the tool to our empirical data, we extracted instances of 17 symbolic and 7 socio-symbolic theoretically proposed patterns to manually confirm their ethnographic relevance. In what follows, we provide illustrations of such manual evaluation using one symbolic pattern, *loose coupling,* and one socio-symbolic pattern, *contagion.*

[3] The semantic networks were mapped based on co-occurrence of signs within the window of 9 (i.e. separated by 7 signs) in the texts. Signs from a customized stop list as well as those with part of speech other than noun, verb, or adjective, were not included in the networks. Semantic networks for local symbolic structures contain signs and their associations used at least two times. Semantic networks for expert symbolic structures contain signs and their associations used at least eight times.

As described in Sect. 2, the pattern *loose coupling* reflects how a sign from an 'independent' symbolic structure is reproduced within a 'dependent' symbolic structure being associated with a pre-existing sign specific to the 'dependent' structure. Following the procedure described in Sect. 3, we started with extracting all the instances of the pattern from the combined semantic network. Then, we visualised all the instances of the pattern in a network plot. For illustrative purposes, we focus on the instances of the loose coupling pattern involving one expert sign reproduced by the locals, 'plan' (see Fig. 8).

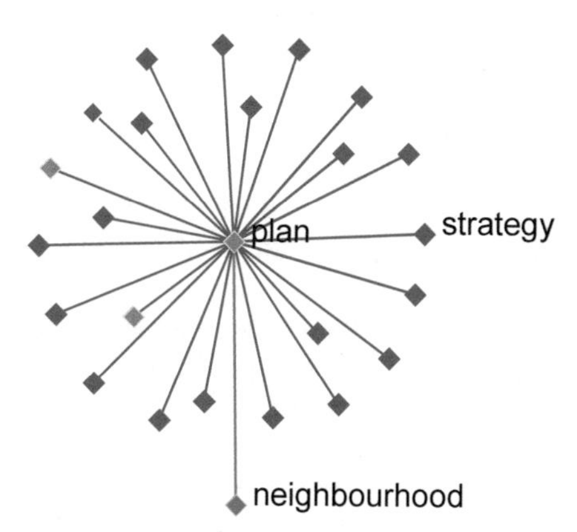

Fig. 8. Instances of the pattern 'loose coupling' containing the sign 'plan'. Red diamonds and lines = signs and their associations used only by the experts; blue diamond and line = sign and association between signs used only by the locals; orange diamonds = signs used by the experts and locals. For illustrative purposes, only labels of signs that appear in the example are shown. Some signs are hidden to reduce visualisation complexity. Signs' part-of-speech is hidden.

The visual representation demonstrates that the sign 'plan' used by the experts and locals has many more associations in the expert symbolic structure than in the local one. This reveals that the meaning of 'plan' is more elaborated for the experts than for the locals. It is, therefore, likely that the associations containing this sign are imposed on the locals. For example, the association between 'plan' and 'strategy' is an expert-specific association. Meanwhile, the association between 'plan' and 'neighbourhood' as well as the latter sign itself are used only by the locals. This means that the locals, reproducing the sign 'plan', discard the association used in the expert's symbolic structure with the same sign ('plan–strategy') and embed this sign in the local symbolic structure by creating a new association with their pre-existing sign 'neighbourhood'. This can be preliminarily interpreted as the locals' re-appropriation of the expert symbolic structure through the association between the sign 'plan' with the more locally relevant sign, 'neighbourhood'.

To put our interpretation of the pattern to test, we extracted and manually examined all textual contexts that contain the associations between the sign 'plan' with the signs 'management' and 'neighbourhood' from the original official documents (N = 240) and the interviews with the local flood group members (N = 9).

The following quote is illustrative of the meaning of the association 'plan–strategy' for the experts:

> The Department's capacity-building support for lead local flood authorities has been well received. It has provided funding support to train staff from across all local authorities to improve their knowledge and expertise of flooding. In addition, the Agency has seconded staff to the local authorities to provide additional resource to complete **strategies** [henceforth, emphasis ours] and develop sustainable urban drainage system **plans.** This is a reciprocal arrangement where some local authority staff have also come into the Agency to improve their understanding of surface water issues. (Expert document, an excerpt)

The experts strive for a systematic and coordinated approach to flood risk management. 'Plans' and 'strategies' are the instruments they use to ensure flood management activities of different stakeholders are aligned and exercised in a timely manner. Hence, the association between 'plan' and 'strategy' is firmly ingrained in the experts' vocabulary.

The analysis of textual contexts extracted from the interviews with members of the LFG II reveals a local meaning of the association between 'plan' and 'neighbourhood' that is best illustrated with the following quote:

> I suppose… the other one [issue] which isn't perhaps as major [a problem] but it [is] certainly significant for [the village], is the local developers. The planning permissions are granted on the understanding that certain flood mitigation steps will be taken. They're… not necessarily everything that was promised initially happened. Developers are only allowed to develop in line with the **neighbourhood plan**. If there's no late **neighbourhood plan,** then they can come in and develop more or less all they want. So [the local steering group] set up to develop a **neighbourhood plan** and it was very successful in doing that and worked very well. Now our group fed into the **neighbourhood plan**. (Informant G, a member of the LFG II)

On the one hand, the locals do reproduce the experts' sign 'plan'. This happens when the LFG II accepts the idea of organising and coordinating flood management stakeholders' activities in accordance with a certain scheme of action. Hence, the sign 'plan' becomes reproduced in the symbolic structure of the flood group. On the other hand, as opposed to a general document that regulates collaboration between stakeholders across a wide range of flood management activities and contexts, the local flood group uses the sign 'plan' when it speaks about coordination of activities between stakeholders involved in local land planning and development. This is done to ensure that new buildings and infrastructure do not adversely impact drainage increasing flood risk in the village. The necessity for the developers to account for the flood risk in the village is outlined in a 'neighbourhood plan' – a local document guiding planning and development in the parish that the LFG II often refers to when it speaks about flood-related problems in the local area. Hence, the reproduced expert sign 'plan' becomes appropriated in the local symbolic structure through the association with the locally relevant sign 'neighbourhood'.

To sum up, our initial interpretation of the loose coupling pattern is supported by the manual inspection of the textual data, supplemented with our ethnographic knowledge of the field.

The socio-symbolic pattern *contagion* involves reproduction of signs and/or associations between them used by one individual in a group by his or her social network alter, who has not used them before, so that such signs and/or associations between them become shared as a result of direct interaction. We extracted all instances of the pattern from the socio-semantic network of the local flood groups and visualised them. The visual representation of all the instances of the contagion pattern for a specific pair of interacting individuals is provided in Fig. 9. We focus on one of the instances of the contagion pattern involving the sign association 'multi-agency–meeting' shared by informants D and E.

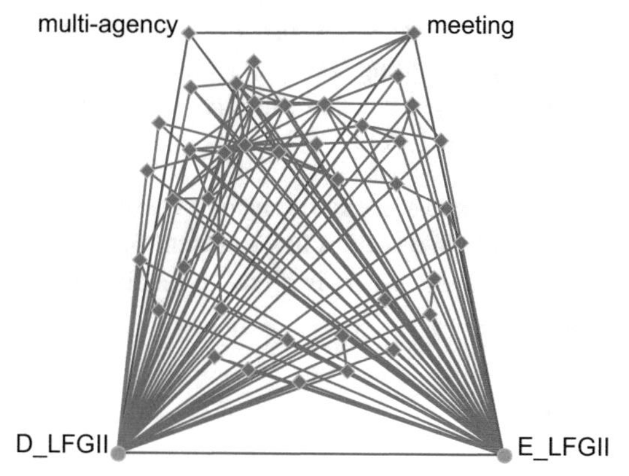

Fig. 9. Instances of the pattern 'contagion' for individuals D and E from LFG II. Blue diamonds = signs; green circles = individuals; blue lines = associations between signs; red line between individuals = social tie; grey lines = sign usage. Links in a single instance of the pattern highlighted in red. For illustrative purposes, only labels of signs that appear in the example are shown. Individuals' labels contain the codename of the group to which they belong.

The figure shows all signs used by informants D and E who are members of the LFGII, as well as corresponding sign associations in the local symbolic structure. Note that not all sign associations are necessarily used by D or E. We identify sign associations used specifically by D and E by looking at sign association attributes indicating their (non-)usage by individuals in a group (see Sect. 3.4). For instance, this way we confirm that D uses the signs 'multi-agency' and 'meeting' and associates them into the term 'multi-agency meeting'. Related to D with a direct social tie, the informant E also uses the same two signs and associates them with each other in the same way as D does. We assume that in the process of interaction with D, the informant E reproduces the association between the signs 'multi-agency' and 'meeting' as used by the informant D.

The extraction of textual contexts for each instance of the association 'multi-agency–meeting' in the original interviews with the informants D and E allows us to verify if the meaning of this association for each informant is similar, and hence, that our interpretation that this association is shared by both informants is correct. For instance, for the informant D, multi-agency meeting is the format that the flood group uses to work with the official flood risk management authorities, which is best showcased by the following quote:

> Arcadis [an engineering consulting company] will now be concepting that drawing out, but each member of the group at the **multi-agency meetings**… will say that this is a problem here and that's a problem there. Like the area at the back of Beech Drive, I know from my childhood… it's been an area of flooding. (Informant D, a member of the LFG II)

Interacting with D, the informant E – who is a newcomer to the flood group – learns the very idea of getting the agencies around the table at multi-agency meetings and reproduces the sign association 'multi-agency–meeting':

> I've had quite a few individual meetings, just me and [another member], we just had a chat about how things are going and where we need to push things generally. Before **multi-agency meetings,** me and [another member] would have a chat. (Informant E, a member of the LFG II)

Thus, the analysis of the textual contexts confirms our expectation that the association 'multi-agency–meeting' part of the pattern has a similar meaning for D and E, which, as we know from the ethnographic work, is likely to result from contagion between a more senior member and a novice.

5 Conclusion

This paper has dealt with the lack of techniques and tools to analyse the interplay between symbolic and social structures at the micro level by inferring symbolic and socio-symbolic patterns that reflect the usage of signs and their associations by socially tied individuals in specific practical contexts. Such patterns allow for examining the co-evolution of symbolic structures of different groups while controlling for the effect of intra-group social network structures on this process. We introduced a technique and a software tool for semi-automatic location, extraction, storage, and visualisation of instances of theoretically derived symbolic and socio-symbolic patterns in (socio-)semantic network data. We illustrated the technique and the tool by analysing two patterns of interplay between expert and local symbolic structures in the context of local social structures using empirical data from our 2019 study of two local flood management groups in England. We conducted a subsequent qualitative analysis of the two instances of the patterns to ensure that our interpretation corresponds to the ethnographic knowledge of the field (see [14]). The limitation of the present illustration is that these data are cross-sectional. Tests based on more extensive longitudinal datasets are to follow.

Acknowledgements. This work was supported by the Russian Science Foundation (grant 19-18-00394 'Creation of knowledge on ecological hazards in Russian and European local communities,' 2019–ongoing). The authors would like to thank two anonymous reviewers for providing valuable comments on an earlier draft of the paper.

References

1. Basov, N., Breiger, R., Hellsten, I.: Socio-semantic and other dualities. Poetics **78** (2020). https://doi.org/10.1016/j.poetic.2020.101433
2. Fuhse, J., Stuhler, O., Riebling, J., Martin, J.L.: Relating social and symbolic relations in quantitative text analysis. A study of parliamentary discourse in the Weimar Republic. Poetics **78**. https://doi.org/10.1016/j.poetic.2019.04.004
3. Godart, F.C., Galunic, C.: Explaining the popularity of cultural elements: networks, culture, and the structural embeddedness of high fashion trends. Organ. Sci. **30**(1), 151–168 (2019). https://doi.org/10.1287/orsc.2018.1234
4. Mohr, J.W., White, H.C.: How to model an institution. Theory Soc. **37**(5), 485–512 (2008). https://doi.org/10.1007/s11186-008-9066-0
5. Padgett, J.F., Prajda, K., Rohr, B., Schoots, J.: Political discussion and debate in narrative time: the Florentine Consulte e Pratiche. Poetics **78**, 1376–1378 (2020). https://doi.org/10.1016/j.poetic.2019.101377
6. Schoots, J., Rohr, B., Prajda, K., Padgett, J.F.: Conflict and revolt in the name of unity: florentine factions in the Consulte e Pratiche on the cusp of the Ciompi Revolt. Poetics 101386 (2020). https://doi.org/10.1016/j.poetic.2019.101386
7. Bródka, P., Chmiel, A., Magnani, M., Ragozini, G.: Quantifying layer similarity in multiplex networks: a systematic study. R. Soc. Open Sci. **5**, 171747 (2018). https://doi.org/10.1098/rsos.171747
8. Roth, C., Cointet, J.-P.: Social and semantic coevolution in knowledge networks. Soc. Netw. **32**(1), 16–29 (2010). https://doi.org/10.1016/j.socnet.2009.04.005
9. Basov, N., Lee, J.-S., Antoniuk, A.: Social networks and construction of culture: a socio-semantic analysis of art groups. In: Cherifi, H., Gaito, S., Quattrociocchi, W., Sala, A. (eds.) Complex Networks & Their Applications V, pp. 785–796. Springer, Cham (2017). https://doi.org/10.1007/978-3-319-50901-3_62
10. Linzhuo, L., Lingfei, W., James, E.: Social centralization and semantic collapse: hyperbolic embeddings of networks and text. Poetics 101428 (2020). https://doi.org/10.1016/j.poetic.2019.101428
11. Blumer, H.: Symbolic Interactionism: Perspective and Method. University of California Press, Berkeley (1986)
12. Mead, G.H.: Mind Self and Society from the Standpoint of a Social Behaviorist. Chicago University Press, Chicago (1934)
13. Basov, N.: The ambivalence of cultural homophily: Field positions, semantic similarities, and social network ties in creative collectives. Poetics **78**. https://doi.org/10.1016/j.poetic.2019.02.004
14. Basov, N., de Nooy, W., Nenko, A.: Local meaning structures: mixed-method sociosemantic network analysis. Am. J. Cult. Sociol. (2019). https://doi.org/10.1057/s41290-019-00084-9
15. R Core Team: R: A language and environment for statistical computing. (2013)
16. Carley, K.: Extracting culture through textual analysis. Poetics **22**(4), 291–312 (1994). https://doi.org/10.1016/0304-422X(94)90011-6
17. Hardy, C., Maguire, S.: Organizing risk: discourse, power, and "riskification". Acad. Manag. Rev. **41**(1), 80–108 (2016). https://doi.org/10.5465/amr.2013.0106
18. Mohr, J.W.: Soldiers, mothers, tramps and others: discourse roles in the 1907 New York City charity directory. Poetics **22**(4), 327–357 (1994). https://doi.org/10.1016/0304-422X(94)90013-2
19. Fine, G.A.: Group culture and the interaction order: local sociology on the meso-level. Ann. Rev. Sociol. **38**(1), 159–179 (2012). https://doi.org/10.1146/annurev-soc-071811-145518

20. Puzyreva, K., Basov, N.: Local knowledge in Russian flood-prone communities: a case study on living with the treacherous waters. In: Babu, G., Qamaruddin, M. (eds.) International Case Studies in the Management of Disasters, pp. 47–60. Emerald Publishing Limited (2021). https://doi.org/10.1108/978-1-83982-186-820201004
21. Hoetker, G., Agarwal, R.: Death hurts, but it isn't fatal: the postexit diffusion of knowledge created by innovative companies. Acad. Manag. J. **50**(2), 446–467 (2007). https://doi.org/10.5465/amj.2007.24634858
22. Binder, A.: For love and money: Organizations' creative responses to multiple environmental logics. Theor. Soc. **36**(6), 547–571 (2007). https://doi.org/10.1007/s11186-007-9045-x
23. Fiol, C.M., O'Connor, E.J.: Waking Up! Mindfulness in the face of bandwagons. AMR **28**(1), 54–70 (2003). https://doi.org/10.5465/amr.2003.8925227
24. Schilke, O.: A micro-institutional inquiry into resistance to environmental pressures. Acad. Manag. J. **61**(4), 1431–1466 (2018). https://doi.org/10.5465/amj.2016.0762
25. Nenko, A., Khokhlova, A., Basov, N.: Communication and knowledge creation in urban spaces: the tactics of artistic collectives in Barcelona, Berlin and St. Petersburg. In: Aiello, G., Tarantino, M., Oakley, K. (eds.) Communicating the City. Peter Lang, Austria (2017)
26. Hallett, T.: The myth incarnate: recoupling processes, turmoil, and inhabited institutions in an urban elementary school. Am. Sociol. Rev. **75**(1), 52–74 (2010). https://doi.org/10.1177/0003122409357044
27. Joseph, J., Ocasio, W., McDonnell, M.-H.: The structural elaboration of board independence: executive power, institutional logics, and the adoption of CEO-only board structures in U.S. corporate governance. Acad. Manag. J. **57**(6), 1834–1858 (2014). https://doi.org/10.5465/amj.2012.0253
28. Bolton, C.D.: Some consequences of the meadian self. Symb. Interact. **4**(2), 245–259 (1981). https://doi.org/10.1525/si.1981.4.2.245
29. Etzrodt, C.: The foundation of an interpretative sociology: a critical review of the attempts of George H. Mead and Alfred Schutz. Hum. Stud. **31**(2), 157–177 (2008). https://doi.org/10.1007/s10746-008-9082-0
30. Basov, N., Brennecke, J.: Duality beyond dyads: multiplex patterning of social ties and cultural meanings. In: Groenewegen, P., Ferguson, J.E., Moser, C., Mohr, J.W., Borgatti, S.P. (eds.) Research in the Sociology of Organizations, pp. 87–112. Emerald Publishing Limited (2017)
31. Fuhse, J.: The meaning structure of social networks. Sociol. Theory **27**(1), 51–73 (2009). https://doi.org/10.1111/j.1467-9558.2009.00338.x
32. Godart, F.C., White, H.C.: Switchings under uncertainty: the coming and becoming of meanings. Poetics **38**(6), 567–586 (2010). https://doi.org/10.1016/j.poetic.2010.09.003
33. Rawlings, C.M., Childress, C.: Emergent meanings: reconciling dispositional and situational accounts of meaning-making from cultural objects. Am. J. Sociol. **124**(6), 1763–1809 (2019). https://doi.org/10.1086/703203
34. White, H.C.: Identity and Control: A Structural Theory of Social Action. Princeton University Press, Princeton (1992)
35. Antonyuk, A., Puzyreva, K., Basov, N.: Principles and patterns of interaction between expert knowledge and local community knowledge. Manuscript in preparation, Centre for German and European Studies, St. Petersburg (2020)
36. Antonyuk, A., Kretser, I., Basov, N.: Creation of local knowledge in interaction. Unpublished manuscript, Centre for German and European Studies, St. Petersburg (2019)
37. Vadera, A.K., Aguilera, R.V.: The evolution of vocabularies and its relation to investigation of white-collar crimes: an institutional work perspective. J. Bus. Ethics **128**(1), 21–38 (2014). https://doi.org/10.1007/s10551-014-2079-x
38. Burt, R.S.: Social contagion and innovation: cohesion versus structural equivalence. Am. J. Sociol. **92**(6), 1287–1335 (1987)

39. Carley, K.: Knowledge acquisition as a social phenomenon. Instr. Sci. **14**(3–4), 381–438 (1986). https://doi.org/10.1007/BF00051829
40. Coleman, J.S.: Social capital in the creation of human capital. Am. J. Sociol. **94**, S95–S120 (1988). https://doi.org/10.1086/228943
41. Monge, P.R., Contractor, N.S.: Theories of Communication Networks. Oxford University Press (2003)
42. Zhou, D., Ji, X., Zha, H., Giles, C.L.: Topic evolution and social interactions: how authors effect research. In: Proceedings of the 15th ACM International Conference on Information and Knowledge Management, New York, NY, USA, pp. 248–257. Association for Computing Machinery (2006)
43. Cucchiarelli, A., D'Antonio, F., Velardi, P.: Semantically interconnected social networks. Soc. Netw. Anal. Min. **2**(1), 69–95 (2012). https://doi.org/10.1007/s13278-011-0030-z
44. Benoit, K., Watanabe, K., Wang, H., Nulty, P., Obeng, A., Müller, S., Matsuo, A.: quanteda: an R package for the quantitative analysis of textual data. J. Open Source Softw. **3**(30), 774 (2018). https://doi.org/10.21105/joss.00774
45. Straka, M., Straková, J.: Tokenizing, POS Tagging, lemmatizing and parsing UD 2.0 with UDPipe. In: Proceedings of the CoNLL 2017 Shared Task: Multilingual Parsing from Raw Text to Universal Dependencies, pp. 88–99. Association for Computational Linguistics, Vancouver (2017)
46. Csardi, G., Nepusz, T.: The igraph software package for complex network research. InterJournal Complex Systems (2006)

Analysis of Directed Signed Networks: Triangles Inventory

Elizaveta Evmenova[1(✉)] and Dmitry Gromov[2]

[1] Faculty of Applied Mathematics and Control Processes, St. Petersburg State University, St. Petersburg, Russia
e.evmenova.o@gmail.com

[2] National Research University Higher School of Economics, St. Petersburg, Russia
dv.gromov@gmail.com

Abstract. Signed networks form a particular class of complex networks that has many applications in sociology, recommender and voting systems. The contribution of this paper is twofold. First, we propose an approach aimed at determining the characteristic subgraphs of the network. Second, we apply the developed approach to the analysis of the network describing the Wikipedia adminship elections. It is shown that this network agrees with the status theory if one does not consider strongly tied vertices, i.e., the vertices that are connected in both directions. At the same time, the strongly connected vertices mostly agree with the structural balance theory. This result indicates that there is a substantial difference between single and double connections, the fact that deserves a detailed analysis within a broader context of directed signed networks.

Keywords: Signed graphs · Weighted graphs · Structural balance · Status theory · Probabilistic methods

1 Introduction

Within just two decades, the theory of complex networks has become a well-established area of research. There have been published dozens of monographs, see, e.g., [1–3] and a large number of papers devoted to this topic (see, for instance, [4, 5] for a good overview and introduction). The scope of the classical complex networks theory is mostly restricted to the study of statistical properties of undirected and/or unweighted graphs. However, there are numerous applications that cannot be considered within this framework and require the use of more general graph structures. In particular, in the last decade there has been a growing interest in studying signed networks, i.e., the networks where two nodes are connected by a signed (or, more generally, weighted) relation.

Although the notion of signed networks was introduced almost 70 years ago by Frank Harary in [6], until recently, the application of signed networks had been mostly confined to sociological studies. Within this context, one associates friendship with a positive relation and antagonism with a negative one. This line of research resulted in the theory of structural balance, see, e.g., [7: Ch. 9, 8] and references therein.

A. Antonyuk and N. Basov (Eds.): NetGloW 2020, LNNS 181, pp. 120–132, 2021.
https://doi.org/10.1007/978-3-030-64877-0_8

In the last two decades, a vast amount of data related to different signed interaction networks have appeared, which has generated a new wave of interest in studying signed networks. We mention highly influential papers [9, 10], where the authors analyze and systematize different approaches to the interpretation of the structure and evolution of such networks, including the status and sentiment analysis (see also [11]). There are also many papers exploring other aspects of signed networks, see, e.g., [12–15] to mention just a few.

When studying signed graphs from the balance-related point of view, it is essential to recognize whether a given graph exhibits a particular structure, which expresses itself in the prevalence of triangles of a specific type. So, if a graph obeys the structural balance law, the signs of the edges should multiply in a positive number. On the other hand, if the graph is governed by the status principle, more complicated cases can appear. Consider, for instance, the triangle shown in Fig. 1. While the theory of balance implies that the unmarked edge should be positive to ensure that the product of all signs is positive as well, the theory of status predicts that the edge from C to A should be negative since C has a higher status than A. On the other hand, if the edge between A and C changes its orientation, both theories would predict the same result: a positive weight. This observation stresses the importance of considering the orientation of the edges.

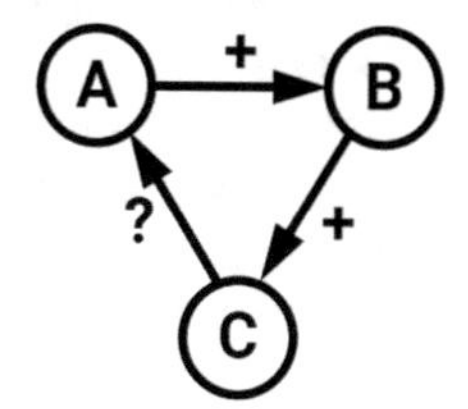

Fig. 1. A triangle illustrating the difference between the theories of balance and status

Typically, the researcher postulates a certain underlying structure and tests this postulate by checking whether the frequencies of particular triangles follow the corresponding rule. In this work, we plan to develop another, problem-centric approach to the analysis of the graph structure. Namely, we suggest that the frequencies of all triangles present in the overall network are to be compared to the theoretical frequencies computed for a probabilistic graph model that has the same distribution of edges as the original one. In this way, we can determine the triangles whose frequencies deviate most strongly from the theoretical frequencies.

The suggested approach is similar in spirit with that used for determining motifs in complex networks as pioneered in [16, 17] (also see a very instructive critical discussion of the first paper in [18, 19]). However, our approach is different in several respects. First, we take a different route when considering graph models. Second, instead of estimating the theoretical frequencies from generated graphs, we compute them (semi-)analytically using the developed algorithm.

We apply the devised approach to the structural analysis of the signed network describing the relationships between Wikipedia users that participated in administrator elections. We carry out a detailed analysis of the network's structure and show that it mostly agrees with the status theory if we do not consider the strongly tied vertices, i.e., the vertices that are connected in both directions. Such vertices mostly tend to be of (+, +)-type. However, there is a considerable proportion of pairs that do not follow this pattern. This problem deserves a separate investigation that would go beyond the framework of triangles and consider more general structures.

This paper is organized as follows. In Sect. 2, we present basic facts from graph theory that will be used later on; in Sect. 3, we describe the developed approach to determine characteristic subgraphs of a graph; an application of the suggested method is presented in Sect. 4, where a detailed analysis and classification of triangles with 3 and 4 edges is carried out. The discussion section concludes the paper.

2 Graph Theory Background and Terminology

A directed weighted graph (weighted digraph) G is defined as a tuple $G = (V, E, W)$, where $V = \{v_1, v_2, \ldots, v_{N_V}\}$ is a finite ordered set of vertices, $E \subseteq V \times V$ is a set of edges, and $W : E \to \{w_1, \ldots, w_n\}$ is a weight function that assigns a unique weight from a finite set to each edge $e \in E$. Sometimes we will employ a shorthand notation e_{ij} to denote the edge (v_i, v_j). For a directed graph, $(v_1, v_2) \in E$ does not imply that $(v_2, v_1) \in E$. If both edges $e_{ij} = (v_i, v_j)$ and $e_{ji} = (v_j, v_i)$ belong to E, we say that the edge e_{ji} is reciprocal of e_{ij} (and vice versa).

We let N_V denote the number of the vertices and N_E the number of edges in the graph, i.e., $N_V = |V|$ and $N_E = |E|$. We will occasionally use notation vert(G) and ed(G) to denote the number of vertices, resp. edges of the graph G.

In this paper, we consider only simple (weighted) digraphs, i.e., directed weighted graphs without self-loops, i.e., $e_{ii} \notin E$ for all $i = 1, \ldots, N_V$. A simple digraph is said to be fully connected if $E = (V \times V)\setminus\{e_{ii}\}_{i=1,\ldots,N_V}$. In this case, $N_E = N_V(N_V - 1)$, which is the maximal possible number of edges, denoted by $N_E^{\max}$.

Let $S \subset V$ be a subset of vertices V. We say that $G_S = (S, E_S, W_S)$ is a subgraph of G generated by S if $E_S = E \cap (S \times S)$ and $W_S = W|_{E_S}$. Note that we assume throughout the text that the vertices in the set S are ordered. With this convention, there are at most $\binom{N_V}{n}$ different subgraphs generated by sets of vertices consisting of exactly $n \leq N_V$ elements. Finally, we let $\Gamma_n(G)$ denote the set of all subgraphs of G generated by ordered subsets $S \subset V$ containing exactly n elements. From now on, we will use small Greek letters, say, γ when specifically referring to subgraphs of the graph G.

In practice, when considering different subgraphs in a graph, we are interested in identifying the structures that are invariant with respect to the way we denote or enumerate the vertices. Such structures are closely related to the notion of graph isomorphism. We say that two graphs $G' = (V', E', W')$ and $G'' = (V'', E'', W'')$ are *isomorphic*, denoted $G' \sim G''$, if there exists a bijection $\Xi\colon V' \to V'$ such that for all $v_1, v_2 \in V'$, $(v_1, v_2) \in E' \Rightarrow (\Xi(v_1), \Xi(v_2)) \in E''$ and for all $(v_1, v_2) \in E'$, $W'(v_1, v_2) = W''(\Xi(v_1), \Xi(v_2))$. Two isomorphic graphs have the same number of vertices and edges and share a number of properties. Put simply, two graphs are isomorphic if one graph can be transformed into another one without breaking or reweighting edges. In Fig. 2, a complete set of signed isomorphic digraphs is shown for illustration.

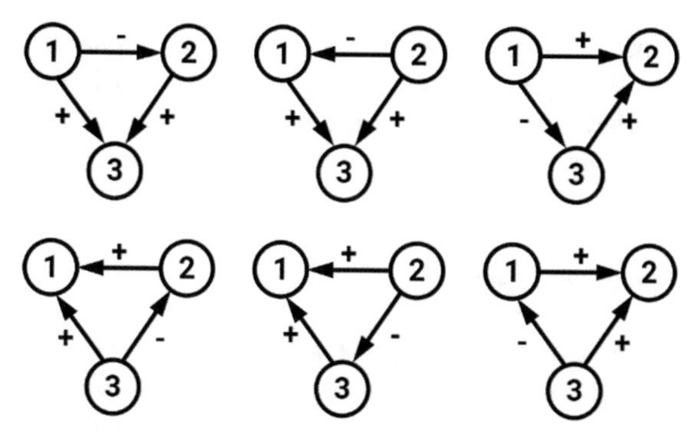

Fig. 2. A set of isomorphic graphs with three vertices. The set of weights consists of two elements: $\{+, -\}$.

The isomorphism relation is an equivalence relation. Hence, for each $\gamma \in \Gamma_n(G)$. it defines the equivalence class $[\gamma] = \{\gamma' \in \Gamma_n(G)|\gamma' \sim \gamma\}$. The set of all equivalence classes defines a partitioning of the set $\Gamma_n(G)$. An important question consists in determining the cardinality of a given equivalence class γ, denoted $\#[\gamma]$, i.e., the number of graphs isomorphic to γ. To start with, we note that in a graph with n vertices, the vertices can be permuted in $n!$ ways; so, there are $n!$ possible bijections of the set of vertices to itself. It remains to determine which of these permutations generate distinct graphs. To do so, we use the notion of symmetry of a graph. Let $\tau(\gamma)$ denote the number of transformations that map a graph to itself. Obviously, any graph has at least one symmetry that coincides with the identity transformation; thus, $\tau(\gamma) \geq 1$. Figure 3 shows a digraph γ such that $\tau(\gamma) = 3$. While for simple graphs like that shown in Fig. 3 the number of symmetries can be determined by mere observation, for more complex graphs this can pose a serious computational challenge. We refer the interested reader to [20, Ch. 14] for a thorough discussion of graph symmetries and related problems within a group-theoretic context.

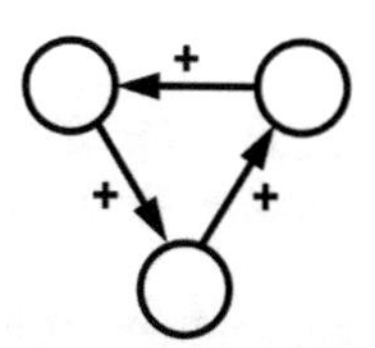

Fig. 3. A digraph with 3 vertices, 3 edges, and 3 symmetries: identity map, rotation by $2\pi/3$ and rotation by $4\pi/3$. Note that the composition of the last two symmetries yields the identity map.

Finally, the cardinality of the equivalence class $[\gamma]$ is computed as

$$[\gamma] = \frac{\text{vert}(\gamma)!}{\tau(\gamma)}.$$

We can easily observe that the graph shown in Fig. 3 has only two unique isomorphic forms, whereas the second form corresponds to the reflection of this graph.

Specific Notation. In this paper, we use specific notation adjusted to the needs of the considered problem. For a given graph γ, we let $\text{ed}(\gamma, w_i)$ denote the number of edges with weight w_i. Similarly, $\text{ed}(\gamma, \emptyset)$ denotes the number of missing edges in the graph γ. When computing these characteristics, we consider the edges (v_i, v_j) and (v_j, v_i) separately.

3 Triangles Classification

When analyzing the structure of a network, it is often instructive to take a closer look at the elementary building blocks of this network. In the following, we will concentrate on the analysis of different oriented signed triangles that appear in a network. We will call a *triangle* the set of three vertices such that any two vertices are connected by an edge, whereas the direction of the edge is not substantial.

Specifically, we will be interested in determining which triangles are *characteristic* for a given network. The notion of characteristic requires some additional elaboration. One might consider a triangle characteristic if it appears in the whole network more often compared to other triangles. However, we know that distinct triangles may have different a priori probabilities of occurrence. To compute these probabilities, we suggest considering a *probabilistic theoretical graph model* that can be used to generate *theoretic frequencies*, in contrast to the observed frequencies of triangles, which are referred to as *empirical frequencies*. The ratio of the theoretical, resp. the empirical frequency of a specific triangle to the total number of all triangles is called the *theoretical*, resp. *empirical probability* of the occurrence of the triangle. In the following, we will slightly abuse notation and speak of the probability of a triangle.

We will consider a specific triangle as characteristic if it possesses two properties:

- The triangle occurs in the network sufficiently often to not be considered as a mere fluctuation.
- The empirical probability of the triangle is sufficiently larger than its theoretical probability.

Admittedly, this definition is somewhat vague as we do not define what means "sufficiently often" or "sufficiently larger". We will address this issue when considering a specific example. As for now, we concentrate on computing theoretical frequencies of triangles.

The key issue in computing theoretical frequencies consists in considering an appropriate theoretic graph model. One possible approach consists in merely considering realization(s) of random graphs. However, it is rather clear that no single realization of a random graph can serve as a reference point for all other graphs. Thus, a family of such realizations should be studied instead. When computing reference graphs, one typically considers graphs that share specific statistical characteristics with the studied graph. Hence, the question is which particular properties are to be retained. To provide a context, we refer the interested reader to a particularly informative discussion in [18]. In there, the authors argue that a random graph should be generated as a random specimen of the class, which the studied graph belongs in. Therefore, it is important to understand what are the characteristic features of the class of graphs, rather than a specific graph: e.g., whether there is a restriction on the vertex degree or the distance between the connected vertices, etc.

On the other hand, for a random graph that possesses sufficiently many statistical parameters, one has to resort to numerical simulations in which a (large) family of random graphs is generated, and the theoretical characteristics are obtained by computing an ensemble average of the quantities of interest. This approach has an inherent problem that the estimations computed in this way do not provide the actual values of the theoretical characteristics, but only tend to them as the number of elements in an ensemble goes to infinity. This means that obtaining accurate estimations for the theoretical frequencies can be rather time-consuming.

In this work, we will pursue a different course and concentrate on the analytical computation of theoretical frequencies of triangles. To do so, we restrict ourselves to considering a theoretic random graph that has the same probabilities of edges as the original one and derive the quantities of interest straight from that theoretic model. Note that no assumptions beyond those imposed on the probabilities of weights are made. First, there is no reason to assume some a priori assumptions on, say, the degree distribution. Second, any additional assumption would make the task of computing the probabilities of interest substantially more involved. Finally, we stress the fact that this approach differs from the described above in the sense that we do not generate the realizations of this random graph, but rather analytically compute the theoretical characteristics of the studied model.

Let $\{\mathrm{w}_1, \dots, \mathrm{w}_n\}$ be the weights of the edges. We define the empirical probability that there is an edge with a specific weight $\mathrm{w_i}$ between two randomly chosen vertices as

$$\mathrm{p_i} = \mathrm{P}\{\mathrm{W(e)} = \mathrm{w_i}\} = \frac{\mathrm{ed(G, w_i)}}{\mathrm{N_E^{max}}}.$$

Also, we define the probability that there are no edges between two arbitrary vertices as

$$\mathrm{p_0} = \mathrm{P}\{\mathrm{e} \notin \mathrm{E}\} = 1 - \sum_{i=1}^{n} \mathrm{p_i}.$$

Multiplying the respective probabilities with the maximal theoretical number of edges we obtain empirical frequencies, e.g., $\mathrm{f_i} = \mathrm{p_i N_E^{\mathit{max}}}$.

When considering a theoretical random graph, we assume that between any two vertices (where we count both (v_i, v_j) and (v_j, v_i)), there can be either a single edge with the weight $\mathrm{w_i}$ or no edge. The former event has the probability p_i, while the latter the probability p_0. Now we wish to consider the probability of a specific triangle. We will specify triangles by their signature, i.e., the sequence of weights of edges between any pair of vertices. Consider, for instance, the triangle γ^* shown in Fig. 3. Assume that its vertices are numbered from left to right and from up to down. Furthermore, we assume that the positive relation $\langle + \rangle$ has the weight $w_1 = 1$. Then, the signature of this particular triangle can be written as

$$\sigma(\gamma^*) = (W(e_{12}), W(e_{21}), W(e_{13}), W(e_{31}), W(e_{23}), W(e_{32})) = (0, 1, 1, 0, 0, 1).$$

The probability of a given triangle $\overline{\gamma}$ is the product of probabilities corresponding to the individual weights of the edges:

$$\prod_{i=1}^{n} p_i^{ed(\overline{\gamma},w_i)} \cdot p_0^{ed(\overline{\gamma},\emptyset)}.$$

However, this probability has to be adjusted to take into account the arbitrary vertices numbering. Since we do not distinguish between two isomorphic graphs, the computed *theoretical probability* should be multiplied with the number of elements in the corresponding isomorphism class:

$$\pi_t(\overline{\gamma}) = P\{\gamma = \overline{\gamma}\} = \prod_{i=1}^{n} p_i^{ed(\overline{\gamma},w_i)} \cdot p_0^{ed(\overline{\gamma},\emptyset)} \cdot \#[\overline{\gamma}].$$

On the other hand, the *empirical probability* of a given triangle $\overline{\gamma}$, denoted $\pi_e(\overline{\gamma})$, is computed as the ratio of the frequency of the triangle occurrence to the total number of triangles in the considered graph, multiplied with the cardinality of the respective isomorphism class.

4 Results of the Wikipedia Election Graph Analysis

In this section, we analyze data obtained from the network that describes the results of the Wikipedia administrator election. Wikipedia is a free encyclopedia, whose operation is supervised and controlled by a relatively small number of administrators, i.e., the users with additional rights and access to technical features that aid in maintenance. If a regular user desires to become an administrator, she/he fills a Request for Adminship (RfA). Subsequently, the Wikipedia community publicly votes on whom to nominate for adminship. One person can cast one of three types of votes: for (positive), against (negative), and neutral. Note that a vote can be cast both by existing admins and by ordinary Wikipedia users.

In [21], a complete dump of Wikipedia page edit history from January 3, 2008, was used to extract the information about 2,800 elections. Out of these, 1,200 elections resulted in a successful promotion, while about 1,500 elections did not result in the promotion.

This information about the elections is represented as a graph. The vertices of the graph are the users (both admins and regular users), and the (weighted) edges correspond to the votes cast. Table 1 lists some basic numbers about the graph.

Table 1. Basic statistics

Number of vertices, N_V	Total number of edges, N_E	Number of positive edges	Number of negative edges	Number of neutral edges	Number of triangles, N_Δ
7194	110087	81862	22497	6888	534758

The density coefficient $\delta = 0.0021$ and the triangle density $\delta_{\triangle} = 8.6214 \cdot 10^{-6}$, which implies that the network is rather sparse. However, the relative degree is $\langle k \rangle = 15.3026$, which indicates sufficiently large voting activity.

To conclude the preliminary analysis, Table 2 contains the data describing the triangle structure of the Wikipedia election graph. The average number of edges in a triangle is $\langle e_{\triangle} \rangle = 3.2812$. Furthermore, only 4% of all triangles have 5 and 6 edges, while the majority of triangles have either 3 edges (76.24%) or 4 edges (19.88%). In the following, we will consider the latter two classes.

Table 2. Statistics on triangles

Number of edges	3	4	5	6
Total number of triangles	407677	106297	18203	2581
Fraction of triangles	0.7624	0.1988	0.0340	0.0048
Number of isomorphism classes	38	162	243	130
Number of isomorphism classes in the network	38	159	188	61

4.1 Classification of Triangles with 3 Edges

In Table 3, a selection of the triangles with 3 edges is presented. In total, the listed triangles appear in the network 391338 times, which amounts to 95.99% of all triangles with three edges. This table does not include 24 triangles that amount to only 4% of the total population. Thus, the collection shown in Table 3 is deemed representative.

We start by observing some general patterns that appear in this list. The first observation is that all triangles, except the first one, form a transitive structure (as opposed to a circular one). That is to say, there are two branches that can be schematically represented as $A \to B \to C$ and $A \to C$. This observation contrasts sharply with the theoretical estimation. Indeed, there at most 10 circular triangles, of which only one is present in the list. Furthermore, this single triangle has the smallest ratio among all triangles. Such a pattern indicates that there is a strong tendency to subordination in the considered triads; whereas the person (denoted above as C), at whom both branches converge, plays a distinct role.

This explanation strongly correlates with the predictions given by the theory of status that suggests that a positive edge $A \to B$ indicates that A considers B as having a higher status and the opposite if the edge is negative. Within the context of the considered problem, it might be more accurate to speak of *respect* or *merit* rather than status: when voting pro or contra a candidacy, the users evaluate the qualities of that person, which may or may not correlate with her/his status. However, while having this interpretation in mind, we will stick to conventional notation.

To extend this model to our case, we will assume that two people connected by a neutral edge consider themselves as having the same status. Now we are ready to check if the triangles shown in the table obey this rule.

It turns out that six most characteristic triangles (9–14) satisfy exactly the status balance rule. For instance, triangle 11 depicts the situation when person A considers

Table 3. List of triangles with three edges. The columns are denoted as follows: N is the number of the graph in the list, f_i^e is the empirical frequency of the ith triangle, and ρ_i is the probability ratio: $\rho_i = \frac{\pi_e^i}{\pi_t^i}$.

N	f_i^e	ρ_i	Triangle	N	f_i^e	ρ_i	Triangle
1	3691	15.13	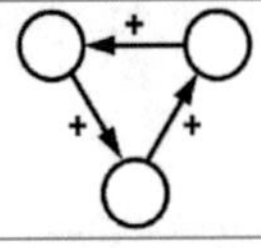	8	32706	487.76	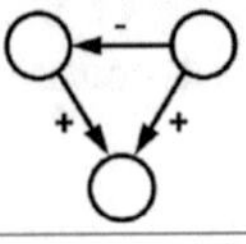
2	12839	191.47	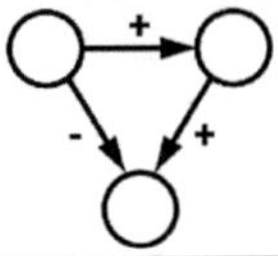	9	11207	545.88	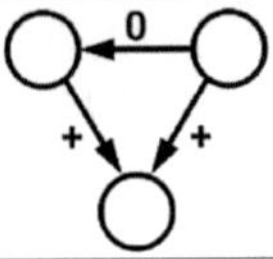
3	3780	205.13	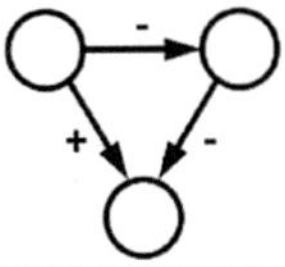	10	11716	570.67	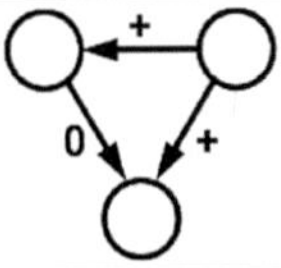
4	6707	363.97	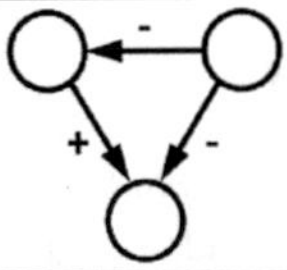	11	3360	595.53	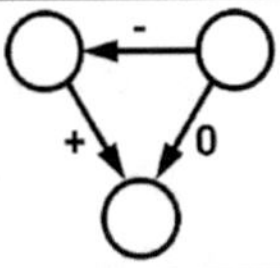
5	7458	404.72	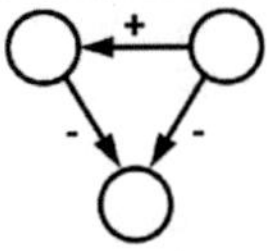	12	3415	605.28	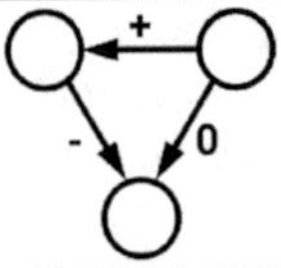
6	30562	455.78	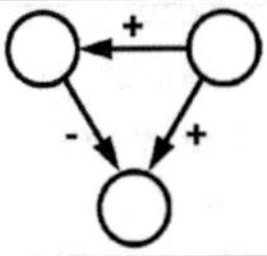	13	3841	758.47	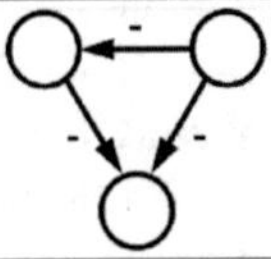
7	9602	467.70	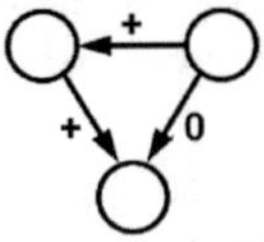	14	250454	1026.47	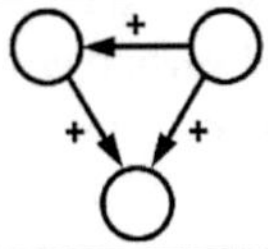

person B as having lower status, whereas person B sees person C as having higher status. This agrees with the fact that person A considers person C as having the same status, i.e., a negative and a positive evaluation cancel.

As we go up the table (i.e., toward less characteristic triangles), we observe triangles that violate the status rule by different extent.

To look at the problem from a different side, we consider whether the shown triangles agree with the structural balance theory. First, note that the structural balance theory is not actually capable of interpreting the neutral edges. This stems from the fact that a neutral edge can potentially be considered either as a weak positive or a weak negative relation. However, within the context of the considered problem, it is clear that in our case, the neutral edge should be interpreted as a weak positive relation. Following this assumption, we see that there are in total 8 triangles (1, 3–5, 7, 9, 10, 14) that *formally* agree with the structural balance theory. However, among those triangles, only three belong to the top 7 most characteristic triangles.

Thus, we conclude that the status theory offers a more consistent description to the observed triangles. However, to make the status theory even more compatible with the observed results we should allow for certain freedom in interpreting the status levels. Since we consider only three weights, any type of subordination has to fit into this rather narrow framework. It is clear though, that often it is difficult to uniquely qualify the subordination relation. Say, person A can consider person B as having lower status, while person B considers person C as having higher status. How should person A qualify person C? It depends. If person C has a much higher status, the subordination of A and B may turn out to be irrelevant as in triangle 8 (a similar situation is with triangle 6).

There is a different aspect that does not fit into the structural balance theory. The point is that the majority of triangles do not have reciprocal edges, although the theory suggests that the triangles with closer ties should prevail. To take a closer look into the problem, we consider the class of triangles with 4 edges, which form one-fifth of the whole population.

4.2 Classification of Triangles with 4 Edges

In Table 4, a selection of the triangles with 4 edges is presented. The listed triangles appear in the network 90495 times (85.13% of all triangles with four edges). While there are still about 15% of all triangles that are not represented in this table, these have rather small frequencies. In total, the left-out 15% correspond to 145 different triangle classes (recall that we do not distinguish between two isomorphic graphs and count only isomorphism classes).

Again, we start by observing some general patterns that appear in this list. And similarly to the previous case, we see that there is a dominating pattern: a pair of closely tied vertices and a third vertex either superordinate or subordinate to these two. It is remarkable that in 8 out of 10 cases, the edges between the tied pair and the third vertex are either both positive or positive and neutral, whereas the sub-, resp. super-ordination is expressed by the direction of the edges. The prevalence of such a pattern can be seen as an argument in favor of the status theory.

On the other hand, when considering the tied pairs of vertices, one can see that the theory of status can explain only a smaller fraction of all such pairs. For if person A considers person B as having higher status, we would expect that person B considers person A as having a lower status. In our collection, we observe exactly the opposite: there are eight pairs of ties of type (+, +), one of type (−, −), two pairs of ties of type (+, 0), and only three pairs of type (+, −). Only the last pair of ties, and partially the second to the last one, agrees with the status rule.

An alternative approach to analyzing oriented subgraphs from the viewpoint of either structural balance or status theory consists in considering such triangles as a superposition of several elementary triangles. In the case of triangles with 4 edges, each such

Table 4. List of triangles with four edges

N	f_i^e	ρ_i	Triangle	N	f_i^e	ρ_i	Triangle
1	1130	10629.66		8	3531	33215.32	
2	1185	11147.03		9	1333	45627.71	
3	1580	14862.70		10	1953	60003.21	
4	2658	25003.21		11	2002	61508.66	
5	10777	27859.98		12	2142	65809.10	
6	3115	29302.10		13	2156	66240.10	
7	3477	32707.36		14	53456	138190.90	

triangle can be uniquely represented as a superposition of 2 elementary triangles of the same type as those studied in Sect. 4.1. Hence, we can extend the results presented in that section to the analysis of 4-edge triangles. Of the listed triangles, six triangles, of which 5 are the most characteristic ones (triangles 5, 10–14), agree with the structural balance rule, while the remaining 8 violate this rule. Even fewer triangles satisfy the status law when analyzed from this perspective.

This apparent difficulty can be resolved if we consider the tied pairs as a single unit. In this case, we have only two actors: a pair and a third person. The relations between these two actors boil down to two scenarios: either both members of the pair consider the third person as having a higher or equal status, or the opposite. Ten out of

14 triangles can be interpreted along this line. The interaction structure within a tied pair is formed according to the different rules that cannot be adequately described by either the structural balance or status theory.

This result stresses the need to distinguish between single edges and pairs of reciprocal edges as was already mentioned in [22].

4.3 Triangles with 5 and 6 Edges

To conclude the performed analysis, we briefly consider the cases of triangles with either five or six edges. In total, such triangles constitute less than 4% of the total population of triangles. Of these triangles, about 60% are the triangles with 5 or 6 positive edges. This result does not require a detailed interpretation, but instead suggests that there is a number of strongly connected groups that support each other. A different question would consist in analyzing the size and structure of such groups, but this lies beyond the scope of this research.

5 Discussion

The main contribution of this paper is twofold. First, we presented a particularly useful approach to determining characteristic subgraphs of a given network. In this paper, we studied only triangles, but the described methodology can be easily (at least conceptually easily) extended to more general classes of subgraphs. Second, we carried out an analysis of a specific directed weighted network and provided a detailed interpretation of the obtained results from the viewpoint of two different theories: the structural balance theory and the status theory. The results are quite remarkable: the considered network mainly agrees with the status theory if we exclude from consideration strongly tied vertices, i.e., vertices that are connected in both directions. The strongly tied vertices tend to be of (+, +) type. However, there exist a considerable proportion of other types of connections that indicates the need for a more detailed analysis of such cases.

The latter result is particularly significant as it points out at the importance of separately considering single and double connections. This fact was already suggested in [22], albeit from a different perspective. We plan to carry out a detailed analysis of this phenomenon within a broader context of directed signed networks. As a concluding remark, we wish to mention that the carried out analysis cannot bring in an ultimate explanation of the inherent structure of the considered network. Indeed, most phenomena observed in real networks cannot be interpreted from a purely mathematical standpoint, but require joint work of mathematicians, computer scientists and sociologists.

References

1. Newman, M.: Networks, 2nd edn. Oxford University Press, Oxford (2018)
2. Chkhartishvili, A.G., Gubanov, D.A., Novikov, D.A.: Social networks: models of information influence, control and confrontation. Studies in Systems, Decision and Control, vol. 189. Springer (2018)

3. Easley, D., Kleinberg, J.: Networks, Crowds, and Markets. Cambridge University Press, Cambridge (2010)
4. Newman, M.E.: The structure and function of complex networks. SIAM Rev. **45**(2), 167–256 (2003)
5. Bolouki, S., Nedich, A., Başar, T.: Social networks. In: Handbook of Dynamic Game Theory, pp. 907–949. Springer (2018)
6. Harary, F.: On the notion of balance of a signed graph. Michigan Math. J. **2**(2), 143–146 (1953). https://doi.org/10.1307/mmj/1028989917
7. Roberts, F.S.: Graph theory and its applications to problems of society. In: Regional conference series in applied mathematics, vol. 29. SIAM (1978)
8. Kunegis, J.: Applications of structural balance in signed social networks. arXiv: 1402.6865 (2014). https://arxiv.org/abs/1402.6865
9. Leskovec, J., Huttenlocher, D., Kleinberg, J.: Predicting positive and negative links in online social networks. In: Proceedings of the 19th International Conference on World Wide Web, pp. 641–650 (2010). https://doi.org/10.1145/1772690.1772756
10. Leskovec, J., Huttenlocher, D., Kleinberg, J.: Signed networks in social media. In: Proceedings of the SIGCHI Conference on Human Factors in Computing Systems, pp. 1361–1370 (2010). https://doi.org/10.1145/1753326.1753532
11. Pang, B., Lee, L.: Opinion mining and sentiment analysis. Found. Trends R Inf. Retr. **2**(1–2), 1–135 (2008)
12. Marvel, S.A., Kleinberg, J., Kleinberg, R.D., Strogatz, S.H.: Continuous-time model of structural balance. Proc. Nat. Acad. Sci. **108**(5), 1771–1776 (2011). https://doi.org/10.1073/pnas.1013213108
13. Traag, V.A., Van Dooren, P., De Leenheer, P.: Dynamical models explaining social balance and evolution of cooperation. PLoS ONE **8**(4), 1–7 (2013). https://doi.org/10.1371/journal.pone.0060063
14. Shi, G., Proutiere, A., Johansson, M., Baras, J.S., Johansson, K.H.: The evolution of beliefs over signed social networks. Oper. Res. **64**(3), 585–604 (2016)
15. Jia, P., Friedkin, N.E., Bullo, F.: The coevolution of appraisal and influence networks leads to structural balance. IEEE Trans. Netw. Sci. Eng. **3**(4), 286–298 (2016). https://doi.org/10.1109/TNSE.2016.2600058
16. Milo, R., Shen-Orr, S., Itzkovitz, S., Kashtan, N., Chklovskii, D., Alon, U.: Network motifs: simple building blocks of complex networks. Science **298**(5594), 824–827 (2002)
17. Itzkovitz, S., Milo, R., Kashtan, N., Ziv, G., Alon, U.: Subgraphs in random networks. Phys. Rev. E **68**(2), 026,127 (2003)
18. Artzy-Randrup, Y., Fleishman, S.J., Ben-Tal, N., Stone, L.: Comment on "Network motifs: simple building blocks of complex networks" and "Superfamilies of evolved and designed networks". Science **305**(5687), 1107 (2004). https://doi.org/10.1126/science.1099334
19. Milo, R., Itzkovitz, S., Kashtan, N., Levitt, R., Alon, U.: Response to comment on "Network motifs: Simple building blocks of complex networks" and "Superfamilies of evolved and designed networks". Science **305**(5687), 1107 (2004). https://doi.org/10.1126/science.1100519
20. Harary, F.: Graph Theory. Addison-Wesley, Boston (2018)
21. Leskovec, J.: Wikipedia adminship election data. Stanford Network Analysis Project. https://snap.stanford.edu/data/wiki-Elec.html
22. Evmenova, E., Gromov, D.: Structural analysis of directed signed networks. In: Smirnov, N., Golovkina, A. (eds.) Proc. IV Stability and Control Processes Conf. Lecture Notes in Control and Information Sciences. Springer (forthcoming)

Emotional Street Network: A Framework for Research and Evidence Based on PPGIS

Aleksandra Nenko(✉), Marina Kurilova, and Maria Podkorytova

ITMO University, St. Petersburg, Russia
al.nenko@itmo.ru

Abstract. In this paper we study subjective perception of the city space represented through geography of emotions. In particular, we analyze the continuity of user experience in the city, considering user emotions as connected states. To do this, we develop a concept of an "emotional street network" and analyze the integrity of human emotional experience through availability and connectivity of its emotional network, as well as its valence. To explore the relevance of the emotional street network concept, we use data from a public participation geoinformational system (PPGIS) Imprecity, where users can leave emoji and comments on their feelings in public spaces of 6 types – joy, anger, sorrow, fear, disgust, and surprise. Dataset consists of more than 2000 emotional marks from 600 unique users in open public spaces of St. Petersburg, Russia. Two networks of positive and negative emotions were built: the locations less than 500 m length from each other (a classic measure for pedestrian accessibility) marked with emoji were connected with Points to path algorithm in QGIS software, afterwards collated with the street-road network. The connectivity of the final networks was calculated through axial connectivity index using QGIS Space Syntax plug-in. Both resulting emotional street networks have hierarchical structure – more connected areas in the city center and less connected in the periphery. The negative emotional network is more dispersed, reflecting a more localized and geographically distanced character of negative emotions, covering more peripheral areas than the positive one.

Keywords: Emotions · Emotional street network · Public space · PPGIS

1 Emotional Street Network: The Concept

Emotions and feelings of urbanites are important for city governors and urban planners while they illustrate the real experience of the city, its excellences and problems in the view of citizens. Scholars show that emotions are embodied in urban environments. City places and streets have features that draw positive and negative reactions [1–3]. Spatialized emotions are often represented as moods in specific locations (e.g., [4]). However, emotions are not localized in particular city spots but are a durable condition felt in certain "milieus" of the city [5–7]. In this paper we argue that to analyze real-life

A. Antonyuk and N. Basov (Eds.): NetGloW 2020, LNNS 181, pp. 133–143, 2021.
https://doi.org/10.1007/978-3-030-64877-0_9

urban structure and continuity of user experience of the city space we need to consider emotions as connected states.

The street has always been regarded as a major public space and "livable streets" have been proclaimed an important principle for urban well-being and urban design [8, 9]. Rich social life at the city streets, availability of points of attraction – local shops, cafes, bars, green streets have been proved as prerequisites of positive life experience, social tolerance, local economic prosperity [10]. Pedestrian traffic-free streets and plazas with the right ingredients (shops, cafes, benches, etc.) offer the best settings for sociability and thus build and strengthen social bonds, which is the city's greatest achievement [11]. The more the city is walkable and attracts pedestrians to the streets, the more it is healthy and productive [12, 13].

In spite of the amount of theoretical literature on the emotional side of the street life, produced by the urban planners, architects and environmental psychologists based on their natural observations and case-studies, empirical studies of emotions in the streets on the city scale are rare. There are valuable exceptions, which belong mostly to the field of study based on geolocated user generated data in social media, rating platforms and other geoinformation systems, such as research engines. However, it is of interest to link the emotional experience with the structure of the built environment and to define which of the features of the built environment correlate with the positive or negative experience. It is also quite important to find reliable sources of data, other than single cases, to illustrate that. Here we elaborate on this issue and present the concept of "emotional street network", i.e., a system of interconnected streets, not only linked by their intersections (as in the classical concept), but also by the similarity of the emotional states people feel in them. Applying this concept, we show how emotional experience is structured by the built environment used by people in their daily life. City dwellers are actually perceiving the city space while using the streets. The nature of emotions in the streets differs and can be interpreted by the nature of activities relevant for them, i.e., more dynamic spatial activities, like walking and sightseeing, and more static spatial activities, such as recreation in or exploration of a particular spot. The correspondent feelings are different – excitement, amusement, joy and relief, peacefulness, enjoyment. To receive rich experience of the city, a person needs to feel aroused and filled with positive emotions while walking the streets during stationery and the transit activities in space. To analyze the integrity of human emotional experience in the city, we trace availability and connectivity of the emotional street network as well as its valence.

2 Constructing an Emotional Street Network of St. Petersburg

To demonstrate the construction and interpret the city through the concept of urban emotional network, we are using data from public participation geoinformational system (PPGIS) Imprecity, created by the authors of the paper [14]. Imprecity is an online interactive map where users can leave emoji and comments on their feelings in public spaces. The emotions used in Imprecity correspond to the list of basic emotions introduced by Paul Ekman – joy, anger, sorrow, fear, disgust and surprise [15]. In this paper we analyze more than 2000 emotional marks from 600 unique users who have expressed what they feel in more than 700 public spaces of St. Petersburg, Russia. Most of the users are women (852 emoji belong to them) compared to men (520 emoji belong to them), 770 emoji were not defined. Emotion of joy is the most popular within the users, it itself is approximately equal to the sum of the rest of the emotions. In this paper we compare positive emotions (joy) with negative ones (anger, sorrow, fear, and disgust) to reduce the disproportion of the subsamples of emotions in the general dataset. The interpretation of emotional state is derived from the comments people leave together with the emoji.

GIS data from the Imprecity platform can be shown in a classic geographical layout – a heatmap of points (emotional marks) depicted over the city map [16]. Heatmaps of positive and negative emotions (Fig. 1a and 1b) are designed in QGIS software with its built-in functions [17]. Such a representation reveals the overall distribution of emotions, as well as the hottest and the coldest spots of emotions in the city space. The distribution of the positive emotion of joy in St. Petersburg shows its concentration in the city center in a triangle shape (ᗅ) and in a handful of more distant areas (which are parks) (Fig. 1a). Joy is mostly concentrated in the city center where people enjoy beautiful views and architecture, places to go and other people to see, and also appears in more distant parks, where people spend their vacation time. The heatmap of negative emotions shows that their distribution has a less compact form in the city center with an L-shape (L), is more dispersed, reaches distant locations in the periphery and is high in concentration there (Fig. 1b). The negative emotions are concentrated in the city center as well, yet for other reasons – people do not like noise and crowd, as well as in remote areas in dense many-storied sleeping quarters, where people feel frustrated. The level of concentration of positive emotions in the city center is almost 4 times higher than of the negative ones.

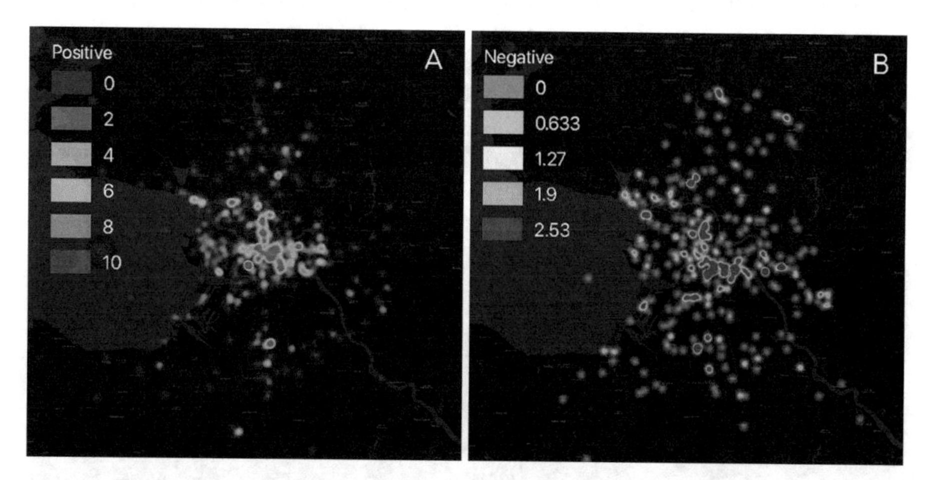

Fig. 1. Heatmaps of positive (A) and negative (B) emotions in St. Petersburg

Though heatmap representation is good for generalization of the spatial distribution, it does not give any understanding of the way emotions correlate with the city streets people generally move through and which form their experience. Our point is to try to show the structure of the city behind a heatmap of objects using the network as a concept and as a tool. Here we present a way to transform the heatmap representation of emotional experience into a network representation. For this we use the Space Syntax approach, which has been developed since the 1970s by Space Syntax Lab, The Bartlett, UCL, for measuring the structure of the city space and its relation to the pedestrian and traffic flows and the functional load of the city space [18].

The algorithm of "network making" evolves as follows. Firstly, the raw geolocated data coming in separate points (Fig. 2a) is transformed into linear data with the "points to path" algorithm built in QGIS. The locations less than 500 m length from each other marked with emoji were connected with this algorithm, thus giving a map of linear emotional data (Fig. 2b). The assumption behind this procedure was to connect points, which might occur on the way of a city dweller and make up her continuous emotional experience. 500 m were taken as a classical proxy for pedestrian accessibility [13, 19]. Figure 2b shows a cluster of connected emotional locations, however this is a mere geographical connection, not yet a real-life structure, which a person might face in the city. Secondly, the linear emotional data was collated with the street-road network layer derived from OpenStreetMap, an open crowd-sourced bank of geolocated urban data highly reliable for St. Petersburg (Fig. 3a) [20]. Network lines connected with the streets were left in the network, others were deleted, thus resulting in an emotional network of the city public space (Fig. 3b). The connectivity of the final network was calculated with the axial connectivity index using the Space Syntax plug-in in QGIS [21]. Connectivity index measures the number of spaces immediately connected to a space of origin [22]. The segments colored in red are the most connected, the segments colored in blue are the least connected (the maximum and minimum connectivity is relative for each network). The connectivity for emotional street networks is interpreted as the extent to which one emotionally charged street connects with other emotionally charged ones. The mostly connected emotional streets are the ones strolling through which a person might face even more positive (or negative) feelings coming from the surrounding streets. Actually, this measure stands for a probability to "meet" emotional situations in the streets. Along with the connectivity index the integration index is calculated. Integration is a normalised measure of distance from any space of origin to all others in a system [22]. It is a static

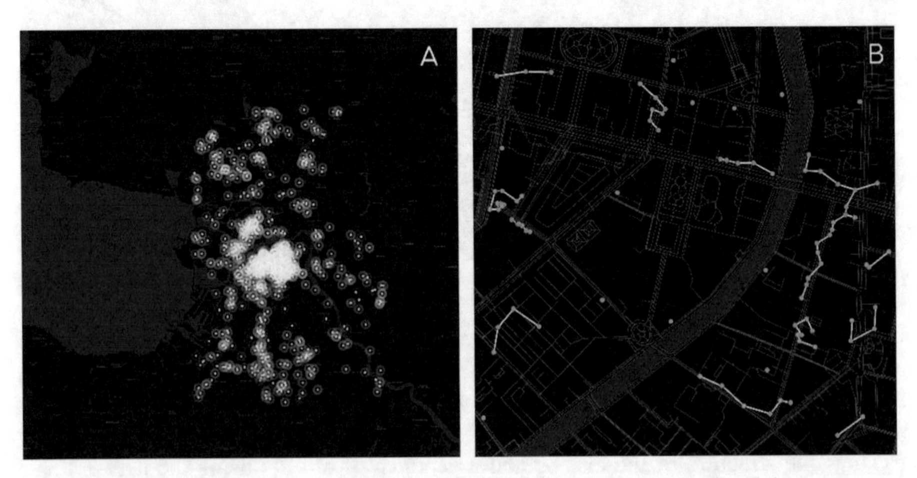

Fig. 2. A: Raw data – points of joy; B: Transformed linear emotional data

global measure and describes the average depth of a space to all other spaces in the system. The spaces of a system can be ranked from the most integrated to the most segregated. In general, it calculates how close the origin space is to all other spaces. In our case this measure has a contextual meaning: the most integrated streets are the most deep in accordance to other streets providing a certain type of emotional experience (negative or positive).

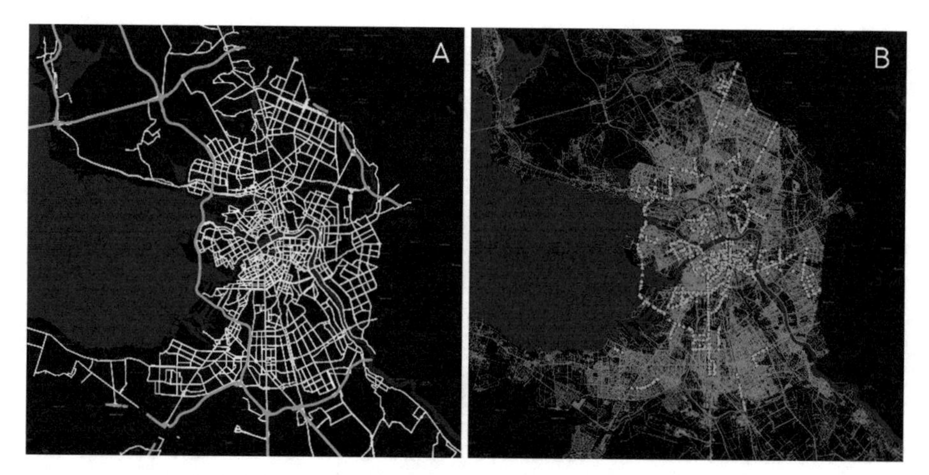

Fig. 3. A: Linear road network data; B: Resulting emotional network

3 Results: Negative and Positive Emotional Networks in St. Petersburg

The resulting maps of positive and negative street networks are presented at Figs. 4a and 4b. Both emotional street networks have a center-periphery hierarchical structure – more connected and emotionally charged streets in the city center and less connected and enclaved in the periphery. The positive street network is more densely connected and integrated (Fig. 4a). It has an evident core of interconnected emotional streets with maximum connectivity in the city center. The "heart" of joy connects central city streets – Spit of Vasilevsky island embankments (Makarova and Universitetskaya), Dvortsovaya embankment, Nevsky avenue and Griboyedova channel; several streets inside the triangle integrate it as arteries (e.g., Italyanskaya street). These streets and linear objects (embankments) are the most touristic, architecturally beautiful, filled with points of attraction locations in the city according to the data from OpenStreetMap (see Fig. 5b). They also have the highest evaluation marks and the biggest number of reviews for local tourist attractions, hotels, nightlife according to TripAdvisor [23]. Besides the core the positive street network contains the axes going toward each of the main geographical directions: South (Moskovsky avenue), North (Medikov avenue, Kantemirovskaya street, Engelsa avenue and Svetlanosky avenue), West (e.g., Sredny avenue, Beringa Street, Smolenka embankment and Korablestroiteley street) and East (e.g., Bolsheokhtinsky Bridge and

Sverdlovskaya embankment). They are less connected, however they are locally important for these areas. An important fact is that these streets are adjacent to the green areas (such as Sosnovka park in the North next to the Svetlanosky avenue), Park Pobedi in the South, next to the Moskovsky avenue. There are only two pendants in the positive street network, which are quite peripheral in comparison to the city center (blue clots on the map).

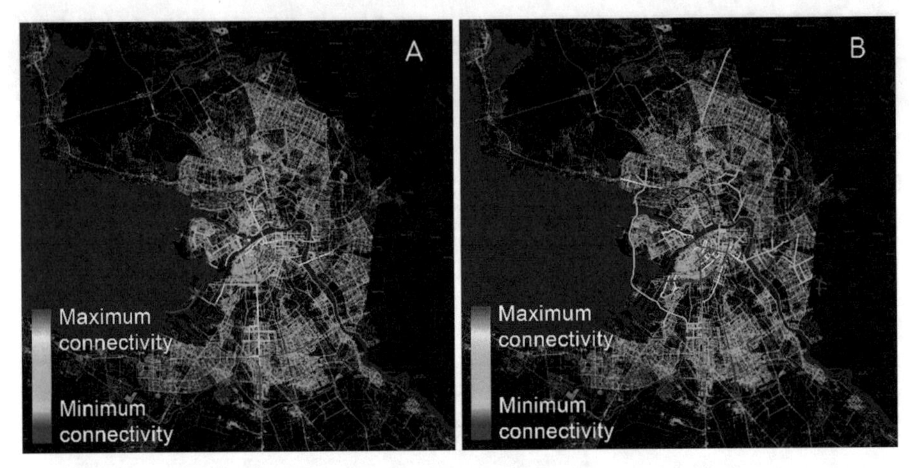

Fig. 4. Positive (A) and negative (B) emotional networks

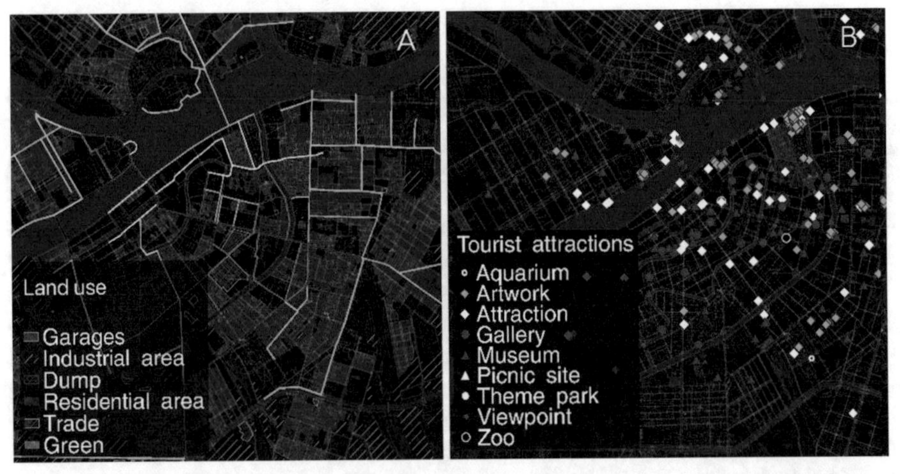

Fig. 5. A: Core of the positive street network; B: Location of points of interest in the city

The negative street network is less connected, more dispersed, reflecting a more localized character of negative emotions, as well as more geographically distanced and covering more peripheral areas than the positive one (Fig. 4b). This character of the negative street network structure can be interpreted the way that negative emotions are "blocked" in certain areas of the city – patches of streets or their intersections in the city periphery and adjacent locations. This might have a good application for the city administration: the hotbeds of bad emotions can be localized and cured with measures of city beautification, optimized transport, pedestrian regulation and others.

It is visible from the comparison of the resulting maps that the same streets turn out to be parts of both positive and negative street networks, and the localization of the most connected negative streets is also the same as in the positive street network. In particular, Nevsky avenue and Moskovsky avenue reappear in both networks. Such a finding can be interpreted as a tendency of the most public, heavily used streets to provide affordances for both good and bad experience. Analysis of the commentaries of Imprecity users, referred to earlier, shows that people are frustrated by the noise, crowd and heavy traffic, therefore Nevsky and Moskovsky avenues, having these features, are perceived as negative streets. However, at the same time, they lead people to beautiful architecture, points of interest and other pedestrian attractions, providing positive experience.

Another finding is that there are streets much more negatively perceived than positively, Obvodny channel embankments, for instance, which are one of the most connected negative streets in the network, though it is partly included into the positive network as well (Fig. 6a). The cityscape of Obvodny channel is illustrated at Fig. 6b. It is a former industrial artery of the city, with nowadays stopped factories, gloomy concrete embankments, dirty water of the channel, heavy traffic, dusty air and lack of greenery. At the same time, it has beautiful redbrick industrial architecture, which is appealing to the pedestrians.

Fig. 6. A: Obvodny channel in the negative street network; B: View on the Obvodny channel

Considering the character of the most connected streets with positive experience, we see a tendency of embankments, which are quite a few in St. Petersburg city center, to become the top ones (see Table 1). The city is known as Northern Venice and historically was built similarly to the Amsterdam model with water arteries, natural (rivers flowing into Neva basin) and artificial (channels) ones. This background might reveal itself in the mental life of the city today, when rivers and embankments become the pathways of joy and symbols of positive feelings. The most connected positive streets are also mostly touristic walkable ones, located in the city center along the main city attractions and public spaces. The most connected negative streets, where people can easily run into bad emotional experience, are those with heavy traffic (Nevsky avenue, Moskovsky avenue), streets in the former industrial belt of the city (Obvodny channel), streets going along messy market (*bazar*) areas with lots of migrants living and working (Sadovaya street, Grivtsova street).

Table 1. Rating of the most connected streets in positive and negative street networks

Index	Positive street network	Negative street network
10-8	Nevsky avenue Makarov embankment Dvortsovaya embankment Fontanka embankment Universitetskaya embankment Bolshaya Konyushennaya street 1st line of Vasilievsky island Rubinstein street	Nevsky avenue Sredny Avenue of Vasilievsky island Obvodny channel Ligovsky avenue Sadovaya street Grivtsova Street Kadetskaya line
7-4	Grivtsova street Moskovsky avenue (the beginning) Sadovaya street Italian street Kamennoostrovsky avenue Zanevsky avenue Millionnaya street Moyka embankment Pestel street Griboyedov embankment Mikhailovsky street	Fontanka embankment Zagorodny avenue Suvorovsky avenue Bakunina avenue Zanevsky avenue Novocherkassky avenue Makarov embankment 15th line of Vasilevsky island Mitrofanyevsky highway Blagodatnaya street Kuznetsovskaya street Basseynaya street Kubinskaya street

In the Table 2, the list of the most integrated streets is provided. The major tendency remains: Nevsky avenue is still the most integrated street both in the positive and negative networks in St. Petersburg, as it is the most connected. The avenue is in the core of the city's cultural life and is integrating other city streets with cultural attractions into a landscape of urban vibe. At the same time, it is one of the most important public and transportation hubs, integrating other streets into a system of the city's heavy infrastructure, however provoking negative emotions of insecurity and aggression.

Table 2. Rating of the most integrated streets in positive and negative street networks

Index	Positive street network	Negative street network
10	Nevsky avenue Kamennoostrovsky avenue Dvortsovaya embankment Makarov embankment	Nevsky avenue
9	–	Obvodny channel
8	–	Moskovsky avenue
7	Fontanka embankment	Ligovsky avenue
6	–	Sadovaya avenue
5	Bolshaya Konyushennaya street	Grivtsova Street
4	1st line of Vasilievsky island	University embankment
3	Moyka embankment	Yefimova street
2	Rubinstein street	Kadetskaya line
1	Grivtsova street Moskovsky avenue (the beginning)	Rubinstein street

4 Conclusion

The study of the positive and negative street networks in St. Petersburg illustrates the continuity of emotional experience of the city space. Both positive and negative street networks have a center-periphery hierarchical structure, with the most connected streets of both emotional valence located in the city center. However a positive street network is more centrally localized and connected, while a negative one is more dispersed and has quite a few negative clusters in the city periphery. Positive streets are the ones containing beautiful architecture, sights and venues as points of interest. Negative streets are devoid of these and are mostly transit areas with heavy traffic, bad pedestrian regulation, noise and crowd. Streets, which are the main public arteries of the city, are included into both networks, while they are providing reasons for joy and for negative feelings simultaneously. Nevsky avenue is the most connected and integrated street both in positive and negative street networks, which shows its particular importance for subjective perception and the mental city map of St. Petersburg citizens.

The study of emotional networks gives a new perspective on the connectivity of the city space, tendencies of spatial distribution of negative and positive subjective experiences, nature of aggregation of different emotions in the main and tertiary, central and peripheral city streets. Knowledge of emotional networks in the city can be used to develop the structure of the city space for the emotional well-being of the citizens. There are several applications for the city administration and urban planners to use this knowledge.

1. Existing positive street network of the city space can be material to analyze appreciated routes for pedestrians and tourists. The existing positive street network can be used in touristic place-making of the city, of which the key part is route-making. Additionally, the negative networks can show which ways better not to take if the tourist is not well acquainted with the city.
2. The structure of the urban emotional network can demonstrate items of the built environment which make an input into intensifying emotional experience along the path walks of people. For example, the streets passing next to the green spots of the city themselves receive attractivity (which is especially important in the case of the remote city areas, where beautiful architecture historically was not built, and sights or venues are few because of the lower demand). Shops, cafes, bars, theaters and museums located at the front of the streets create their inviting atmosphere. Creating more items like that in the city streets will improve their "emotional profile".
3. The structure of the positive street network can show the streets and pathways in need of more connections for better emotional flow. For example, it can show the shortcuts to make in-between the streets to preserve the mood.

5 Discussion

Developing the concept and methodology of the emotional street network requires further steps.

Firstly, for a more detailed account on the graph analysis, the axial and segment levels of the emotional network could be considered using Space Syntax metrics. This could give evidence on the consistency of an emotional rate of a street at different structural levels.

Secondly, to avoid the risk of the missing data, the original OpenStreetMap network has to be completed with non-official paths, such as footpaths, not to lose items of the pedestrian routes, which can be quite tactical and spontaneous. Otherwise, only official streets will eventually appear in the emotion street network. To build specific recommendations for the city streets, an initial street network needs to be quite precise.

Thirdly, the more data on emotions is gathered, the more disaggregated the analysis could be. For instance, more local tendencies could be traced on the level of separate administrative districts, not the whole city.

Acknowledgments. This paper was supported by Russian Foundation for Basic Research grant № 20-013-00891 A "Emotional perception of urban environment as a factor for urban resilience".

References

1. Lynch, K.: The Image of the City. MIT press, Cambridge (1960)
2. Gehl, J.: Cities for People. Island Press, Washington (2010)
3. Ellard, C.: Places of the Heart: The Psychogeography of Everyday Life. Bellevue Literary Press, New York (2015)

4. Quercia, D., O'Hare, N.K., Cramer, H.: Aesthetic capital: what makes London look beautiful, quiet, and happy? In: Proceedings of the 17th ACM Conference on Computer Supported Cooperative Work & Social Computing, pp. 945–955. Association for Computing Machinery, New York (2014). https://doi.org/10.1145/2531602.2531613
5. Debord, G.: Introduction to a critique of urban geography. Praxis (e) press (1955/2008)
6. De Certau, M.: The practice of everyday life, eng. Trans. By S. Rendall. University of California Press, Berkeley and Los Angeles (1980)
7. 'Quercia, D., Schifanella, R., Aiello, L.M.: The shortest path to happiness: recommending beautiful, quiet, and happy routes in the city. In: Proceedings of the 25th ACM Conference on Hypertext and Social Media, pp. 116–125. Association for Computing Machinery, New York (2014). https://doi.org/10.1145/2631775.2631799
8. Appleyard, D.: Livable Streets. University of California Press, Berkeley (1981)
9. Lennard, S.H.C., Lennard H.L.: Livable Cities, Social and Design Principles for the Future of the City. Gondolier Press, Southampton (1987)
10. Jacobs, J.: The Death and Life of Great American Cities. Random House, New York (1961)
11. Montgomery, C.: Happy City: Transforming Our Lives Through Urban Design. Macmillan, New York (2013)
12. Rudofsky, B.: Streets for People: A Primer for Americans. Doubleday, Garden City, New York (1969)
13. Spek, J.: Walkable City. How Downtown Can Save America, One Step at a Time. North Point Press, Konnur (2013)
14. Nenko, A., Petrova, M.: Emotional geography of St. Petersburg: detecting emotional perception of the city space. In: Alexandrov, D.A., Boukhanovsky, A.V., Chugunov, A.V., Kabanov, Y., Koltsova, O. (eds.) Digital Transformation and Global Society, vol. 859, pp. 95–110. Springer, Cham (2018)
15. Ekman, P.: Basic emotions. In: Dalgleish, T., Power, M. (eds.) Handbook of Cognition and Emotion, 45-60. Wiley, Sussex (1999)
16. De Smith, M., Goodchild, M., Longley, P.: Geospatial Analysis: A Comprehensive Guide to Principles, Techniques and Software Tools. The Winchelsea Press, Matador (2011)
17. Graser, A.: Learning QGIS 2.0. Packt Publishing, Birmingham (2013)
18. Space Syntax. spacesyntax.com Accessed 01 June 2020
19. Talen, E.: Pedestrian Access as a Measure of Urban Quality. Plann. Pract. Res. **17**, 257–278 (2002). https://doi.org/10.1080/026974502200005634
20. Open Street Map, Saint-Petersburg. https://www.openstreetmap.org/relation/337422. Accessed 01 June 2020
21. QGIS Space Syntax toolkit. https://plugins.qgis.org/plugins/esstoolkit/. Accessed 09 Feb 2019
22. Hillier, B., Hanson, J.: The Social Logic of Space. Cambridge University Press, Cambridge (1984)
23. TripAdvisor, activities in St. Petersburg. https://www.tripadvisor.ru/Attractions-g298507-Activities-St_Petersburg_Northwestern_District.html. Accessed 01 June 2020

Application of Semantic Networks and Enterprise Architecture Approaches for Creation of Digital Twins of Organizations

Miroslav Kubelskiy(✉)

Peter the Great St. Petersburg Polytechnic University, St. Petersburg, Russia
mirqube@gmail.com

Abstract. Creation of digital twins of various objects and processes relevant to the enterprise activity for management tasks is becoming increasingly relevant in the era of digital transformation of enterprises, which is part of the fourth industrial revolution. It can be both individual products, production facilities and processes, or entire enterprises. The creation of the last type of digital twins, i.e. a digital twin of organizations (DTO), is the least developed topic. This is largely due to the low degree of formalization of many aspects relevant to enterprises, as well as the presence of a significant gap between the physical and information worlds in the case of an insufficient degree of the organization digitalization. The main topic of this research is a conception of methodology for DTO creation based on ontologies, knowledge graphs and enterprise architecture. Enterprise Architecture has generated many modeling-based methods for description and formalization of the activities and structure of enterprises in terms of business – and IT-architecture, but by nature, most models are qualitative and static and thus not applicable for the task of digital twin creation. Presentation of various architectural models in the form of semantic networks for structuring the subject domain, as well as using as a data structure for graph databases, allows developing a methodology for building digital twins of organizations. Usage of the methodology can provide enterprise data integration, continuous data acquisition, data warehousing and prepare data for further analysis to support managerial decision-making.

Keywords: Digital twin of organization · Semantic networks · Knowledge graph

1 Introduction

Thc basis of Industry 4.0 [1] is the process of digital transformation of enterprises [2]. Within the enterprise, there can be many directions of business activity and information systems supporting them, creating and using ever-growing data arrays, which greatly increases the complexity of the control object, i.e. enterprise. This trend makes it impossible for a manager to perceive all the information necessary for making an effective managerial decision. In turn, the ability of organizations to use (receive, store,

A. Antonyuk and N. Basov (Eds.): NetGloW 2020, LNNS 181, pp. 144–161, 2021.
https://doi.org/10.1007/978-3-030-64877-0_10

analyze and manage) these great arrays of data and information becomes a new competitive advantage and provides them with an unprecedented opportunity to use artificial intelligence methods for management tasks.

According to the systems approach to management [3], the main method for studying, analyzing and predicting the behavior of any system, i.e. combinations of interacting elements organized to achieve one or more of the set goals [4, 5], is modeling. A model is a simplified description of a system, reflecting only those properties that are essential for modeling purposes [6]. During the evolution of management decision-making support tools, a large number of modeling methods of varying degrees of formalization and complexity have been developed that allow describing both the qualitative and quantitative parameters of enterprises [7]. Most of them do not meet the challenges of modern time, because of their insufficient accuracy, static, complex construction and use, the need for constant updates.

According to [8], the following components of Industry 4.0 can be distinguished: Cyber physical systems (CPS), Cloud systems, Machine to machine (M2M) communication, Smart factories, Augmented reality and simulation, Data mining, Internet of things, Enterprise resource planning (ERP) and business intelligence, Virtual manufacturing, Intelligent robotics. The subject of the research is the key technology for constructing cyber physical systems, namely the creation of digital models of real-world objects and ensuring the updating of these models in real time, i.e. creation of digital twins of real-world objects. A digital twin is a set of informational constructions that fully describe the potential or actual real produced product from the microscopic to the geometric macroscopic level. The description provided by the digital twin should be "virtually indistinguishable from its physical counterpart" [9].

This technology is a modern trend in the development of industry in accordance with the fourth industrial revolution and is quite mature. Technological solutions are widely discussed in the literature, which make it possible to build digital twins of such objects as production processes [10], complex technical systems, for example, products [11] or buildings [12]. The use of digital twin technology can significantly reduce costs at all stages of the modeling object life cycle and improve its functional characteristics. The interest in this technology is also due to the fact that its widespread use can make it possible to change the world industry and make the transition to a circular economy, which will lead to a significant reduction in the amount of waste produced by mankind [13].

The problem of creating digital twins of organizations is not covered enough in the literature, and there are no ready-made solutions that can fully realize the "Digital Twin of Organization" (DTO) technology, what represents a research gap. The construction of DTO is complicated by the following factors:

1. An enterprise is not a physical object with a constant structure, but is a complex dynamically changing sociotechnical system that exists in the real world. Many of its elements and relationships between them are difficult for formal description in contrast to, for example, purely technical systems. The problem is relevant to the field of enterprise architecture.

2. A significant part of concepts used in organizational management may not have clear definitions or have many different definitions, which also complicates the construction of formal enterprise models. The problem belongs to the field semantic technologies, namely ontologies and knowledge graphs.
3. The data and information that should be integrated into the DTO have a different nature (qualitative, quantitative and multimedia), a heterogeneous structure, differ greatly in the degree of structuredness, are in various, often unrelated, sources. This is a problem of integration, governance and analysis of data and information.

The synthesis of achievements in the field of enterprise architecture, knowledge engineering (semantic networks – ontologies and knowledge graphs) and data science allows to overcome the problems identified above and to develop a methodology for creation of Digital Twins of Organizations, which is the purpose of the research. The proposed methodology is based on the building of an organization's knowledge graph. This approach is strictly relevant to organizational and social computational sciences [14], according to which it is possible to represent the organization as a meta-network, i.e. a network of linked agents, knowledge, resources and tasks [15], and perform further analysis using computational models.

At the current stage, research can be classified "research in progress", because the stage of practical implementation and testing of the methodology is the subject of further research; therefore, the article presents the conception of the proposed methodology. The practical implementation of DTO technology, which has significant potential for application in the corporate and public sectors, can significantly increase the effectiveness of managerial decisions at all levels, which justifies the relevance of this topic.

The paper has the following structure: the second section of the paper describes the basic concepts necessary to create a methodology for constructing a DTO, the following describes the research methodology and provides a description of the methodology conception for building a DTO proposed by the author. The paper concludes with further steps to develop the methodology.

2 Foundations

2.1 Digital Twins

Michael Grieves, as described in his white paper [16], firstly introduced the term "digital twin" in 2003. This technology gained wide popularity in 2012, when NASA and USA Air Force Vehicles identified it as one of the key ones and gave it the following definition: "A Digital Twin is an integrated multiphysics, multiscale, probabilistic simulation of a […] system that uses the best available physical models, sensor updates, […] history, etc., to mirror the life of its corresponding […] twin." [17]. The main goal of creating digital twins for NASA and USA Air Force Vehicles was the need to analyze the behavior and current state of aircraft during operation by building their digital models, based on building mathematical models, collecting and analyzing historical and current data. Since 2012, the number of scientific publications and examples of the application of this technology in other industries has been steadily growing [18].

In [16] three main elements of a digital twin were identified for the first time, which are independent of the modeling object: a physical component, a virtual component, and a connection between the two. In a later publication [19] the number of constituent elements was expanded to five – the data was separated into a separate element, and a service system appeared, which an interface for interacting with the technology user is. Later in [20] it was concluded that only the organization of bidirectional data exchange between a physical object and its virtual copy allows the full implementation of a digital twin technology. Thus, we can conclude that data plays an important role in this technology.

The digital transformation of enterprises, taking place as part of the fourth industrial revolution, concerns not only products, technical devices and production processes, but also involves the digitization of other objects relevant to enterprises – people (employees and consumers), infrastructure, organizational processes. That will generate huge amounts of data that are potentially valuable, revealing the interconnections between which, it is possible to build a holistic, coherent and comprehensive digital model of the organization, namely the digital twin of the organization.

One of the possible definitions of DTO is presented in [21]: "This digital twin will be a living digital simulation model of the organization that updates and changes as the organization evolves. And upon becoming available, the digital twin will allow scenarios to be thoroughly tested to predict the performance of potential tactics and strategies". In [22] another definition is presented: "A digital twin of an organization is a set of information that allows you to track a change in a company's situation when modeling various impacts on it: managerial, disturbing environmental influences, etc. For this, such a set of information should take into account all the essential causal relationships, as well as contain the necessary and sufficient set of data to simulate the behavior of an enterprise in a market environment."

Based on the foregoing, it is possible to conclude that one of the main objectives of the DTO is the integration of constantly updated data from disparate sources, including databases of information systems used at the enterprise, with the possibility of subsequent analysis to support management decisions. Hence, the following requirements can be formulated for the DTO technology in the field of operations with data, the DTO should provide:

- Data modeling and mapping (i.e. the ability to compare digitized data with the objects of the real world to which they relate, as well as modeling the data structure);
- Data acquisition (i.e. collection of various types of data (qualitative, quantitative, text, media, etc.) from heterogeneous sources, as well as manual data entry, if they are not in the database of enterprise information systems);
- Data integration (i.e. integration of disparate data in a single space with a sufficient degree of their quality and purity);
- Data warehousing (i.e. storage of received and integrated data with the ability to read, write, change);
- Data analysis (i.e. the use of various data analysis methods to search for significant dependencies, including visual analysis, as well as modeling methods – simulation, factor, predictive).

In addition, it should be noted that the possibility of building a DTO for a particular enterprise depends heavily on what stage of the digital transformation it is in. If most of the data is still in analog storage, for example, in the form of paper documents, then until they are digitized it will be impossible to build the DTO, which is beyond the scope of the proposed methodology.

2.2 Enterprise Architecture

Enterprise Architecture [23] is an interdisciplinary field of research that allows for a holistic understanding of the structure of the enterprise, gaining particular relevance in the era of the fourth industrial revolution. The focus of Enterprise Architecture is the design, management and transformation of a modern enterprise as a complex system that includes business and IT components. A distinctive feature of the application of enterprise architecture for management is the active use of concise and clear visual models that describe the activity, structure, i.e. subjects and objects of activity, as well as the goals of the enterprise. It carries additional value for key stakeholders in the process of strategic and tactical decision-making, because minimizes the loss and misunderstanding of information [24, 25].

Enterprise Architecture is based on the formal documentation of various aspects relevant to the enterprise, including corporate strategy, business activities, organizational structure, information technology, including application, data and information architecture, with a special emphasis on the interaction and interdependence of these elements. The main result of this formalization is the creation of artifacts that can take the form of lists/registers, classifiers, tables/matrices, and mentioned earlier visual diagrams. Recently, in modeling notations and methodologies of enterprise architecture management more attention has been paid to modeling physical infrastructure, which is necessary to create full-fledged digital twins of organizations [26].

Over the thirty-year history of enterprise architecture, many methodologies, frameworks, and modeling languages have been created, for example:

- Zachman Architectural Framework [27];
- Architecture of Integrated Information Systems (ARIS) [28];
- The Open Group Architecture Framework (TOGAF) [29];
- ArchiMate modeling language [26].

In addition, in the enterprise architecture modeling, methods originating in the field of computer science are used, for example:

- Unified Modeling Language (UML);
- Entity-Relationship Model (ER-model) [30].

In [18] enterprise architecture is singled out as one of the most relevant management disciplines for creating digital twins of organizations. This is due to the fact that the digital twin of the organization cannot function separately from the information systems already existing in the enterprise because they contain data and information related to various aspects of the enterprise, to which the DTO must have access for its continuous updating. In turn, the enterprise architecture allows to describe the IT landscape of the enterprise, the available data and information architecture, and also understand how it is all interconnected with the business aspects of the enterprise. It allows integrating the DTO into the structure of the enterprise harmoniously and ensuring its coordinated interaction with other enterprise information systems.

The main disadvantages of the architectural models used in enterprise architecture are:

- the relative high cost and complexity of building and updating;
- static nature of models – once built a model can only be changed manually;
- minimal integration with real enterprise data;
- limited opportunities for quantitative calculations.

Despite this, the tools developed in this discipline allow to perform object-functional modeling of the enterprise with sufficient accuracy, which will be shown later.

2.3 Ontologies and Knowledge Graphs

A knowledge graph is a collection of objects, their attributes, and typed relationships between them. The connections between the objects, the objects themselves and their attributes are defined in the models of the subject area, for which a knowledge graph is built, which are called ontologies [31]. Ontology is a formal, explicit specification of the conceptualization of the subject area [32], which includes a hierarchy of concepts used in the subject area, the types of relations between them and the imposed restrictions. The types of relationships between objects can strongly depend on the domain being modeled using an ontology, but the most common are relations such as "a class of" (used to create taxonomies), "consists of" (used in partonomies), "refers to", "performs", etc.

Building ontologies, or ontological engineering, is the main method in such a discipline as knowledge engineering, which is the direction of artificial intelligence associated with the construction of knowledge bases and knowledge-based systems. Building ontologies allows you to get a formal description of various subject areas available for processing by computers, for example, knowledge from some scientific discipline – physics or biology, any types of systems, for example, enterprises.

"Ontology" and "knowledge graph" are semantic networks, i.e. networks describing the logical (semantic) relationships between objects of the subject area, and are related to each other. Exist two types of ontologies: ABox and TBox. TBox ontologies focus on classes, attributes and axioms of the domain. The focus of ABox ontologies are the instances of classes. Both types are used in knowledge graph, because in order to move from an ontology (TBox) to a knowledge graph, it is necessary to populate the ontology with instances (ABox).

Ontologies are often used as a precise model of data circulating in information systems, which makes its use in the building of digital twins of organizations especially relevant [33], because when implementing the DTO technology it is necessary to ensure interaction, i.e. exchange of data, between all elements of the system – people, information systems, equipment, etc. Using ontological data models can provide the required quality and consistency of data, as well as minimize the number of errors and failures when reading and writing.

3 Research Methodology

The framework of "design science research methodology" [34] is a methodological basis of the research. It is widely applied in the field of Enterprise Architecture, Business Informatics and Information Science Research. According to [35] four basic principles should be met by the research: "generalization – applicability to the class of problems; originality – the expansion of existing knowledge about the subject; validity – supported by arguments and ensuring the possibility of validation; usefulness – the ability to benefit (either in the near future or in the future) to interested parties."

Artifacts or instantiations are the main results of the research: "Design science research aims at the development of artifacts, namely constructs (e.g. concepts, terminologies, models, methods), and instantiations (i.e. concrete solutions implemented as prototypes or production systems)" [35].

According to [36], design science research includes six main steps: 1) problem identification and motivation; 2) define the objectives for a solution; 3) design and development; 4) demonstration; 5) evaluation; 6) communication.

Description of research problem and motivation for developing the methodology of creation of digital twins of organizations are included in the paper introduction part. The main objectives for a proposed methodology are described at the end of the part dedicated to digital twins. The logic of designing and developing is presented in the section that describes the conception of methodology for creation of digital twins of organizations. The focus of this article is to communicate the intermediate result of the study, i.e. the methodology conception. Its demonstration and evaluation are the subject of further research.

4 A Conception of Methodology for Creation of Digital Twins of Organizations

As a first step in enterprise replication in digital space, it is necessary to describe the real object, i.e. enterprises with a sufficient degree of formalization, for which an ontological approach is best. The advantage of this approach is the possibility of reusing ontologies created earlier with their subsequent adaptation to better match the current task, as well as the possibility of varying the degree of formalization of the model.

The presence of the enterprise ontological model allows comparing the available data with the objects of a real enterprise – people, information systems, equipment, processes, i.e. to make modeling and mapping of data. Therefore, it is necessary to identify the main classes of objects, their properties, and the relationship between objects, the totality of which is an enterprise.

Since ontological modeling of enterprises is quite a mature discipline, several ontologies of the enterprise have been developed in the academic community:

- Toronto Virtual Enterprise ontology (TOVE) [37];
- Business Entity Ontological Model (BEOM);
- Enterprise ontology [38];
- Organization from W3C [39].

Unfortunately, most of these ontologies are not widely used among management practitioners because of their complexity, and not all of them are kept up to date.

Enterprise architecture practitioners often use the modeling language ArchiMate [26], which is the modeling standard within the TOGAF EAM methodology [29]. ArchiMate has a rich notation that contains about 60 types of objects and 11 relationships, which allows to describe various aspects of the enterprise with sufficient accuracy. In addition, starting with version 3.1, the modeling language includes classes of objects related to the physical infrastructure, i.e. equipment, facilities, materials that are necessary to create a full-fledged DTO.

As in any modeling language, ArchiMate is based on its metamodel, i.e. a set of all existing classes of objects, denoted by a unique symbol, and valid relationships between them. This allows to use the ArchiMate metamodel as the initial ontology needed to build the DTO. An abridged conceptual model of the ArchiMate metamodel, created by the author, is presented in Fig. 1.

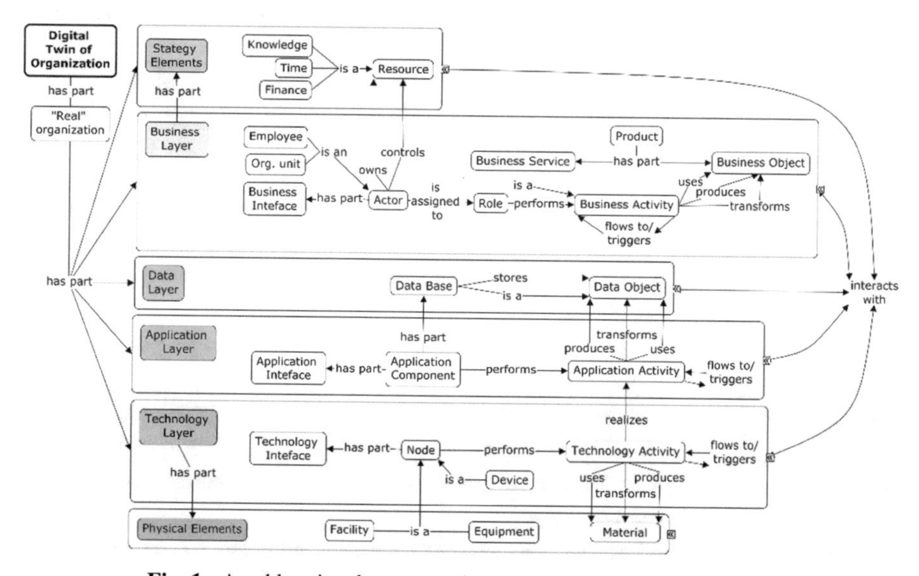

Fig. 1. An abbreviated conceptual model of ArchiMate metamodel

As part of the TOGAF enterprise architecture management methodology, it is assumed that any enterprise can be represented as a combination of 3–4 layers, each of which relates to certain types of objects, which is reflected in ArchiMate. Thus, in ArchiMate there are three main types of objects [26], which belong to three layers: Business Layer, Application Layer, Technology Layer.

A separate group of Strategy Elements belongs to the Business Layer and Physical Elements to the Technology Layer. The Motivation Elements group also stands out separately in the modeling language, but they are not shown in the figure. The 4th layer is sometimes allocated in enterprise architecture, which is associated with the data architecture, which is important for the purpose of creating a DTO, therefore, the Data Layer is present in the figure.

In accordance with the ArchiMate language, elements of different levels can interact with each other according to certain rules, but for clarity, all inter-level interaction is represented by the "interacts with" relationship.

ArchiMate allows to describe various activities inside and outside the enterprises, performed by various actors, in the form of processes, functions or services, which are combined in the figure into Activity-elements, although initially such an element is not in the notation. The definitions of the layers, groups of elements and the elements themselves from the notation used in the illustration are presented in Table 1.

Table 1. Definitions of ArhiMate concepts. Source: [26]

Concept	Definition
Application component	An application component represents an encapsulation of application functionality aligned to implementation structure, which is modular and replaceable
Application interface	An application interface represents a point of access where application services are made available to a user, another application component, or a node
Application layer	The Application Layer elements are typically used to model the Application Architecture that describes the structure, behavior, and interaction of the applications of the enterprise
Business actor	A business actor represents a business entity that is capable of performing behavior
Business interface	A business interface represents a point of access where a business service is made available to the environment
Business layer	Business Layer elements are used to model the operational organization of an enterprise in a technology-independent manner
Business object	A business object represents a concept used within a particular business domain
Business role	A business role represents the responsibility for performing specific behavior, to which an actor can be assigned, or the part an actor plays in a particular action or event
Business service	A business service represents explicitly defined behavior that a business role, business actor, or business collaboration exposes to its environment
Data object	A data object represents data structured for automated processing

(*continued*)

Table 1. (*continued*)

Concept	Definition
Device	A device represents a physical IT resource upon which system software and artifacts may be stored or deployed for execution
Equipment	Equipment represents one or more physical machines, tools, or instruments that can create, use, store, move, or transform materials. Equipment comprises all active processing elements that carry out physical processes in which materials (which are a special kind of technology object) are used or transformed
Facility	A facility represents a physical structure or environment. It represents a physical resource that has the capability of facilitating (e.g., housing or locating) the use of equipment. It is typically used to model factories, buildings, or outdoor constructions that have an important role in production or distribution processes
Material	Material is typically used to model raw materials and physical products, and also energy sources such as fuel and electricity
Node	A node represents a computational or physical resource that hosts, manipulates, or interacts with other computational or physical resources
Physical elements	The physical elements are included as an extension to the Technology Layer for modeling the physical world
Product	A product represents a coherent collection of services and/or passive structure elements, accompanied by a contract/set of agreements, which is offered as a whole to (internal or external) customers
Resource	A resource represents an asset owned or controlled by an individual or organization
Strategy elements	The strategy elements are typically used to model the strategic direction and choices of an enterprise, as far as the impact on its architecture is concerned. They can be used to express how the enterprise wants to create value for its stakeholders, the capabilities it needs for that, the resources needed to support these capabilities, and how it plans to configure and use these capabilities and resources to achieve its aims
Technology interface	A technology interface represents a point of access where technology services offered by a node can be accessed
Technology layer	The Technology Layer elements are typically used to model the Technology Architecture of the enterprise, describing the structure and behavior of the technology infrastructure of the enterprise

For the task of building a DTO, it is important to focus on the fact that there are two types of elements that describe the enterprise structure – active structure element (able to do activity, i.e. it is an activity subject) and passive structure element (unable to do activity, i.e. it is an activity object). Activity inside or outside the system is presented by behavioral elements. Passive structure elements are used, produced and transformed during the activity performed by the subject. For example, this applies to data, which includes data objects that may change during the activity of information systems. An actor can change something in the data only through the application interface. The same applies to resources and materials, which is a reflection of the resource footprint of activities that is described in [22].

Thus, the wide range of elements presented in this modeling language allows to describe almost any enterprise in a first approximation. The modeling language allows to customize elements and relationships as well. It is possible to produce a specification and determine the attributes of the elements. Specification means the ability to subclass element classes with the ability to inherit their properties. For example, the parent element "Actor" can spawn elements such as "individual", "Organizational Unit" and "Organization". The ability to set attributes is carried out by creating profiles of elements and relationships: "A profile is a data structure which can be defined separately from the ArchiMate language but can be dynamically coupled with elements or relationships" [26], what is shown in Fig. 2.

To demonstrate this approach, an architectural model was constructed for the fictitious situation of a manufacturing company, shown in Fig. 3.

At the top of the figure are strategic elements, namely the Value Flow and Resources. The Value Stream consists of two stages, which are the first two stages of the product life cycle – Research and Development and Production, each of which has two attributes – duration and total cost. For the implementation of each of these stages, a specific set of resources is required, including finance, necessary knowledge in various subject areas, physical and information technology resources.

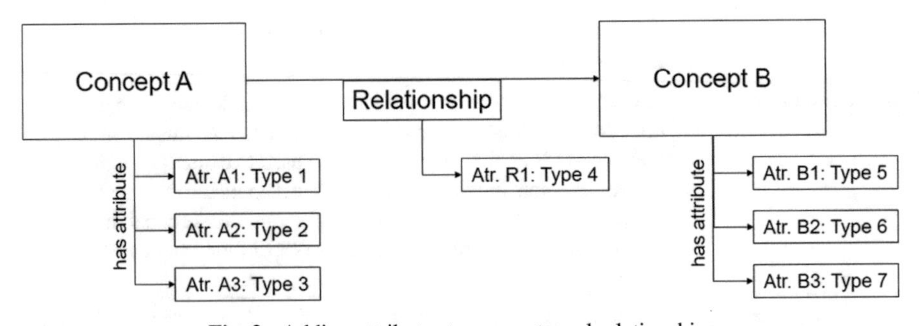

Fig. 2. Adding attributes to concepts and relationships

At the bottom of the model are elements related to all three architectural layers. To implement the first stage of the product life cycle, it is necessary to implement a business process dedicated to R&D, in which the roles of a designer and a marketer take part, belonging to specific actors with certain knowledge and a specific salary. An information system with a paid monthly license is used for design as well. The result of this process is the design of the planned product, presented as a data object – CAD-model of the product.

The second stage of the life cycle is implemented using another business process, "Production process", which receives the previously developed product design. To implement this business process, two other employees are involved. As part of the business process, the technological process is being implemented, consisting of two sub-processes "Manufacturing" and "Assembly". Two units of equipment of a specific model with a monthly rent are used to implement the technological process "Manufacturing", i.e. "3D-printer" and "CNC machine", which require the previously obtained CAD-model of the product, as well as materials – Plastic, Metal, Electricity, also having its value and needed quantity. At the output of this process, two separate parts are obtained, which are transformed into a ready product after passing through the "Assembly" process.

Presence of attributes and their values for each individual element makes it possible to evaluate the financial assets required for the implementation of two stages of the product life cycle. Unfortunately, most modeling tools do not allow this to be done automatically, as well as displaying the attributes of elements in a similar form. The author to illustrate the approach did it manually.

It is important to pay attention to the fact that in the demonstrated example there are both elements related to classes, for example, roles, and specific instances of classes, for example, actors having specific attribute values, i.e. related data. In this case, the presented architectural model can be considered as an ontological knowledge graph, i.e. a set of nodes and semantic edges that describes a separate part of the enterprise. If we remove the attribute values and class instances from the model, we get the ontology of a particular enterprise, which can play a role of the enterprise data structure.

There is the possibility of presenting the resulting architectural model, i.e. knowledge graph, in common ontology formats – OWL or RDF. This makes the model available for machine analysis, which allows to use graph analysis methods and to save the model in a database. The use of SQL databases for these tasks does not seem appropriate since this model has a network structure, because graph and table data types are very different from each other. This can greatly complicate the structure of the database, the result of which will be a decrease in speed and complication of the formation of queries to the database management system. The most suitable for the task of storing this kind of data is the graph type of the database [40]. The structure of such a database will largely coincide with the TBox ontology of the enterprise.

A system of adapter needs to be built between the enterprise IT landscape and DTO to ensure continuous data acquisition. Adapters perform data parsing from enterprise information systems and convert them to the required form in accordance with the ontology, after which the data is written to the corresponding part of the knowledge graph. This process is illustrated in Fig. 4, where the main types of enterprise elements that are parts of the ontology are highlighted, and the main types of information systems that contain data relevant to these elements are identified. Data flows pass through adapters from information systems to the corresponding ontology objects.

The main aspects of the methodology for creating a DTO can be integrated into a single conceptual model, presented in Fig. 5.

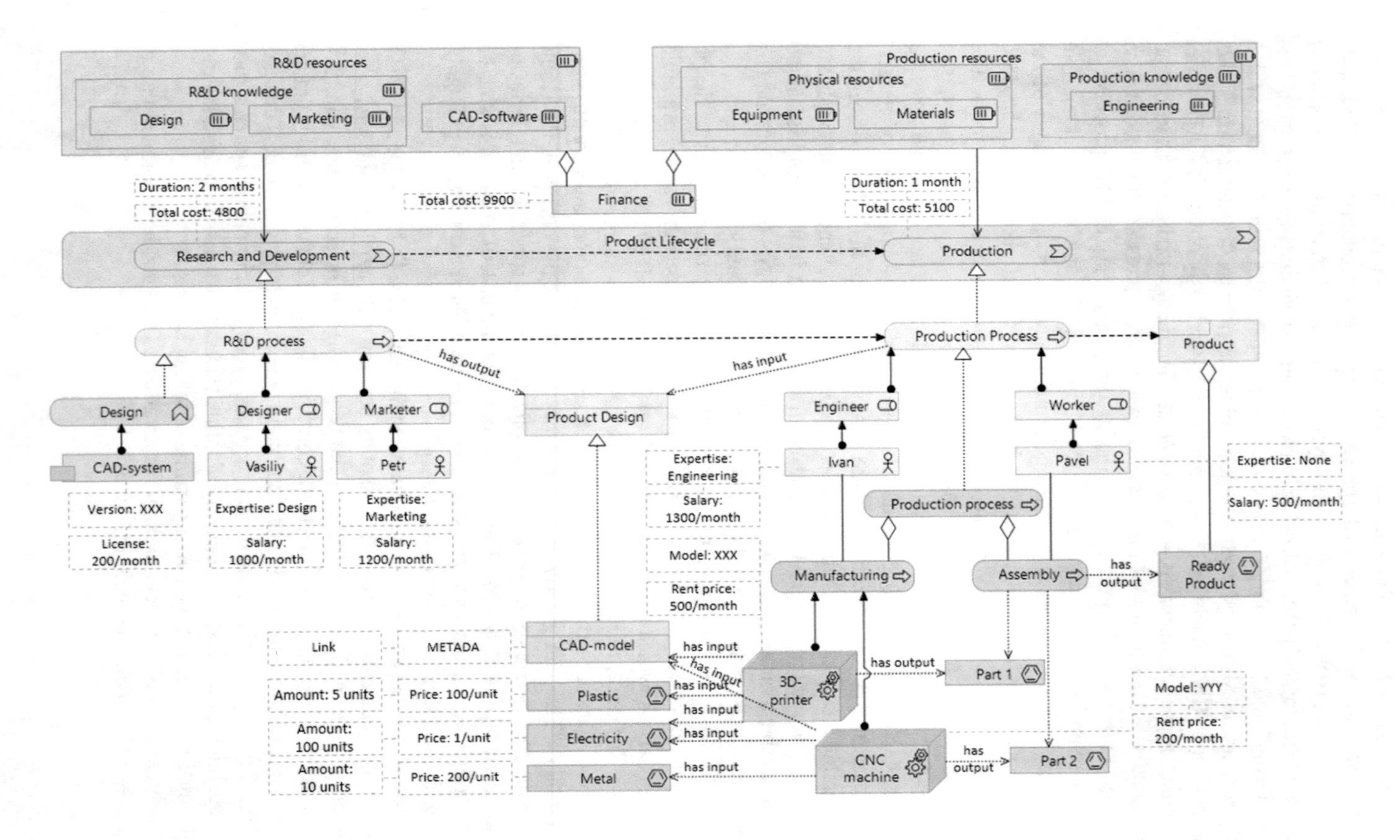

Fig. 3. An architectural model for the example case with added attributes

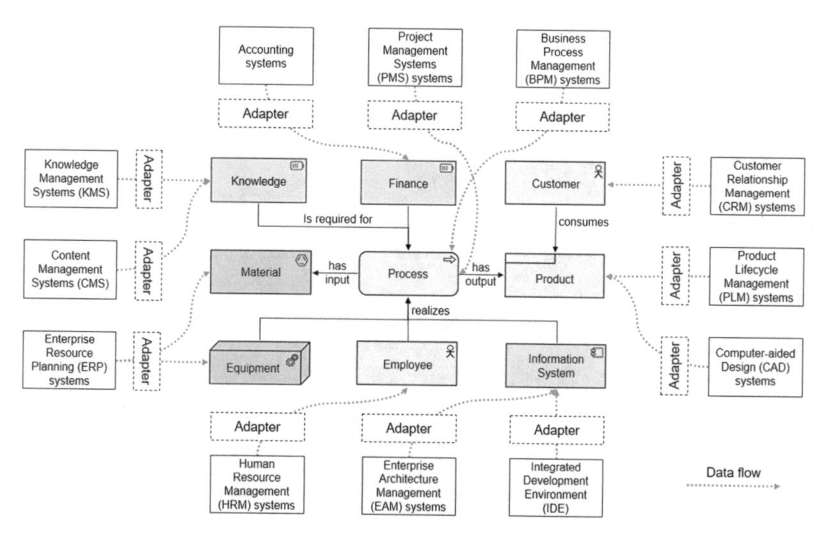

Fig. 4. A process of data acquisition

The model distinguishes five components of the DTO – the real organization, its virtual copy, the relationship between them, data and the service system. A color codification has been introduced for each element and aspects of the methodology relevant to it.

The main element of the DTO is the data that can exist in the organization both in digital and non-digital forms. The possibility to build an organization's DTO is highly dependent on what stage of the Digital Transformation it is at. The key point of the proposed methodology is to ensure interaction with the enterprise information systems, which ensures constant updating of the DTO. If the organization does not actively use information systems, then for building a DTO it will have to digitize the available data, which can significantly complicate the task.

A real organization is a collection of architectural layers, each of which includes certain types of objects, their attributes and the relationships between them. Digital data contained in enterprise information systems are data objects related to the architectural data layer.

A virtual organization includes an enterprise architectural model based on the ArchiMate modeling language, in which prepared and transformed data are integrated. Changing this model, i.e. creating, deleting, changing the configuration of objects, attributes, relationships, allows to manage the enterprise ontology. The enterprise ontology serves as a description of the global enterprise data structure and as a schema for the enterprise knowledge graph. When the ontology is enriched with data, it transforms into the enterprise knowledge graph, which performs the function of semantic integration of enterprise data.

Adapters play a role of "Connection between two" element and they perform the data acquisition from enterprise information systems. Adapters perform parsing and transformation of the received data in accordance with the global data structure described by the enterprise ontology.

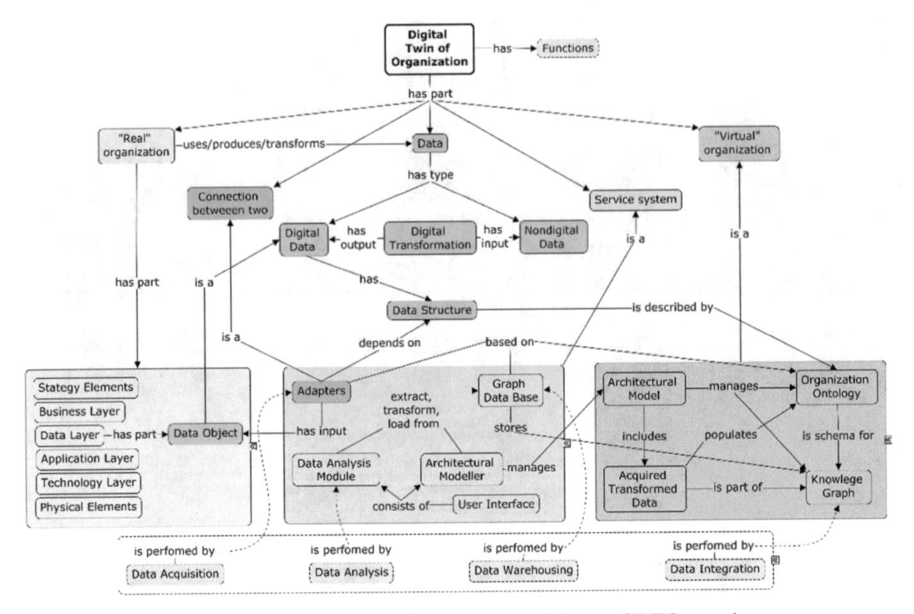

Fig. 5. A conceptual model of the methodology of DTO creation

Adapters are part of a service system, which also includes a graph database that performs the function of data warehousing, the structure of which is also determined by the ontology. Interaction with the service system occurs through the user interface which includes an architectural modeler and allows the construction of architectural models and a description of a real enterprise, as well as a data analysis module. The data analysis module allows to apply various management decision support methods, including data visualization, machine learning methods, as well as various modeling methods (imitation, factorial, predictive) using quality integrated data.

5 Conclusion

The concept of a methodology for constructing digital twins of organizations (DTO) based on semantic technologies – ontologies and knowledge graphs, as well as enterprise architecture was presented in the course of the research paper. The presented methodology includes all the basic elements of digital twins described in the academic literature and allows carrying out the entire cycle of work with enterprise data, also necessary for the DTO viability: acquisition, integration, warehousing and analysis of data.

Potential advantages of the presented methodology include:

- Synthesis of knowledge from different disciplines, namely the use of visual modeling from the field of enterprise architecture to manage the enterprise knowledge graph;
- The potential opportunity to build a DTO for almost any organization that has stepped into a digital transformation;

- Representing an enterprise in the form of semantic networks enables network analysis and graph analytics to be applied, for example, to find the most critical nodes by determining its centrality [41];
- A holistic view of the enterprise is provided with the help of an architectural model that can be easily changed and adapted during the organization life;
- Implemented semantic data integration, opening up new possibilities for data analysis through the ability to take into account semantic relationships between different data;
- Reduces time and costs, increases the availability of data analysis and modeling methods by providing quick access to high-quality and structured data, as well as through integrated decision support methods;
- The proposed methodology for building a DTO provides the ability to operate not only with data, but also with the knowledge of the enterprise.

A significant limitation of the presented research is the lack of practical testing and evaluation of the results of both the proposed methodology in particular and the entire technology of the DTO in general. Moreover, in the case of a large and complex enterprise, the process of building and updating architectural ontological models can be very costly. To overcome it, it is necessary to automate this process using machine learning and process mining methods, which can also allow the discovery of new objects and relationships that pretend to enrich architectural models to make the digital twin dynamic. The second option is to model the most critical areas of the enterprise. Another potential disadvantage of the methodology is the cognitive overload that occurs when building complex visual models containing a large number of objects. To overcome this disadvantage, the development of new user interfaces is required.

Further research involves the development of a service system that implements the methodology, as well as testing of the presented methodology.

Acknowledgments. The reported study was funded by the Russian Foundation for Basic Research (RFBR), project number 20-07-00854.

References

1. Lasi, H., Fettke, P., Kemper, H.G., Feld, T., Hoffmann, M.: Industry 4.0. Bus. Inf. Syst. Eng. **6**(4), 239–242 (2014)
2. Christian, M., Hess, T., Benlian, A.: Digital transformation strategies. Bus. Inf. Syst. Eng. **57**(5), 339–343 (2015)
3. Volkova, V., Emelyanova, A.: Teoriya sistem i sistemnyy analiz v upravlenii organizatsiyami [Theory of Systems and System Analysis in the Management of Organizations]. Finansy i Statistika, Moscow (2006). (in Russian)
4. ISO/IEC 15288:2008 (IEEE Std. 15288–2008): Systems and software engineering—system life cycle processes. International Organization for Standardization (2008)
5. Volkova, V.: A comparison of definitions of systems and system research and design approaches. Problemy Upravleniya v Sotsial'nykh Sistemakh **4**(6), 36–50 (2012)
6. Volkova, V., Chernenkaya, L., Mager, V.: Klassifikatsiya modeley v sistemnom analize [Classification of models in system analysis]. Comput. Sci. Telecommun. Control **3**(174), 33–43 (2013). Scientific and technical statements of SPSTU. (in Russian)

7. Gavrilova, T., Kubelskiy, M., Kudryavtsev, D., Grinberg, E.: Modeling methods for strategy formulation in a turbulent environment. Strateg. Change **27**(4), 369–377 (2018)
8. Oztemel, E., Samet, G.: Literature review of Industry 4.0 and related technologies. J. Intell. Manuf. **31**(1), 127–182 (2020)
9. Michael, G.: Origins of the digital twin concept. https://www.researchgate.net/publication/307509727_Origins_of_the_Digital_Twin_Concept. Accessed 09 May 2020
10. Uhlemann, T.H.-J., Schock, C., Lehmann, C., Freiberger, S., Steinhilper, R.: The digital twin: demonstrating the potential of real time data acquisition in production systems. Procedia Manuf. **9**, 113–120 (2017)
11. Tao, F., Cheng, J., Qi, Q., Zhang, M., Zhang, H., Sui, F.: Digital twin-driven product design, manufacturing and service with big data. Int. J. Adv. Manuf. Technol. **94**(9–12), 3563–3576 (2018)
12. Azhar, S.: Building information modeling (BIM): trends, benefits, risks, and challenges for the AEC industry. Leadersh. Manag. Eng. **11**(3), 241–252 (2011)
13. Kjaer, L., Pigosso, D., Niero, M., Bech, N., McAloone, T.: Product/service-systems for a circular economy: the route to decoupling economic growth from resource consumption? J. Ind. Ecol. **23**(1), 22–35 (2019)
14. Carley, K.: Computational organizational science and organizational engineering. Simul. Model. Pract. Theory **10**, 253–269 (2003)
15. Carley, K.: Smart agents and organizations of the future. In: Lievrouw, L., Livingstone, S. (eds.) The Handbook of New Media: Social Shaping and Social Consequences of ICTs, pp. 206–220. Sage, London (2002)
16. Grieves, M.: Digital twin: manufacturing excellence through virtual factory replication. White paper (2014)
17. Glaessgen, E., Stargel, D.: The digital twin paradigm for future NASA and U.S. Air Force vehicles. In: 53rd Structures, Structural Dynamics and Materials Conference, pp. 1–14 (2012)
18. Wache, H., Dinter, B.: The digital twin–birth of an integrated system in the digital age. In: Proceedings of the 53rd Hawaii International Conference on System Sciences (2020)
19. Tao, F., Zhang, M., Nee, A.Y.C.: Digital Twin Driven Smart Manufacturing. Academic Press, London (2019)
20. Kritzinger, W., Karner, M., Traar, G., Henjes, J., Sihn, W.: Digital Twin in manufacturing: a categorical literature review and classification. IFAC-PapersOnLine **51**(11), 1016–1022 (2018)
21. Rashik, P., Aija, L., Llewellyn, T.: Building an organizational digital twin. Bus. Horiz. **63**, 725–736 (2020)
22. Samosudov, M.: Resource footprint of activities as an element of the digital twin of the enterprise. E-Management **2**(3), 38–47 (2019)
23. Kudryavtsev, D., Arzumanyan, M.: Enterprise architecture: transition from IT-infrastructure design to business transformation. Russ. Manag. J. **15**, 193–224 (2017)
24. Buckl, S., Schweda, Ch.M.: On the state-of-the-art in enterprise architecture management literature. Technical report. Fakultät für Informatik der Technischen Universität, München (2011)
25. Lankhorst, M.: Enterprise Architecture at Work, 3rd edn. Springer, Berlin (2013)
26. ArchiMate® 3.1 specification. https://pubs.opengroup.org/architecture/archimate3-doc/toc.html. Accessed 09 May 2020
27. Zachman, J.: The Zachman framework: the official concise definition. Zachman International (2008)
28. Scheer, A.-W., Schneider, K.: ARIS architecture of integrated information systems. In: Bernus, P., Mertins, K., Schmidt, G. (eds.) Handbook on Architectures of Information Systems, pp. 605–623. Springer, Berlin (2006)

29. TOGAF: The Open Group Architectural Framework. http://www.opengroup.org/subjectareas/enterprise/togaf. Accessed 09 May 2020
30. Chen, P.: The entity-relationship model—toward a unified view of data. ACM Trans. Database Syst. (TODS) **1**(1), 9–36 (1976)
31. Villazon-Terrazas, B., Garcia-Santa, N., Ren, Y., Srinivas, K., Rodriguez-Muro, M., Alexopoulos, P., Pan, J.Z.: Construction of enterprise knowledge graphs (I). In: Exploiting Linked Data and Knowledge Graphs in Large Organisations, pp. 87–116. Springer, Cham (2017)
32. Gruber, T.: Toward principles for the design of ontologies used for knowledge sharing. Int. J. Hum Comput Stud. **43**, 907–928 (1995)
33. Negri, E., Fumagalli, L., Macchi, M.: A review of the roles of digital twin in CPS-based production systems. Procedia Manuf. **11**, 939–948 (2017)
34. Hevner, A., Chatterjee, S.: Design science research in information systems. In: Design Research in Information Systems, pp. 9–22. Springer, Boston (2010)
35. Österle, H., Otto, B.: Consortium research. Bus. Inf. Syst. Eng. **2**(5), 283–293 (2010)
36. Peffers, K., Tuunanen, T., Rothenberger, M.A., Chatterjee, S.: A design science research methodology for information systems research. J. Manag. Inf. Syst. **24**(3), 45–77 (2007)
37. Gruninger, M., Fox, M.: An activity ontology for enterprise modelling. Submitted to AAAI-1994, p. 321. Department of Industrial Engineering, University of Toronto (1994)
38. Uschold, M., King, M., Moralee, S., Zorgios, Y.: The enterprise ontology. Knowl. Eng. Rev. **13**(1), 31–89 (1998)
39. World Wide Web Consortium (W3C). https://www.w3.org/. Accessed 05 May 2020
40. Robinson, I., Webber, J., Eifrem, E.: Graph Databases. O'Reilly Media, Inc. (2013)
41. Borgatti, S., Everett, M.: A graph-theoretic perspective on centrality. Soc. Netw. **28**(4), 466–484 (2006)

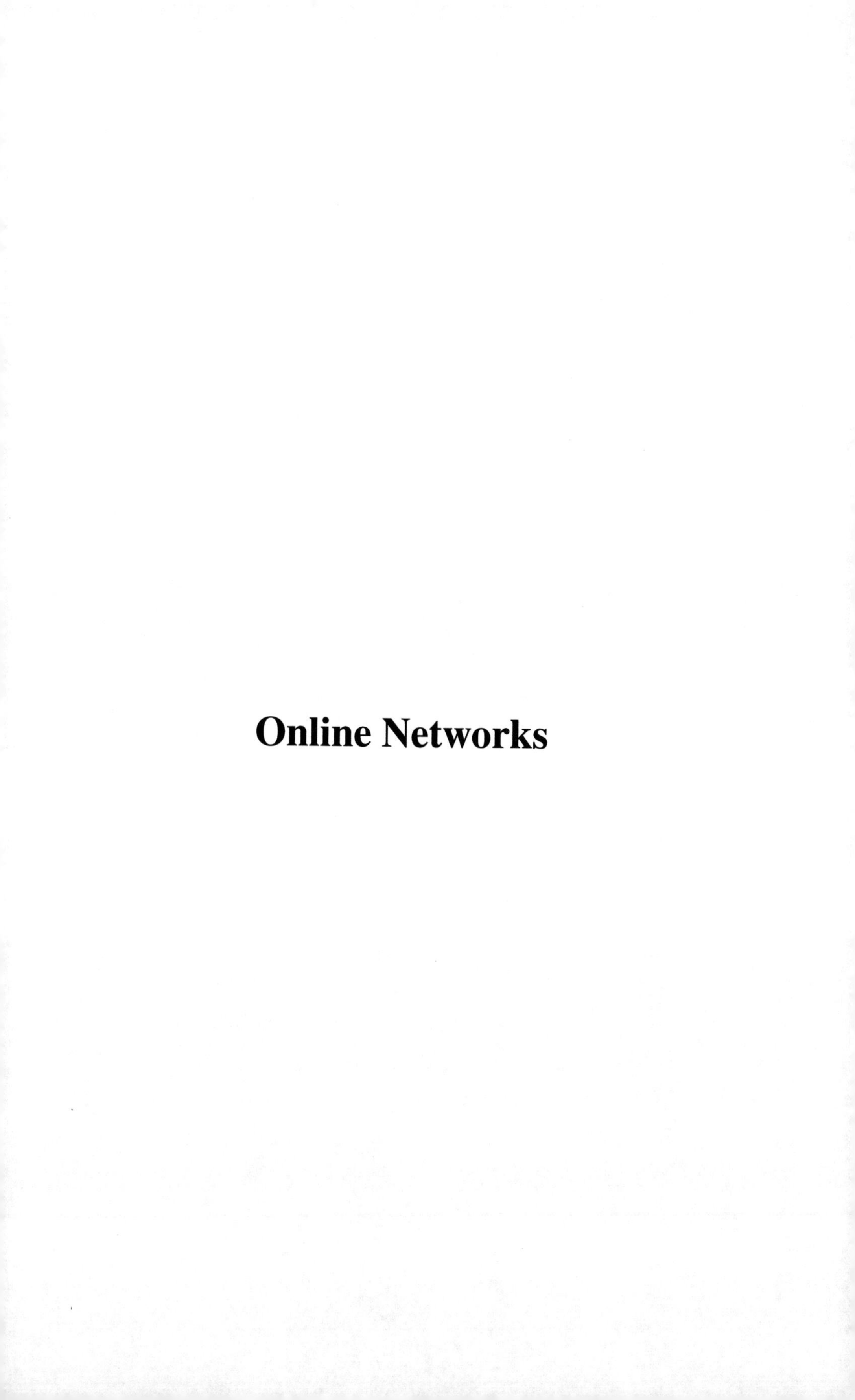

Online Networks

The Boundaries of Context: Contextual Knowledge in Research on Networked Discussions

Svetlana S. Bodrunova(✉)

St. Petersburg State University, St. Petersburg, Russia
s.bodrunova@spbu.ru

Abstract. Today's studies of networked discussions may be divided into theory-driven and data-driven, but both lines of research neglect the role of contextual knowledge in assessment of real-world public discourse. As scholars note, without context, data lose meaning and value; however, there is a striking vacuum of scholarly discussion on how to delineate the relevant context for network discussion studies, as well as what procedures of its description in academic publications should be employed. As a mediator between theories and data-driven results, context has a potential of eliminating the opposition between theory- and data-driven research designs. In an attempt to conceptualize context, we suggest to adapt the long-term experience of cognitive linguistics and critical discourse analysis for developing rigorous procedures of selection, assessment, and explicit description of relevant context(s). We bring attention to the paradox that, in online discussion studies, scholars extract sociologically relevant conclusions from the data of non-sociological nature (that is, either text or network structures), and argue it might be fruitful for selection of appropriate contextual background. After meta-reviewing the conceptual papers on online discussion research and using our own experience in such studies of over 7 years, we suggest three types of contexts for network discussions: cognitive, platform-technological, and media/communicative contexts – that need to be taken into account in network discussion studies.

Keywords: Context · Social networks · Networked discussions · Research design · Cognitive context · Technological context · Communicative context · Data-driven science

1 Introduction

The appearance and rapid proliferation of social networking sites around the world has brought to life unprecedented amounts of written data that could be subjected to various types of social, political, linguistic, and cultural assessment, often with the help of statistics, probabilistic instruments of textual analysis, and social network analytics. Accordingly, a whole industry of *networked discussion studies* (NDS) has evolved; it has been using textual data in combination with the network data to explain both structure (and its dynamics) and socially relevant essence of networked discourses.

A. Antonyuk and N. Basov (Eds.): NetGloW 2020, LNNS 181, pp. 165–179, 2021.
https://doi.org/10.1007/978-3-030-64877-0_11

Very roughly, the body of research that uses textual/metadata/network data may be divided into two streams that can be called "theory-driven" and "data-driven" (or, rather, "method-driven") research [1–4]. Despite the seemingly big difference between the two approaches in how the research designs are constructed, both might be helpless enough in explaining relations between individuals or social groups without a third pillar of the research design, which is context [5].

The role of contextual knowledge has been perceived both positively and negatively within social science as well as within computer science, at least in the part that works with textual datasets. Its positive role is linked to its explanatory power, as well as to the power to correct and ground abstract research questions and hypotheses; its negative role is especially evident in comparative research, as divergent contexts undermine comparability and reproducibility of research results. However, despite the obvious high impact of contextual knowledge upon both setting the hypotheses and interpreting the results, most NDS authors still use contextual knowledge intuitively and often set no clear boundaries for themselves on what context they should consider relevant and how the relevance is established. The discussion on the role of context in the research on networked discussions began in the late 2000s but wound down too soon.

At the same time, cognitive linguistics, critical discourse analysis, and other research areas have developed deep and detailed accounts on what context is and which mental procedures need to be used to reconstruct the speech situation, as well as wider social meaning of the speakers' messages [6]. The long history of studies of context provides instruments applicable for networked discussion research; but these instruments need to be used explicitly and re-thought for networked platform-based user interactions.

The gap between cognitive science and big textual data research might be narrowed by introducing rigorous and explicit procedures that describe relevant context for both posing the hypotheses and explanation of results.

In the remainder of this paper, we show that contextual knowledge may be regarded as a mediator between theoretical assumptions and results received by automated data processing; it may serve to eliminate the opposition between theory-driven and data-driven approaches in big social data studies, including NDS. We reconstruct this gap and suggest two additional types of context relevant for NDS. We argue that, in addition to cognitive, social, political, cultural, and historical context [7], technological limitations, like platform affordances and platform policies, should be regarded as contextual. We also argue that media/communicative context, different from cognitive context of communication, needs to be taken into consideration in NDS. We draw attention to the paradox that, in online discussion studies, scholars extract sociologically relevant conclusions from the data of non-sociological nature (that is, either text or network structures), which makes context especially relevant for networked discussion studies.

2 Theory-Driven vs. Data-Driven Big Textual Data Research: The Role of Context as a Mediator in Both Research Designs

2.1 Theory-Driven or Data-Driven? Two Paradigms of Big Textual Data Studies

As stated above, today, two approaches may be outlined in the studies of big textual data collected from networked discussions that take place, first and foremost, on social media platforms. These approaches are of paradigmatic nature, as they shape all the parts of research designs. As Strasser puts it, "[p]roponents of 'data-driven' and 'hypothesis-driven' science argue over the best methods to turn massive amounts of data into knowledge" [1: 85].

In the theory-driven (or hypothesis-driven [1]) research designs, the classic structure of academic enquiry is preserved: existing theories are juxtaposed to observed new phenomena, either in the very focus of the research or around it, and are used to pose the research questions; then, the received results are put against the background of the previous conceptual knowledge to amplify it. This classic design allows for continuity of the scientific worldview, as it interlinks older and newer research. However, the disadvantage of such an approach is that, in working with largely new phenomena, it may fail with posing adequate research questions and miss the point with the hypotheses.

In the data-driven research, another logic may be traced. Such research designs are based on the assumption that scholars may allow "data speak for themselves" [8]. The essence of "big data" in comparison with "small data" is that no sampling in its traditional sense is needed. More precisely, sampling is done based on criteria far from social representativity, such as in hashtag-based data collection. The resulting collections, in a sense, consist of *all available* data, and use of automated methods makes it unnecessary to sample further. Sampling of data is sometimes substituted by "narrowing" the data by narrowing the query put toward the dataset [2] – that is, by adapting the research question to the nature of the data. This crucially changes the research design, as research questions are posed based on the researches' expectations from preliminary reading of parts of datasets, while theories "change the place" within the research designs and are used for interpretation of the results received after "running data." Often, the place of theory is here taken by method – running data with the help of various types of automated analysis brings the results that not only vary substantially but may be interpreted only within the boundaries of a certain automated method. However, this limitation is seen as natural, which does not prohibit the data "speak for themselves"; for us, it is legitimate to call this approach also "method-driven", as not only queries need to be adapted to both the dataset and the applied method but also, often enough, the very possibility of a study depends on whether the method and instruments are available.

Indeed, data-driven research can, at least to an extent, help overcome the inertia in theory-driven science and allow to see from the processed data some phenomena previously unseen, unexpected, and/or impossible to discover without big data. At the same time, within the last three decades the research community has moved from universalist optimism of seeing the world as possible to translate into the common language of coding [9, 10] to the understanding that the feeling of "all-seeing, infallible God's view" [11: 4] the large-scale data provide might be highly misleading. Also, the advantages of theory-driven science have become the disadvantages of the method-driven science: using theory only to interpret the processed data without posing research questions out of the previous theories potentially breaks the continuity of science, as well as lowers the chances of the results to suit for further development of theories.

Echoed in NDS [12], the gap between theory-driven and method-driven research designs will continue shaping this research area for at least some near future.

2.2 The Importance of Context

However, sometimes researchers run into situations when neither theories nor methods can provide meaningful explanations to what one sees in the data. In our earlier paper [4], we described a networked discussion where the cleavages between the conflicting social groups were conceptualized in the way unpredicted by theories of inter-group relations. At the same time, the data themselves could not provide a clue that the three groups massively described by Twitter users as "vegetables", "zombies", and "cosmonauts" were immigrants who traded vegetables, the local inhabitants who stood up against them, and the helmet-wearing special police troops who stood in between the two conflicting communities, respectively. What provided the explanation, though, was contextual knowledge, or, in short, *context*.

In NDS, the question of context has gradually risen as one of the sharpest. A scholar who consistently brought context to the forefront of big social data studies has been danah boyd. Starting from her PhD dissertation, she has insisted that, "[i]n mediated environments, technology helps shape context, but technology alone does not define the context" [13: 143]. In a later influential paper, boyd and Crawford stated that "without context, data lose meaning and value" [5: 670] and that "retaining context remains critical" [5: 671] for the quality of academic enquiry. Manovich, in his turn, has brought attention to the difference between "surface data" and "deep data" as dependent on context in differing ways. He also underlined that, according to the very definition of social computing given by Third IEEE International Conference on Social Computing in 2011, "social computing refers to 'computational facilitation of social studies and human social dynamics as well as design and use of information and communication technologies that consider social context'" [14: 464]. In early to mid-2010s, both works are echoed in the papers by Bruns and Burgess who speak of, more narrowly, networked discussions, not all big social data [3, 12]. Several more papers of that time period [2] have posed questions regarding use of context for coping with the large amounts of data and creating appropriate queries.

By today, though, the discussion has nearly stopped at stating the importance of context – that is, without elaborating on what "context" actually is or should be. What one considers as context, what is the place of context within research designs, how exactly scholars select contextual knowledge and which hazards it may bring on, has not yet been properly discussed. Perhaps this is due to the fact that, at some point in the 1990s, context has become "considered too chaotic and idiosyncratic to be systematically characterized" [15: 11, 16]. But it might also be the case that the importance of context is either under-estimated in data-driven science or seen as unnecessary to regulate.

2.3 Studies of Context in Cognitive Linguistics, Critical Discourse Studies, and Computational Science

Meanwhile, in cognitive science, or, more precisely, in cognitive linguistics and critical discourse studies, the studies of context of message, dialog, discourse, and communication on the whole have a half-a-century history reconstructed well in [6].

The authors of the book draw attention to the complexity of the phenomenon of context but, in general, see context in a two-fold way. In a wider sense, context is background for comprehension; in a narrower sense, though, it is seen as presupposed common ground among the participants in a conversation [17]. The authors note that three different visions on the term "context" may be traced in textual linguistics, cognitive science, and computational social studies.

Thus, in linguistics, context is regarded as a tool for interpretation of linguistic utterances, as well as their pragmatic and cognitive elements such as frames, maxims, or "common ground." Context seen as "co-text", cognitive environment for discursive manifestations, and as a precondition of speech. According to Van Dijk, "language users not only need to have general 'knowledge of the world', and not only knowledge about the current communicative situation, but of course also mutual knowledge about each other's knowledge" [7: 72].

In cognitive science, including cognitive linguistics, context is mostly understood as situational. In most cases, the communicative situation is seen as evolving on-line, simultaneously to speech, and includes the immediate circumstances of the discourse [18]; here, the role of context is definitive, as "meaning… does not exist without context" [15: 11]. This, to an extent, blurs borders between the context *of* a discussion and the context *behind* a discussion, and the situational context is limited to the physical location, or "place" in a wider sense (e.g. an online milieu). For social scientists, we argue, the importance of the context behind the discussion might be much higher than that of the discussion. We see the context behind a discussion as a complex of external factors that shape the discussion but lie beyond the immediate spatial, time, and discursive boundaries of the discourse.

In computational science, the authors claim, context of whatever nature is mostly seen as an instrument to restrict the space within which a problem can be solved, as well as a means of reducing dimensionality of large-scale datasets, to easier perform automatic inferences, human-like reasoning, and analysis of human-computer relational constraints [15]. For computer scientists, context is knowledge external to the task or data that is subjected to formalization and systematization; it cannot be vague or based on expert opinion.

All the three approaches unite under one "umbrella" thought: whether situational or not, the word "context" is understood as a *limiting circumstance* (or a set of circumstances). In the discourse studies, context narrows down communicative possibilities [19] – and, in its limiting capacity, lowers the likelihood of one scenario of a discussion and rises that of another. Moreover, and counter-intuitively, context, however diverse, reduces the dimensionality of the research design by providing constraints on both the possible hypotheses and interpretation of results. It is not occasional that many studies use the word "context" within a phrase "in the context of…": the limiting notion provides for necessary parameters to contextualize the query to data. Thus, "putting results in broader context" should not mislead: the better context is reconstructed, the lower the number of possible interpretations.

Thus, by today, context in connection to analysis of human discussion is mostly seen as a limiting circumstance within the fabric of talk – and next-to-never as a wider field of reference that shapes the discussions "from outside."

2.4 Discussion Context as Different from Discursive Context

And yet, despite the extensive reconstruction of context studies and a truly deep academic discussion on fundamentals of contextual knowledge in linguistics and cognitive studies, all the three visions partly miss the point when applied to the extant network discussion studies.

As said above, in a big part of computer science, context is seen as knowledge that needs to be subjected to automation, to help approximate judgments by machine to human judgments. However, not all computational tasks need such approximation; in many cases, it is human interpretation of the results received by computational methods that is needed, as the task comes from social sciences, not from computational linguistics or methodological experiments. Amplification of method by involving contextual parameters (e.g. introducing contextually relevant topic keywords for semi-supervised topic modeling [20]) is, undoubtedly, a separate research task; but we also need to see context in a different, more social-scientific way in computational studies – the one that a researcher employs for interpretation, beyond the method itself. Above, we have called it context behind the discussion. Also, computational scientists suggest that contextual knowledge differs from external (= irrelevant) knowledge, and contextual knowledge might become "proceduralized" – that is, "invoked, assembled, structured and situated according to the given focus" [21: 1]. But they do not set up a clear understanding on how context becomes selected for "proceduralization."

Similarly, unlike in cognitive linguistics, contextual knowledge in NDS might be used for a wide variety of interpretational purposes beyond immediate assessment of user speech utterances. It should be applied not only to inducing socially relevant meanings to the discourse generated by users but to explaining and predicting the whole complexity of the discussion. The latter would include interlinkages between users, appearance and social status of various types of influencers [22], modularity of the discussion graphs including echo chambers [23, 24], discussion dynamics and its relations with the offline processes in the public sphere [25], user inclusion/exclusion patterns, and many more. Thus, in NDS, one needs to speak of the *discussion context*, rather than the situational context of user-generated speech only. Here, interaction patterns matter, and what influences them matters, as the discussion is seen from a perspective of social network analysis, not only from the linguistic perspective. The features of the discussions on social media are, of course, discourse-dependent, but not fully, and the cognitive procedures suggested for analysis of text from the linguistic viewpoint might as well be rethought and re-applied to interpretation of extra-linguistic/network parameters of the discussions.

Of the three understandings of context described in [6] and summarized above, the situational approach has, so far, been the closest to the type of context assessment that is used by the majority of NDS scholars. However, reconstruction of the situation around a discussion is, most often, done intuitively in NDS, without proper reflection on three issues: 1) where exactly within the research design the context behind the discussion needs to be involved; 2) how, by what principles and procedures, this context is selected; 3) how contextual knowledge needs to be described and structurally organized in the paper. Below, we reflect on these concerns.

3 Hard Questions About Use of Context in NDS

3.1 Augmented Research Design: Context as a Mediator Between Theories and Results in Big Social Data Studies

The role of context for theory-driven research seems to have been known for ages: outside knowledge is employed to contextualize – that is, to refine the RQs and hypotheses and make them more related to the observed reality. Many scholars have used this strategy very successfully but intuitively, without rigorous re-assessment of how exactly they drag in contextual knowledge into their study and how selection of relevant context is made. In the data-driven studies, context plays no smaller role: contextual knowledge corrects interpretations and shows how limited the application of theories for explanation of the results might be.

Thus, context plays an intermediary role in two major parts of the research design: 1) between the existing theoretical conclusions and precise research questions and hypotheses; 2) between the received results and the added theoretical value of the research. Context is a principal mediator between theories and data, both at the stage of posing research questions and putting data at the background pf previous knowledge. In this capacity, it has a potential of eliminating the opposition between the hypothesis-driven and method-driven approaches if used in these two crucial parts of the research design.

As a mediator in the "theory – results – theory" chain, context needs to be seen as a major part of NDS research designs. But, unlike for other parts of academic inquiry (literature review, elaboration of hypotheses, conceptual and technical parts of methodologies, and description of results), selection and description of context does not enjoy any analytical procedures of comparable rigor in NDS. Below, we pose some obvious questions that arise in inter-disciplinary NDS groups.

3.2 Hard Questions on Use of Context in Research Designs

First, these questions relate to the rigorous selection of defining, selecting, and assessing relevant contextual knowledge behind the discourse. How do we assess the relevance of bits of context? How do we create a non-controversial picture that mediates theory to hypotheses? How do we prove that the selection is relevant to our study?

Here, again, one needs to look at cognitive studies. Several contributions in [6] reconstruct and assess in detail the procedures that are used to involve contextual knowledge into research. These include accommodation of context dependence, resolution of conflicting contexts, bottom-up activation of schemata vs. top-down activation of context, and complying with the criteria of satisfaction and clarity [17]. Perhaps, for NDS, we need to adapt them and place them, as well as other approaches, within the research designs of both theory- and data-driven nature.

Second, three major issues need to be discussed. The first is subjectivity of the scholar who selects the contextual data and explanations. There is much evidence that subjectivity in big data research is unavoidable [5], as producing quantitative results does not mean producing meanings; data demands interpretation, and interpretation is subjective. But this is exactly why the academic community needs to discuss and, if possible, establish the procedures which would help better assess how relevant the selected context is to a given case. Expert knowledge is one way of dealing with this issue; another is working with contextual variables (see below). The second issue is reproducibility vs. context-dependence. Do we consider a research piece irreproducible if the context around a problem changes? Is methodology enough to ensure a study is reproducible – or is context an imminent part of the case and critically affects the reproducibility of the studies, especially in networked discussions, given their dissipative and ever-changing nature? Or, vice versa, how does a scholar persuade the academic community that his/her study, even if contextualized, remains both reproducible (in formal terms) and relevant for other contexts (cultures, societies, times)? And, linked to this, the third question is case comparability. The more you contextualize the discussions you study, the less comparable they might become. Moreover, it may happen so that *nearly all* the findings in a case seemingly similar to other cases may be explained by its individual context – that is, what you see through the data in compared cases *is* local context that rises to you from your data, as unique as it can be. How not to fail into individual explanations per each case if everything in them is so well explained by local peculiarities, and how – if at all – to use context to generalize?

Third, we need to develop frameworks for organizing descriptions of contextual knowledge in academic papers. Perhaps, if better reflected upon, description of relevant context could take its own place in the paper structure. Appearance of such sections in papers would signal of the use of the employed reflexive procedures to which the contextual knowledge was subjected. In general, what is needed is *explication* of the mental procedures used for selection and structuring of relevant context.

3.3 The Challenge of *Context Collapse*

Another hard question lies in the fact that integrity of context as a part of discursive situation has already been seriously questioned in NDS. Costa [26] reconstructs the line of research that states *context collapse* as a feature of socially-mediated communication. Context collapse as a term originates from the works by Meyrovitz [27] and boyd [28] – who, interestingly, stated the necessity of contextual assessment in one range of works and dismissed context of speech as collapsed in another. Context collapse refers to the necessity of a user to address multiple unrelated contexts with varying normativity and response structures [26]. This was later conceptualized as collapsing of several contexts upon one another [29]. After 2008, a range of works [30, 31] have supported the notion of collapsed contexts, and this view has not been challenged till as late as 2014 [32]. Since then, some evidence has been collected that users cope rather well with confronting social groups of divergent values and beliefs by diverging their platform use and presence [26], even if the necessity to confront such groups may lead to self-silencing in an attempt to avoid exclusion when failing to please everyone [33].

Evidently, the context collapse studies view context from the user-centric and situational perspective: here, contexts are complex pieces of reality relevant for individual users that include platform affordances, social norms, and collective behaviors in interaction with the norms and behaviors of a given user. This view is more complex and closer to the idea of context behind the discussion; its only limitation is that context is seen from the individual-user perspective. Bringing the idea of context collapse to the level of a discussion, we see that the challenge of overlapping contexts may arise just as well for the whole discussion. In addition to the "outer world" contexts, overlapping contexts that originate from various academic disciplines provide their own explanations for what is seen in the data. Integral exchange of meanings inside a discussion viewed through multiple disciplinary lens may receive fractured or contradictory understandings.

However, even if the interpretation remains ultimately subjective and potentially fractured, there are means of ensuring the integrity of interpretational procedures and non-controversial stance of the contextual picture, for which the cognitive operations mentioned above may be used. Also, just as users reproduce the divisions between their social contexts in practices of use of social media [26], researchers are entitled to separate contextualized interpretations and address them individually. A bigger issue here, to our viewpoint, is reconstruction of the full enough contextual continuum and finding the borders of context, after which adding more details becomes unnecessary, – that is, creating a *necessary and sufficient* background of a study.

4 Media and Society as Context: Platform Affordances, Social Variables, and Mediated Public Sphere

Before elaborating on how to shape context in research papers, one needs to systematize what types of context may be suggested for NDS. As shown above, context *of* discussions differs from that *behind* discussions: the former is the evolving situation of speech with the immediate circumstances that surround it, while the latter is background that might spread wide enough to cover "cognitive, social, political, cultural, and historical environments of discourse" [15: 10]. The latter also corresponds to what is conventionally understood as "social context at large." As it is the widest possible option for contextualized cognition and interpretation, we will call it *cognitive context*. Below, we also suggest two other contextual domains for NDS, namely *technological context* and *media/communicative context*.

4.1 Cognitive Context: Social-Scientific Research Questions vs. Textual Data

Situational approaches to context described above are quite promising in their mediating logic; within them, context might include "the space, social situation, and people" [13: 143]. In boyd's study of youngsters' communication in MySpace [13], these dimensions were interpreted in line with the cognitive approach – namely, as the imagined audience group, their online presence, and relations that defined belonging to the group, without formally involving the wider context behind the discussion itself.

Taken more broadly, these dimensions remain important for selecting and structuring "social context." A similar approach may be found in economic studies, where the networks of actors ("people"), arenas ("space"/"location"), and final consumer markets ("situation") were identified as "three qualitatively different dimensions of contextual knowledge" [34: 745]. This three-part scheme is one way of reducing the extremely wide array of contextual options; its transgression from the context of the discussions [11] to the context behind them tells of high potential of working with cognitive contexts.

Arguably, another way of selection of appropriate context comes from the collision between the origin of research questions and the nature of data. In NDS, the research inquiry comes from social science (including sociology, individual or group psychology, political or cultural studies, history, and interdisciplinary studies like public memory or gender research), while the data are textual and mathematical. In social science, a decades-long tradition of variable-based data collection and analysis provides for successful deriving meanings from statistical and "small" qualitative data. But variable-based research designs are not always applied to network analysis and NDS. Thus, many factors that previously served as independent and control variables are pushed out to subjective and almost intuitive assessment in the "context" part of the research design. Formally deprived of capability to affect modeling of social processes due to absence or scantiness of the very modeling, they continue shaping expectations at other stages of research.

Cognitive context is, of course, a major challenge for any NDS researcher; in this paper, we only wish to (re)start the discussion on its involvement into contemporary research on socially-mediated discussions. But we argue that, already complex enough, cognitive context is not the only one that needs to be formally explained in practically any NDS paper.

4.2 Platform Affordances and Platform Policies as Context

One cannot under-estimate the impact of communication technology as context of networked discussions. Here, again, we do not aim at providing exhaustive reasoning or even description of all the platform-related issues that form technological context for NDS; we only point out to the aspects of contextual knowledge that need to be addressed in both individual papers and scholarly debate. In a platform society [35], at least two types of platform-induced limitations crucially matter, namely the platform affordances and the platform policies for data collection and publication.

Platform affordances are qualitative features of platforms explicated via specific properties of interface that crucially shape (and limit) users' communication capacities. In conventional academic view, platform affordances have to be taken into account as limitations of method, especially in data collection. But this view may be amplified at least for three reasons.

First, as Costa [26] argues, platform affordances cannot be viewed as stable features of interfaces. Thus, affordances like persistence, visibility, spreadability, and searchability [26] were accepted and employed in a similar way in Euro-Atlantic world, which created a feeling that the affordances and their shapes were immanent for the platforms. But Costa shows that modes of use of platforms change in other regions of the world and introduces the concept of affordances-in-practice, to underline the cultural variability of the same platform affordances for end-user. Second, platform affordances and changes in them (via the changes in interface features) shape user behaviors, just as user behaviors shape platform affordances as documented by Costa. Third, technical affordances are "socialized", as "age, gender, race/ethnicity, socioeconomic status, online experiences, and Internet skills all influence the social network sites people use and thus where traces of their behavior show up" [36: 63]. A lot more research is needed to better define which of these three features crucially influence NDS and have to be incorporated into the "necessary and sufficient" context reconstructions.

Another puzzle piece of technological context is platform policies. While the platform policies on data access, ownership, and control [37] relate more to methodological limitations, the ethical policies, like in case of Facebook constantly reworking its standards towards malicious content and computational propaganda, are a significant part of contextual knowledge that shape perceptions and behavior of users, e.g. in Russia [38]. Where relevant, also state policies towards platforms may be included into the list, as in case of LinkedIn blocked in Russia due to change of Russian Internet legislation and refusal of the platform to store the Russian users' data within Russia.

4.3 Media/Communicative Context: A Public-Sphere Approach

A contextual domain that is particularly elusive, despite it is the most evident and natural for NDS, is media/communicative context. We argue that, for networked discussions, media and communication structures and group communication patterns constitute a separate part of contextual knowledge that inevitably shapes the discussion in addition to cognitive context. Earlier, we already stated the necessity of developing a view upon media and public sphere as a context for network discussion assessment [4]; here, we elaborate more on this thought.

However contested, the public sphere and media theories provide a valuable contribution to how one assesses public life. On a junction of both, Calhoun [39] argued that (legacy) media remain the structural carcass of public discussion (as a process) and public sphere (as a spatial metaphor of a milieu that unites entities and issues with public status). Media entities, as well as other structural and essential components of the public sphere, may become contextual for networked discussions in several ways.

First, with wide proliferation of socially-networked discussions, much evidence was gathered that legacy media retain their status within the two-step communication flow [40] and gatekeeping/gatewatching [41, 42]. They keep it, *i.a.,* by becoming influencers [22], bridging discussion centers with discussion peripheries, and preserving balance and informational roles within heated conflictual discourse [43], including multi-lingual globalized discussions [44]. Thus, two factors matter for practically any networked discussion: the configuration of the media system (with its leaders, political preferences, and divisions) and representation of the media system on a particular platform, as platforms are distorted mirrors of the hybrid media systems in terms of who of the players is active on a platform and who is not.

Second, the nature of the discussion has to be taken into account. NDS scholars have long ago established differences between, e.g., *ad hoc* and calculated publics [45]. We would expand this dichotomy to the discussions themselves and extend it to *ad hoc* discussion outbursts, intentionally cultivated discussions, and "chronic", or ongoing, topic- or issue-based discussions on social media. Our research shows that discussion outbursts critically differ from the surrounding "tissue" of the discussion in terms of equality of participants [46]. Thus, the communicative nature of a talk needs to be understood as pre-requisite in NDS research – which is more or less already the case for the majority of studies but, perhaps, needs to be explicitly stated within research papers. Further on, the dynamics of discussion growth and dissipation might become a necessary variable in assessing socially-mediated discussions, whatever (inter-)disciplinary research questions be posed to it.

Third, inter-group communication patterns, that are part of the "social" or, more precisely, cognitive context of communication, are reshaped on a particular platform by mere fact that not all social groups use the platform the same way – or even do not use it at all. In our experience, a conflictual discussion on anti-immigrant pogroms in Moscow of 2013 was shaped by absence of immigrants on Twitter [22], and the Twitter flaming on Ferguson had a particular shape due to preferential use of Twitter by African Americans [24]. Despite its evident significance, this factor is next-to-never taken into account in analysis of real-world social networks and discussions on them – perhaps because of unavailability of data on social representation on platforms in general, which is an issue that needs to be raised much beyond this paper. Media diets and selective use of platforms by social groups play a crucial role in under-representation of some of them within online publics, which is problematized by the communicative scholars but sometimes disregarded by computer scientists and linguists.

Fourth, contemporary public spheres are characterized by the growing complexity and dissonance [47] – that is, the consensus ideal [48] on the "earnest Internet" [49] looks increasingly distant, which needs to be embraced by the scholars. Criteria for assessment of the level of dissonance are, perhaps, a matter of future; but, already today, it is clear that the dissonant, dissipative, and disruptive nature of online discussions has to be – again – explicitly introduced into the research designs as a mediating level of assessment, compulsory to have in mind.

5 Conclusion

In this paper, we have argued in favor of introducing assessment of contextual knowledge as an integral and explicit part of research designs in NDS, be they theory-driven or data-driven. We expect the context as a third pillar of the research design (along with theory and method) to eliminate the opposition between theory-driven and data-driven research via the mediator role of context in both posing hypotheses and interpreting results. To make context an efficient mediator, the scholars need to address the questions of proper procedures of context reconstruction, as well as those of subjective impact, reproducibility, and comparability of research results.

In NDS, the nature of data (users' texts and structural relations) does not match the traditions of social sciences that have formed before social networking sites were in place. While in NDS the discussion on involvement of contextual knowledge has nearly stopped, in cognitive science and critical discourse analysis, there is a long and fruitful tradition of studying what contexts are and which procedures one applies to relevantly reconstruct them. These findings need to be adapted to NDS, and a scholarly debate on this adaptation is, without exaggeration, a burning issue.

In this paper, we have argued in favor of several propositions. We have divided the context into the one *of* the discussion (immediate/situational) and that *behind* the discussion. The latter, assessed within a research design, allows for involving the factors beyond the discussion that crucially shape it. We have reminded that context needs to be viewed as a limiting circumstance (or a set of circumstances) that reduces interpretation dimensionality and also helps in posing relevant hypotheses; we have also raised the challenge of context collapse to the level of the whole discussion and argued that it needs to be tackled by NDS scholars. Our main contribution to this debate is that, for NDS, contexts may comprise cognitive contexts ("social-contexts-at-large"), platform-technological context, and media/communicative context. The latter needs to be assessed in literally any research on networked discussions, which is only partly a case today, and is complicated by scarcity of knowledge on platform use and the dissonant and disruptive nature of today's online discourse.

Acknowledgements. This research has been supported in full by Russian Science Foundation, grant 16-18-10125-P (2019-2020).

References

1. Strasser, B.J.: Data-driven sciences: from wonder cabinets to electronic databases. Stud. Hist. Philos. Sci. Part C Stud. Hist. Philos. Biol. Biomed. Sci. **43**(1), 85–87 (2012)
2. Fan, W., Huai, J.P.: Querying big data: bridging theory and practice. J. Comput. Sci. Technol. **29**(5), 849–869 (2014)
3. Bruns, A., Burgess, J.: Researching news discussion on Twitter: new methodologies. J. Stud. **13**(5–6), 801–814 (2012)
4. Bodrunova, S.S.: When context matters. Analyzing conflicts with the use of big textual corpora from Russian and international social media. Partecipazione e conflitto **11**(2), 497–510 (2018)
5. Boyd, D., Crawford, K.: Critical questions for big data: provocations for a cultural, technological, and scholarly phenomenon. Inf. Commun. Soc. **15**(5), 662–679 (2012)
6. Airenti, G., Cruciani, M., Plebe, A.: Context in Communication: A Cognitive View. Frontiers Media, SA (2017)
7. Van Dijk, T.A.: Contextual knowledge management in discourse production. In: Wodak, L., Chilton, P. (eds.) A New Agenda in (Critical) Discourse Analysis, pp. 71–100. John Benjamins Publishing Company, Amsterdam (2005)
8. Gould, P.: Letting the data speak for themselves. Ann. Assoc. Am. Geogr. **71**(2), 166–176 (1981)
9. Haraway, D.J.: Simians, Cyborgs and Women: The Reinvention of Nature. Routledge, New York (1990)
10. Haraway, D.: A manifesto for cyborgs: science, technology, and socialist feminism in the 1980s. In: Seidman, S. (ed.) The Postmodern Turn: New Perspectives on Social Theory, pp. 82–115. Cambridge University Press, Cambridge (1994)
11. Kitchin, R.: Big data, new epistemologies and paradigm shifts. Big Data Soc. **1**(1) (2014). https://doi.org/10.1177/2053951714528481
12. Bruns, A., Burgess, J.: Notes towards the scientific study of public communication on Twitter. In: Tokar, A., Mahrt, M., Peters, I., Weller, K., Keuneke, S., Beurskens, M., et al. (eds.) Science and the Internet, pp. 159–169. Düsseldorf University Press, Düsseldorf (2012)
13. Boyd, D.M.: Taken out of context: American teen sociality in networked publics. University of California, Berkeley (2008)
14. Manovich, L.: Trending: the promises and the challenges of big social data. Debates Digit. Humanit. **2**, 460–475 (2011)
15. Faber, P., León-Araúz, P.: Specialized knowledge representation and the parameterization of context. Front. Psychol. **7**, 9–12 (2016). Article 196
16. Ervin-Tripp, S.M.: Context in language. In: Slobin, D.I., Gerhardt, J., Kyratzis, A., Jiansheng, G. (eds.) Social interactions, social context, and language, pp. 21–36. Lawrence Erlbaum Associates, Hillsdale (1996)

17. Airenti, G., Plebe, A.: Editorial: context in communication: a cognitive view. Front. Psychol. **8**, 6–8 (2017). Article 115
18. Evans, V., Green, M.: Cognitive linguistics: An Introduction. Edinburgh University Press, Edinburgh (2006)
19. Leech, G.N.: Principles of Pragmatics. Routledge, London (2016)
20. Bodrunova, S., Koltsov, S., Koltsova, O., Nikolenko, S., Shimorina, A.: Interval semi-supervised LDA: classifying needles in a haystack. In: Proceedings of the Mexican International Conference on Artificial Intelligence (MICAI 2013), pp. 265–274. Springer, Heidelberg (2013)
21. Brézillon, P.: Role of context in social networks. In: Russell, I., Markov, Z. (eds.) Proceedings of the 18th International Florida Artificial Intelligence Research Society (FLAIRS) Conference, AAAI, Clearwater Beach, FL, pp. 20–25 (2005)
22. Bodrunova, S.S., Litvinenko, A.A., Blekanov, I.S.: Comparing influencers: activity vs. connectivity measures in defining key actors in Twitter ad hoc discussions on migrants in Germany and Russia. In: Proceedings of International Conference on Social Informatics, pp. 360–376. Springer, Cham (2017)
23. Newman, M.E.: Modularity and community structure in networks. Proc. Natl. Acad. Sci. **103**(23), 8577–8582 (2006)
24. Bodrunova, S.S., Blekanov, I., Smoliarova, A., Litvinenko, A.: Beyond left and right: real-world political polarization in Twitter discussions on inter-ethnic conflicts. Media Commun. **7**(3), 119–132 (2019)
25. Smoliarova, A.S., Bodrunova, S.S., Yakunin, A.V., Blekanov, I., Maksimov, A.: Detecting pivotal points in social conflicts via topic modeling of Twitter content. In: Bodrunova, S.S., et al. (eds.) Internet Science. INSCI 2018 International Workshops. LNCS 11551, St. Petersburg, Russia, 24–26 October 2018, Revised Selected Papers, pp. 61–71. Springer, Cham (2019)
26. Costa, E.: Affordances-in-practice: an ethnographic critique of social media logic and context collapse. New Media Soc. **20**(10), 3641–3656 (2018)
27. Meyrowitz, J.: No Sense of Place: The Impact of Electronic Media on Social Behavior. Oxford University Press, Oxford (1986)
28. Boyd, D.: Social network sites as networked publics: affordances, dynamics, and implications. In: Papacharissi, Z. (ed.) Networked Self: Identity, Community, and Culture on Social Network Sites, pp. 39–58. Routledge, New York (2010)
29. Wesch, M.: YouTube and you: experiences of self-awareness in the context collapse of the recording webcam. Explor. Media Ecol. **8**(2), 19–34 (2009)
30. Marvin, C.: Your smart phones are hot pockets to us: context collapse in a mobilized age. Mob. Media Commun. **1**(1), 153–159 (2013)
31. Marwick, A.E., Boyd, D.: I tweet honestly, I tweet passionately: Twitter users, context collapse, and the imagined audience. New Media Soc. **13**, 114–133 (2011)
32. Davis, J.L., Jurgenson, N.: Context collapse: theorizing context collusions and collisions. Inf. Commun. Soc. **17**(4), 476–485 (2014)
33. Attrill, A.: The Manipulation of Online Self-Presentation: Create, Edit, Re-Edit and Present. Springer, Heidelberg (2015)
34. Aspers, P.: Contextual knowledge. Curr. Sociol. **54**(5), 745–763 (2006)
35. Van Dijck, J., Poell, T., De Waal, M.: The Platform Society: Public Values in a Connective World. Oxford University Press, Oxford (2018)
36. Hargittai, E.: Is bigger always better? Potential biases of big data derived from social network sites. Ann. Am. Acad. Polit. Soc. Sci. **659**(1), 63–76 (2015)
37. Puschmann, C., Burgess, J.: The politics of Twitter data. In: Bruns, A., Mahrt, M., Weller, K., Burgess, J., Puschmann, C. (eds.) Twitter and society [Digital Formations, Volume 89], pp. 43–54. Peter Lang, New York (2014)

38. Bodrunova, S.S.: Information disorder practices in/by contemporary Russia. In: Waisbord, S., Tumber, H. (eds.) Routledge Companion to Media Misinformation & Populism. Routledge, New York (in print)
39. Curran, J.: Rethinking the media as a public sphere. In: Dahlgren, P., Sparks, C. (eds.) Communication and Citizenship: Journalism and the Public Sphere, pp. 27–57. Routledge, London (1991)
40. Katz, E.: The two-step flow of communication: an up-to-date report on a hypothesis. Public Opin. Q. **21**(1), 61–78 (1957)
41. Bastos, M.T., Raimundo, R.L.G., Travitzki, R.: Gatekeeping Twitter: message diffusion in political hashtags. Media Cult. Soc. **35**(2), 260–270 (2013)
42. Bruns, A.: Gatewatching: Collaborative Online News Production. Peter Lang, New York (2005)
43. Bodrunova, S.S., Litvinenko, A.A., Blekanov, I.S.: Please follow us: media roles in Twitter discussions in the United States, Germany, France, and Russia. J. Pract. **12**(2), 177–203 (2018)
44. Bodrunova, S.S., Smoliarova, A.S., Blekanov, I.S., Zhuravleva, N.N., Danilova, Y.S.: A global public sphere of compassion? #JeSuisCharlie and #JeNeSuisPasCharlie on Twitter and their language boundaries. Monitoring Obshchestvennogo Mneniya: Ekonomicheskie i Sotsial'nye Peremeny **1**(143), 267–295 (2018)
45. Bruns, A., Burgess, J.: Twitter hashtags from ad hoc to calculated publics. In: Rambukkana, N. (ed.) Hashtag Publics: The Power and Politics of Discursive Networks, pp. 13–28. Peter Lang, New York (2015)
46. Bodrunova, S.S., Blekanov, I.S.: Power laws in ad hoc conflictual discussions on Twitter. In: Proceedings of the International Conference on Digital Transformation and Global Society, pp. 67–82. Springer, Cham (2018)
47. Pfetsch, B.: Dissonant and disconnected public spheres as challenge for political communication research. Javnost-The Public **25**(1–2), 59–65 (2018)
48. Fenton, N., Downey, J.: Counter public spheres and global modernity. Javnost-The Public **10**(1), 15–32 (2003)
49. Hedrick, A., Karpf, D., Kreiss, D.: The earnest Internet vs. the ambivalent Internet. Int. J. Commun. **12**, 1057–1064 (2018)

Modeling Cascade Growth: Predicting Content Diffusion on VKontakte

Anna Moroz(✉), Sergei Pashakhin, and Sergei Koltsov

Laboratory for Social and Cognitive Informatics,
National Research University Higher School of Economics, Saint Petersburg, Russia
aria.ferrero@gmail.com, {spashahin,skoltsov}@hse.ru

Abstract. Online social networks have become an essential communication channel for the broad and rapid sharing of information. Currently, the mechanics of such information-sharing is captured by the notion of cascades, which are tree-like networks comprised of (re)sharing actions. However, it is still unclear what factors drive cascade growth. Moreover, there is a lack of studies outside Western countries and platforms such as Facebook and Twitter. In this work, we aim to investigate what factors contribute to the scope of information cascading and how to predict this variation accurately. We examine six machine learning algorithms for their predictive and interpretative capabilities concerning cascades' structural metrics (width, mass, and depth). To do so, we use data from a leading Russian-language online social network VKontakte capturing cascades of 4,424 messages posted by 14 news outlets during a year. The results show that the best models in terms of predictive power are Gradient Boosting algorithm for width and depth, and Lasso Regression algorithm for the mass of a cascade, while depth is the least predictable. We find that the most potent factor associated with cascade size is the number of reposts on its origin level. We examine its role along with other factors such as content features and characteristics of sources and their audiences.

Keywords: News diffusion · Machine learning · Information cascades · Online social networks · Cascade size prediction

1 Introduction

Today's world is overloaded with an enormous amount of information circulating through various environments. Some environments – emerged from the development of technologies, and their penetration in individuals' daily life – became widely and skillfully exploited by media on the global level [1]. Social networking platforms and services as one of such channels provide individuals with facilities for rapid communication and sharing of texts, photos, videos, links to external resources, or any other digital pieces of information with other users [2].

A. Antonyuk and N. Basov (Eds.): NetGloW 2020, LNNS 181, pp. 180–195, 2021.
https://doi.org/10.1007/978-3-030-64877-0_12

The mechanism of such information spreading, in general, involves a source, that posts information in the first place, and a circle of users exposed to this content. Users could be tied to a source by friendship/followership relations, comprising an audience, or accidentally being exposed to the post. Then, some of this audience may repost the message and become a source for their audience. This chain of sharing actions generates a hierarchical tree of information resharing, usually referred to in the literature as a cascade [3,4]. Cascades capture information spread better than the 'small world' model [5,6], and allow studying social influence in networks [4,7]. Although information cascades are capable of reaching an enormous number of users, they vary in size and rarely become large [8]. It is still unclear what factors contribute to the scope of information cascading and how to predict this variation accurately [9].

Currently, there are two approaches to this task: generative (deductive) and feature-based (inductive) [10]. The generative approach involves characterizing and modeling the process of content becoming popular in a social network. Although it provides excellent interpretability, it predicts poorly the variation observed in real-world cascades and may miss possibly valuable predictors. The feature-based approach formulates this task as a regression/classification problem that could be solved using a learning algorithm and a set of features with varying contributions to the explained variance, providing a framework for both prediction and explanation.

Previous works related to the prediction of cascade growth had been investigating factors connected to it. Although some progress in successful prediction has been achieved, a consensus on what features are the most essential to it is not established. Specifically, Cheng and others in [9], aiming to predict whether the size of information cascade will exceed the predetermined number, discovered that content features of an original (root) post (i.e., attached images and captions of the post), although being weak predictors on their own, affect the influence of structural (friendship/followership networks' properties) and temporal (the speed of reshares) features. Simultaneously, they report that the average connectivity of the first reposters contributes to the increasing accuracy of prediction. An alternative finding concerning the influence of content features was reported in [11], in which features of an author of a tweet along with tweet features appeared to be the most essential for prediction. Another study by Hong and colleagues [12] indicated that the best model performance is achieved when contextual features are used along with temporal ones and that user activity aspects enhance marginal predicting performance. One more work [13] experimented on features sets for prediction by applying a hybrid methodology for feature selection. The results were that channel (information source) features, i.e., the author's followers, and content features (whether a post contains URL and/or image), were repeatedly given the best rank by several feature selection algorithms. In the interim, a study by Tsur and others [14] proved that a hybrid model incorporating several feature types (i.e., contextual, structural, and temporal ones) predicts information dissemination much better than partial models, evidencing that no predicting feature type's influence is concealed by the presence of other predicting feature types. There was also an attempt to propose a model able to integrate all notable findings of the previous research in [15], that was successfully empirically tested on the Twitter resharing cascades afterward. However, the authors' highest result of prediction hardly accounted even for a half of the variation in the cascade size.

Additionally, studies on the topic had reported that complex models outperform simpler models applied for prediction. Precisely, the finding is valid for the work by Cao and co-authors [10] that involved a deep learning extension of the Hawkes processes model in comparison to regularized linear regression, both utilized to predict the influence of a retweet path. The already mentioned work by Cheng and colleagues [9] also proved that non-linear algorithms perform a little better than linear ones when predicting the variation in a cascade's size.

In this work, we seek to perform an accurate prediction of cascade growth using six types of regression machine learning algorithms with a wide set of feature categories. We opt for the regression task for two reasons. The size of a propagation cascade, with no additional transformations implemented, is a numeric value. The classification formulation requires transforming this value either to a binary or multinomial variable, which involves dividing the range of values into categories based on some threshold. As cascade's size is commonly power-law distributed [9,16,17], defining such threshold is problematic and reduces the amount of likely valuable information. Hence, this study aims to determine which algorithms are the best for predicting cascade's growth, and to explore which features are the most strongly associated with the eventual scope of a cascade. Although we consider factors associated with information propagation, we aim for a methodological contribution: to investigate the applicability of several learning algorithms to the prediction of cascades observed in a real online network.

Although the most popular social networking service for studying information propagation is Twitter, we choose VKontakte (VK) to address the lack of studies on Russian-speaking platforms and possible discrepancies connected to local audiences and media. Moreover, VK is the most popular Russian-speaking service, similar to Facebook, with an audience of 73.4 million users per month [18].

We divide a notion of cascade's size into three differentiated metrics: its depth, width, and mass. We apply the notion of a cascade level, considering that there is a hierarchy of reposting from node to node. Thus, we define these metrics as follows:

- **width** is the biggest amount of nodes at one of the levels of a resharing cascade;
- **mass** is the total number of nodes at all levels of a cascade;
- **depth** is the maximum number of levels in a propagation cascade.

The rest of this paper is organized in the following way. The next section discusses the data used in the study. Then we elaborate on the methodological pipeline, after which the results of modeling are summarized. Finally, Sect. 5 provides conclusions and considers the meaning and value of the findings.

2 Data

2.1 Dataset Description

As it was already mentioned, the data needed for modeling and further analysis were retrieved from VK service. The scope of the data was narrowed to the official public pages of leading state-owned Russian television channels (see Table 1).

The data included (1) reposting data – the chains of reposts for top posts from the channels, (2) reposters data – publicly available profile information of users who at least once have reshared one of the top posts at any level of a cascade, (3) channel's summary statistics – total numbers of posts' comments, likes to posts, likes to posts' comments for each channel computed for the top posts, (4) post metadata – the popularity figures for each post (i.e., post's number of comments, likes, and reposts, the latter accounting for the number of nodes on the first level of a resharing cascade), and, finally, (5) topic modeling data, a matrix containing probability distributions of 86 labeled topics over news texts posted on behalf of the channels, so that each studied information cascade has some probability of belonging to all of the topics. The topic modeling procedure was reported in [19].

The full set of features used for prediction can be found in Table 2. The final dataset used for fitting and assessing the models consisted of 4,424 observations. Its detailed description by sampled channels is provided in Table 1.

Table 1. Dataset description

News community	N of posts	N of reposts	N of unique reposters	N of 1st-level reposts	Max cascade's width	Max cascade's depth	Max cascade's mass
RIA News	953	44,703	27,871	40,325	1,048	10	1,151
Russia Today	944	17,661	11,725	15,772	591	6	749
RBC	706	28,030	14,313	24,536	665	8	1,169
Dozhd	593	9,233	5,892	7,593	278	8	405
NTV	441	10,262	7,480	8,448	1,279	12	1,413
Russia 24	252	8,072	6,120	6,733	784	6	907
Russia-Culture	123	3,530	2,050	2,704	109	9	273
MIR24	113	5,412	1,264	4,984	153	8	176
Channel-5	101	2,960	2,312	2,208	126	7	217
Channel 1	72	23,755	18,261	20,656	1,170	6	1,627
TVC (News)	64	392	311	270	10	4	18
Russia-1	51	5,505	3,246	3,281	332	9	614
Monson	8	1,684	1,435	1,500	234	4	255
InoTV	3	15	15	15	6	1	6

Table 2. List of features within each feature category

Feature category	Feature	Data type
Root post features	N of comments to each root post	Numeric
	N of likes to each root post	Numeric
	N of reposts of each root post on the first level of a cascade	Numeric
Channel features	The total N of comments to channel's top posts	Numeric
	The total N of likes to channel's top posts	Numeric
	The total N of likes to comments to channel's top posts	Numeric
	N of channel's followers	Numeric
Channel's audience features	The average age of those subscribed to a channel	Numeric
	The cumulative N of followers of those subscribed to a channel	Numeric
	The prevalent sex ('1' for female and '2' for male) of those subscribed to a channel	Binomial
Content features	Distributions of probabilities of shared information, text of a root post, belonging to each of the established by topic modeling procedure [19] topics. 86 separated variables in total	Numeric

2.2 Training and Testing Datasets

The compiled dataset was split into two samples: the training set for fitting models, and the test set one for assessing their performance, as the standards of predictive machine learning modeling entail [20].

There are several ways of splitting the initial dataset, and the most common strategy is to randomly assign 80% of observations to the training sample and 20% to the testing sample. In the case of our data, we can split the dataset by date of posts' publication. Thus, models fitted on the older data would be extrapolating to more recent cascades with more reliable predictions in terms of external validity. Hence, the data was split as follows: the training set included all posts up to October 2017 (including), and the test set covered the period from November 2017 to February 2018. Ultimately, the convention of 80% by 20% data split was met.

2.3 Normalization

In our case, the target variables – width, depth, and mass of a cascade – were found to be power-law distributed. To comply with the assumptions of the penalized linear regression models – requiring a normally distributed outcome variable – the resulting data was duplicated, and normalization procedure was applied so that two sets of data were obtained, non-normalized and normalized. For normalization, the scaling of matrix-like objects algorithm [21] was applied.

3 Methods

3.1 Algorithms and Hyperparameters

For modeling cascade's metrics, a set of algorithms of increasing complexity with 5-fold cross-validation was chosen. The set included three kinds of penalized linear regressions, i.e., Lasso regression, Ridge regression, and Elastic Net regression, Decision Tree algorithm, and more complex ensemble methods, such as the Random Forest and Gradient Boosting Machine algorithms.

The hyperparameters were set to default for Decision Tree, as it was found that changing hyperparameters does not affect performance significantly on this data after a series of preliminary tests. In the case of Random Forest, the picking-up procedure revealed the most optimal hyperparameters for the algorithm. As for Gradient Boosting, the algorithm automatically chose the hyperparameters out of a specified list of values based on the RMSE metric. The same was done for all three types of penalized linear regressions. The detailed settings of hyperparameters are present in Table 3.

3.2 Fitted Models

Models were built in a three-stage procedure aimed at defining the best-fit for predicting cascades' growth. During the first step, three algorithms, Decision Tree, Random Forest, and Gradient Boosting, were fit to the non-normalized training data, each of the algorithms predicting three cascade's metrics separately. After that, the same was reproduced on the normalized training dataset, with penalized regressions added. Finally, a total of 27 models, with 9 of them predicting each of three target values, were run on the testing sets and compared in prediction accuracy based on the R-squared metric. This metric was chosen as the only one that can be applied to both scaled and non-scaled datasets, disregarding discrepancy in variables' value range. In situations when R-squared values were about the same within one of the cascades' metrics prediction, the RMSE measure, calculated as the square root of a mean squared difference between predicted and observed outcomes, was additionally used.

3.3 Interpretation

After the best-fitting models predicting depth, width, and mass of a cascade were chosen, several methods were applied for the purpose of interpretation. Some of the utilized algorithms, i.e., linear regressions, are easily interpretable on their own, producing coefficients for all of the involved independent variables needed to specify how inputs interact with each other to generate the output.

However, other algorithms – sometimes called "black box" models – are more complex, which hampers the interpretability of results. Considering this fact, for interpretation of the Gradient Boosting models, the relative influence of explanatory features was used, which is the number of times a feature was selected for

Table 3. List of tuned hyperparameters

Algorithm	Parameter name	Parameter description	Range
Decision Tree	*mincriterion*	A value of test statistics or (1-p) that should be exceeded to perform a split	0 (*default*)
Random Forest	*mtry*	The number of features randomly sampled as candidates at each split	50
	ntree	the total number of trees to grow in one run	500
Gradient Boosting Machine	*interaction.depth*	number of splits an algorithm has to perform on a single tree	6 (*Salford default setting* [22])
	n.trees	The total number of trees to grow in one run	500, 800, 1000, 2000
	shrinkage	A learning rate, stands for the amount of penalty that will be applied to reduce the effect of each additional tree built	0.001, 0.005, 0.01, 0.05, 0.1, 0.5
	n.minobsinnode	The minimum number of observations in terminal nodes of each tree	10, 15, 20, 25
Lasso Regression	*lambda*	Determines the amount of shrinkage, the penalty term, to be applied to regularize the effect of predicting features	$10^{seq(-3,3,length=100)}$
	alpha	Determines what type of penalized regression model is fit	1
Ridge Regression	*lambda*	Determines the amount of shrinkage, the penalty term, to be applied to regularize the effect of predicting features	$10^{seq(-3,3,length=100)}$
	alpha	Determines what type of penalized regression model is fit	0
Elastic Net Regression	*lambda*	Determines one of the amounts of shrinkage, the first penalty term, to be applied to regularize the effect of predicting features	0.1, 0.2, 0.3, 0.4, 0.5, 0.6, 0.7, 0.8, 0.9, 1
	alpha	Determines one of the amounts of shrinkage, the second penalty term, to be applied to regularize the effect of predicting features	0.1, 0.2, 0.3, 0.4, 0.5, 0.6, 0.7, 0.8, 0.9, 1

each tree split balanced by the improvement of the SSE (sum of squared errors) value and averaged over all trees [23].

In the case of Random Forest, the variable importance score was used. The scores are computed using MSE (mean square error) first obtained on the subsample of data for each tree that was not used during model construction, and once again with variables reshuffled. The differences are then averaged and divided by the standard error. Lastly, to infer the significance of independent variables of penalized linear regression, the coefficients as the estimators of features' importance, controlled by a penalty term, were obtained. After the order of importance of predicting features in each considered model was acquired, the individual conditional expectation plots (ICE plots) were constructed to decide the directionality of the relationship between the outcome and the most important features. The partial dependence plot displays the marginal effect of one or more features on the predicted outcome, and it was used to assess the direction of the relation between an outcome and a feature [24].

Finally, the whole procedure was repeated without variable with the number of reposts at the first level of resharing to investigate its relative importance for reasons discussed below.

4 Results

Table 4 shows the performance of all nine models predicting the mass of a cascade on both non-normalized and normalized data. Due to differently preprocessed sets of the data, values of RMSE for the first three and the remaining six models fall into different number ranges. However, it is still possible to compare R-squared values among nine models, and RMSE values within models built on the same data. The highest performance on cascade's mass prediction is attributed to the penalized linear regression – the percentage of explained variance in the outcome goes beyond 97%. Among these three models, the most accurate appears to be the Lasso regression, having the smallest root mean square error.

R-squared and RMSE metrics of nine models predicting cascade's width and nine models for cascade's depth are laid out in the same table, respectively, with the very same technical peculiarities mentioned for the mass predicting models' performance. The best models for prediction of cascade's width are Gradient Boosting Machine on non-normalized data and Lasso and Elastic Net regressions on the normalized dataset, achieving the highest R-squared values of approximately 0.99, and the lowest RMSE values. Finally, in the case of depth of a cascade, the best predictive performance is shown by the Gradient Boosting algorithm on both datasets, with 57% explained variance in the outcome variable. Judging by lower RMSE value, this model built on the non-normalized data is slightly more accurate.

Additionaly, it should be noted that width and depth of information cascade are predicted more accurately, with models accounting for almost 100% of the variance in the target variable while cascade's depth was predicted poorly, compared to other metrics of cascade's size, reaching maximum of 57% of explained variance.

Table 4. Models' performance figures

Cascade's Metric	Model	R-squared Value	R-squared Value (with one feature out)	RMSE	RMSE (with one feature out)
Mass	Decision Tree (*non-scaled data*)	0.962	0.247	20.226	89.598
	Random Forest (*non-scaled data*)	0.966	0.252	18.540	90.018
	Gradient Boosting (*non-scaled data*)	0.97	0.242	17.670	88.503
	Lasso Regression (*scaled data*)	0.972	0.230	0.166	0.888
	Ridge Regression (*scaled data*)	0.970	0.178	0.196	0.906
	Elastic Net Regression (*scaled data*)	0.972	0.232	0.168	0.888
	Decision Tree (*scaled data*)	0.960	0.024	0.208	1.861
	Random Forest (*scaled data*)	0.923	0.084	0.320	2.039
	Gradient Boosting (*scaled data*)	0.966	0.059	0.182	2.156
Depth	Decision Tree (*non-scaled data*)	0.539	0.089	0.673	0.957
	Random Forest (*non-scaled data*)	0.550	0.117	0.662	0.949
	Gradient Boosting (*non-scaled data*)	0.571	0.136	0.648	0.921
	Lasso Regression (*scaled data*)	0.316	0.115	0.828	0.950
	Ridge Regression (*scaled data*)	0.334	0.120	0.815	0.971
	Elastic Net Regression (*scaled data*)	0.305	0.103	0.836	0.951
	Decision Tree (*scaled data*)	0.511	0.034	0.70	1.059
	Random Forest (*scaled data*)	0.560	0.113	0.666	0.981
	Gradient Boosting (*scaled data*)	0.572	0.112	0.665	0.956
Width	Decision Tree (*non-scaled data*)	0.969	0.312	15.260	68.407
	Random Forest (*non-scaled data*)	0.980	0.319	11.817	70.180
	Gradient Boosting (*non-scaled data*)	0.984	0.320	11.064	67.911
	Lasso Regression (*scaled data*)	0.984	0.266	0.122	0.867
	Ridge Regression (*scaled data*)	0.983	0.201	0.155	0.893
	Elastic Net Regression (*scaled data*)	0.984	0.270	0.122	0.867
	Decision Tree (*scaled data*)	0.964	0.058	0.196	2.528
	Random Forest (*scaled data*)	0.924	0.107	0.333	2.121
	Gradient Boosting (*scaled data*)	0.981	0.070	0.137	2.405

The following models were considered for interpretation: Lasso regression for cascade's mass prediction and Gradient Boosting built on the non-normalized dataset predicting both width and depth of a cascade. The most prominent feature associated with cascade's mass was the number of reshares on the first level of a cascade. On the relative scale of overall variable importance, assessed on the base of absolute values of regression coefficients, ranging from 0 to 100, this feature scored the maximum. Figure 1 (see in Appendix) depicts the mentioned variation. Further conclusions about features' importance were drawn on the base of the same method, partial dependence plots. The same was observed for other metrics – their increase was strongly associated with the number of reposts on the first level of a cascade. In the model predicting cascade's width, this feature reached 97 out of 100 points on the normalized scale of features' relative influence. In predicting the depth, it scored 70 out of 100, compared to other features that barely exceeded 3 points.

Thus, most of the variation in cascade's size can be explained by the change in the number of reshares on the first level of the information propagation tree. This relation itself appeared to be positive, meaning that the increase in the number of reposts on first-level is attributed to the growth of a cascade in depth, width, and mass (see Fig. 2 and Fig. 3 in Appendix).

Among other features, relatively significant appeared to be content features: probability distributions of several text topics, all of the used channel's audience features, one of the root poster features, and channel features. For prediction of cascade's mass, the number of reposts on the first level of a cascade is followed by prevailing sex of channel's followers, the number of channel's followers, the cumulative number of channel's audience' followers, the number of comments to channel's posts, the number of root post's likes, and average age of channel's audience. According to the coefficients (see Fig. 2 in Appendix), the prevalence of male users among channel's followers and the number of channel's followers are negatively associated with the mass of a cascade, while the rest of the mentioned features – positively.

As for the cascade's width prediction, the number of reposts on the first level is followed by the probability distributions of the following topics: "West-Russia relations", "The Voice Russia", "TV shows: comedy and dancing", and "Weather". They appeared to be positively correlated with the cascade's width, as the marginal effect of each topic exceeds zero. Finally, for the depth prediction the number of comments to channel's posts, the number of channel's audience' followers, and the number of likes to channel's posts are relatively important. Interestingly, only the number of reposts at the first level of resharing, among all mentioned features, is positively associated with the depth of a cascade.

As the number of reposts at the first level of propagation cascade accounted for the overwhelming part of the variance in all three metrics modeled, we propose that the growth of a cascade can be attributed solely to this feature. To investigate this proposition, we repeated the whole procedure without this variable. As a result, the set of best-fit algorithms has changed (see Table 4). Now, the best model for predicting the mass of a cascade is Random Forest trained and assessed on non-normalized data. As for cascade's width and depth, Gradient Boosting Machine on the non-normalized dataset emerged as most accurate. R-squared values of all nine models for each metric substantially decreased, not reaching even half of the value of their counterparts in a situation when the removed feature was present.

Further, scaling of the dataset for prediction of all three metrics gives a substantial fall in R-squared values compared to those resulting from the non-scaled data. As for the most crucial features, the cumulative number of channel's audience followers, content characteristics, and the number of comments to channel's posts explain most of the variation. Surprisingly, the increase in the latter shows a negative association with cascade's depth, while others are positively associated with cascade's metrics.

Additionally, we considered marginal effects of age and sex of a channel's audience on the cascade's size in the models without the number of reposts on the first level. The prevailing sex of a channel's audience indicated by users as "male" is negatively correlated with the size of a cascade. The marginal effect of the average age of a channel's audience on cascade's mass equals zero for most of the predictor values except for the value of 33.84, where it falls below zero. Its marginal effect on the width of a cascade is negative, while on the cascade's depth, it is the exact opposite, indicating positive association. However interesting, we abstain from an in-depth analysis of factors associated with cascade growth and their contribution and limit our interpretations as we aim to investigate the methodological value of the findings.

5 Conclusions and Discussion

In this work, we examine the issue of information spread in online social networks by attempting to find the best-fit model that can be used to predict the growth of propagation cascades. Although there are existing studies that have addressed similar objectives, we contribute to the research by approaching the problem with the data from Russian-speaking social networking service. After obtaining the best-fit models, we look at the features that contributed the most in order to determine the possibly noteworthy constituents of the models able to predict variation in the outreach of an information cascade.

The results showed that non-linear algorithms perform better when predicting the growth of information cascade – which is consistent with [10] and [9] works. Except for the prediction of mass of a cascade with a full set of features, which is most accurately predicted with regularized linear regression. Furthermore, it should be noted that R-squared values for the models with all features included are objectively quite high, indicating that models can explain up to 98% of the variation in cascade's size – in contrast to Martin and his colleagues' study [15] which achieved a maximum of 48% of explained variation.

The interpretation of the best-fit models indicated that features of all categories either way appeared as one of the most contributing to the prediction, similarly to [14]. However, not all features have the same degree of importance for the accuracy of prediction, which is more in line with [9]. The growth of an information cascade in width, depth, and its gain in mass can be attributed entirely to the number of reposts on the first level of a cascade – formally, to the number of edges directly connected to the root of a cascade, while other features have a relatively small effect on the outcome when included in a model altogether. Although logically evident, this finding raises questions whether this connection can be explained by the propagation mechanics and the data specificity, or is it an artifact of the platform where the study was conducted. It should be noted that on VK, there is no way a user can see the number of likes, reshares, or views of the original post at the point when this post was reposted, i.e., from the account of a user who did the repost. In addition to that, when a repost is done, a reposter is able to place a caption to the reshared post – fairly modifying resharing information.

Nevertheless, a sharp decrease in R-squared value by three times for all models, after removing the number of reposts on the first level of resharing from the predicting features, suggests that the eventual outreach of content diffusion can be predicted with high accuracy by this feature solely. Hence, we can conclude that the observed structural variation in depth and width of a propagation tree, and its eventual magnitude highly depends on the activity of the source' audience that serves as an intermediary in letting the information leak beyond where it was initially posted. Note that such an audience includes not only followers of a channel but also users who are not bounded by the followership relations to a channel.

Additionally, the results of [11] study were also partly supported – features related to an author of a post (in our case, channel features) combined with content features benefited the predictive power of the models, yet only when the number of shares on the first level was excluded. This finding lets us suggest that if we assume that there are specific scripting strategies that each channel uses to differentiate itself and the content it posts from other news channels, the eventual reach of information propagation depends on the source's (channel's) specificity. Furthermore, users that make the propagation possible – in line with the selective exposure theory [25] – intentionally choose from what source and what news information to consume depending on the attributes of both source and content.

Content features on their own were consistently important for predicting cascade' s growth with models using all features and models without the number of shares on the first level, compared to other features. This validates conclusions of work by Hong and colleagues [12]. However, predicting models without the number of reposts on the cascade's first level let us conclude that, contrary to Elsharkawy and colleagues' work [13], audience features are more relevant for prediction of cascade's growth than channel's features. Further, if cascade's growth can be relatively successfully predicted by audience features, a question about users' similarity or dissimilarity as a possible driver of information diffusion can be considered.

Finally, channel features and one of its audience features, prevailing sex of channel's audience indicated as male, that showed negative association when reaching a certain value with the depth of a cascade, raise suspicions. In the case of channel features, it appears that the number of followers of the author of the content (a channel) – simply put, channel's popularity – has a negative influence on information propagation depth. We can speculate that such a phenomenon can be linked to the reputation of Russian news channels' representation in the chosen social network. The online channels are assumed to be perceived by users of this social network as not trustworthy enough to 'participate' in spreading its content. Besides, the negative association between the number of channel's followers and the depth of a diffusion tree can be attributed to specifics of the publication sorting mechanism of the VKontakte platform. Still, further research is needed for the explicit connection and rationale of such a pattern to be established. As for male audiences having a negative impact on the cascade's growth, it can be allegedly explained by male users' tendency to less frequently become intermediate recipients (brokers) of information in a network of information diffusion, compared to female users. Yet, to claim it as a fact, more research on the topic should be conducted.

Limitations

It is important to note that the reposting data used in the study is news posts published on official communities of Russian television channels, making the results valid only for such or a similar sample. This study's findings cannot be generalized on reposting patterns of any other content on social media platforms except for news data. Another limitation of our research is connected to metrics used to assess models. In the case of "black box" models, it is debatable whether the R-squared value can be applied for assessment of their predictive power.

Acknowledgements. This work is an output of a research project implemented as part of the Basic Research Program at the National Research University Higher School of Economics (HSE University).

A Appendix

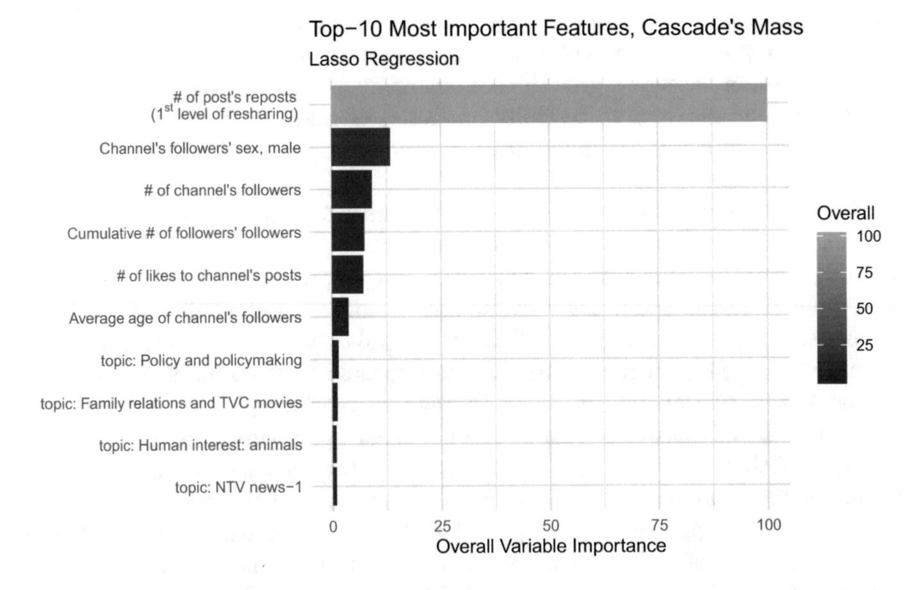

Fig. 1. A graph showing top-10 of the most tangible predicting features used as input by the Lasso Regression algorithm for cascade's mass prediction.

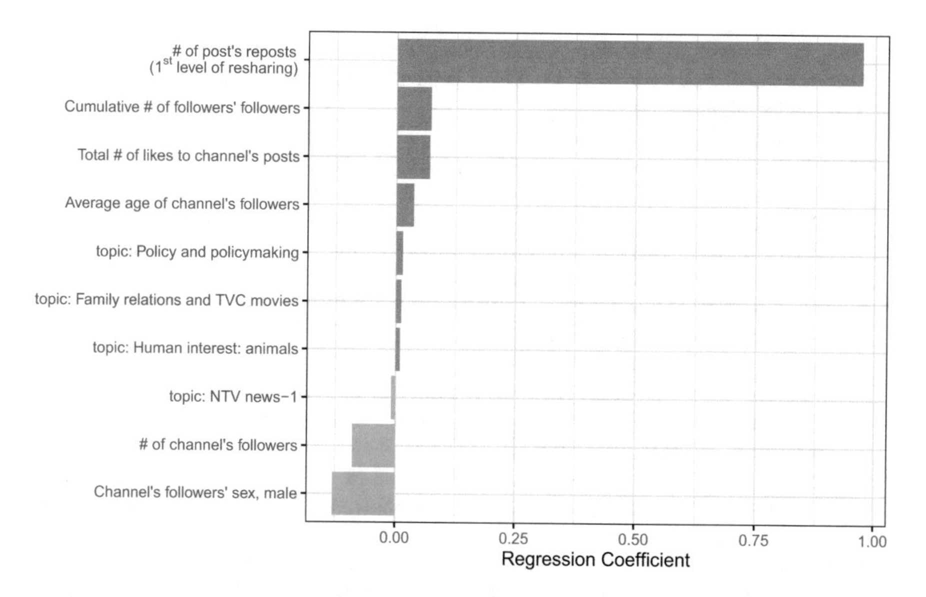

Fig. 2. A plot displaying negative to positive values proportion of Lasso Regression variables' coefficients.

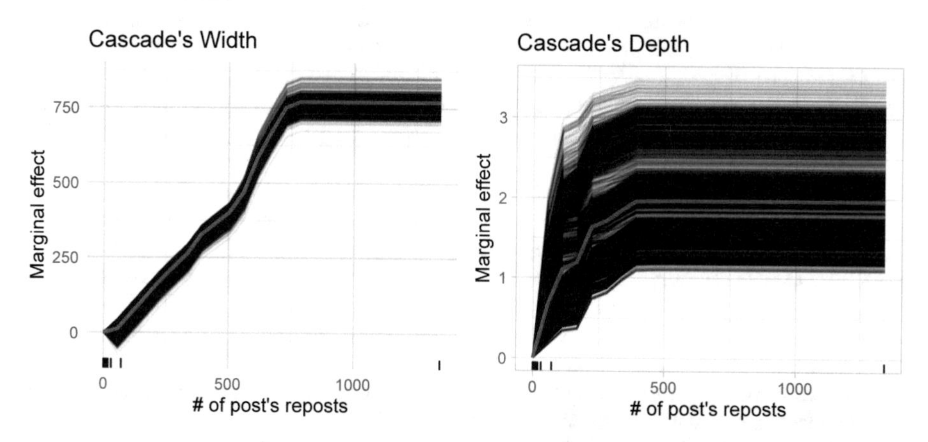

Fig. 3. Plots showing observation-level effects of the number of posts first-level reposts on cascade's depth (on the left) and width (on the right). The multiple black lines are individual conditional expectation (ICE) curves, while the red line stands for the averaged values across all predictions [24] by Gradient Boosting algorithm.

References

1. Thorson, K., Wells, C.: Curated flows: a framework for mapping media exposure in the digital age. Commun. Theory **26**(3), 309–328 (2015)
2. Boyd, D.M., Ellison, N.B.: Social network sites: definition, history, and scholarship. J. Comput.-Mediated Commun. **13**(1), 210–230 (2007)

3. Sun, E., Rosenn, I., Marlow, C.A., Lento, T.M.: Gesundheit! Modeling contagion through Facebook news feed. In: Third International AAAI Conference on Weblogs and Social Media (2009)
4. González-Bailón, S., Borge-Holthoefer, J., Moreno, Y.: Online networks and the diffusion of protest. In: Analytical Sociology, pp. 261–278 (2014)
5. Liben-Nowell, D., Kleinberg, J.: Tracing information flow on a global scale using internet chain-letter data. Proc. Natl. Acad. Sci. **105**(12), 4633–4638 (2008)
6. Gomez-Rodriguez, M., Leskovec, J., Krause, A.: Inferring networks of diffusion and influence. ACM Trans. Knowl. Discov. Data **5**(4), 1–37 (2012)
7. Myers, S.A., Zhu, C., Leskovec, J.: Information diffusion and external influence in networks. In: Proceedings of the 18th ACM SIGKDD International Conference on Knowledge Discovery and Data Mining, KDD 2012 (2012)
8. Bakshy, E., Hofman, J.M., Mason, W.A., Watts, D.J.: Everyones an influencer. In: Proceedings of the Fourth ACM International Conference on Web Search and Data Mining, WSDM 2011 (2011)
9. Cheng, J., Adamic, L., Dow, P.A., Kleinberg, J.M., Leskovec, J.: Can cascades be predicted? In: Proceedings of the 23rd International Conference on World Wide Web, WWW 2014 (2014)
10. Cao, Q., Shen, H., Cen, K., Ouyang, W., Cheng, X.: DeepHawkes: bridging the gap between prediction and understanding of information cascades. In: Proceedings of the 2017 ACM on Conference on Information and Knowledge Management, pp. 1149–1158 (2017)
11. Petrovic, S., Osborne, M., Lavrenko, V.: RT to Win! Predicting message propagation in Twitter. In: Fifth International AAAI Conference on Weblogs and Social Media (2011)
12. Hong, L., Dan, O., Davison, B.D.: Predicting popular messages in Twitter. In: Proceedings of the 20th International Conference Companion on World Wide Web, WWW 2011 (2011)
13. Elsharkawy, S., Hassan, G., Nabhan, T., Roushdy, M.: Towards feature selection for cascade growth prediction on Twitter. In: Proceedings of the 10th International Conference on Informatics and Systems, INFOS 2016 (2016)
14. Tsur, O., Rappoport, A.: What's in a hashtag? Content based prediction of the spread of ideas in microblogging communities. In: Proceedings of the Fifth ACM International Conference on Web Search and Data Mining, pp. 643–652 (2012)
15. Martin, T., Hofman, J.M., Sharma, A., Anderson, A., Watts, D.J.: Exploring limits to prediction in complex social systems. In: Proceedings of the 25th International Conference on World Wide Web, pp. 683–694 (2016)
16. Leskovec, J., Mcglohon, M., Faloutsos, C., Glance, N., Hurst, M.: Patterns of cascading behavior in large blog graphs. In: Proceedings of the 2007 SIAM International Conference on Data Mining (2007)
17. Vicario, M.D., Bessi, A., Zollo, F., Petroni, F., Scala, A., Caldarelli, G., Stanley, H.E., Quattrociocchi, W.: The spreading of misinformation online. Proc. Natl. Acad. Sci. **113**(3), 554–559 (2016)
18. Mail.ru Group Limited Annual Report for FY 2019 and unaudited IFRS results for Q1 2020, April 2020. https://corp.imgsmail.ru/media/files/engq1-2020-results.pdf
19. Koltsov, S., Pashakhin, S., Dokuka, S.: A full-cycle methodology for news topic modeling and user feedback research. In: International Conference on Social Informatics, pp. 308–321. Springer (2018)
20. James, G., Witten, D., Hastie, T., Tibshirani, R.: An Introduction to Statistical Learning, vol. 112. Springer, Heidelberg (2013)

21. Becker, R.A., Chambers, J.M., Wilks, A.R.: The new s language, April 2018
22. TreeNet stochastic gradient boosting: an implementation of the MART methodology. http://docs.salford-systems.com/TreeNetManual_v1.pdf
23. Quan, Z., Valdez, E.A.: Predictive analytics of insurance claims using multivariate decision trees. SSRN Electron. J. (2018)
24. Friedman, J.H.: Greedy function approximation: a gradient boosting machine. Ann. Stat. **29**, 1189–1232 (2001)
25. Sullivan, L.E.: Selective exposure. In: The SAGE Glossary of the Social and Behavioral Sciences, p. 465 (2009)

Friends Network Expansion and Reduction: Investigating the Role of Structural and Psychological Factors

Yadviga Sinyavskaya[1(✉)] and Alexander Porshnev[1,2]

[1] Laboratory for Social and Cognitive Informatics, National Research University Higher School of Economics, St. Petersburg, Russia
ysinyawskaya@hse.ru

[2] Department of General and Strategic Management, National Research University Higher School of Economics, Nizhny Novgorod, Russia

Abstract. Rich data from social networks sites (SNS) attracts attention of psychologists and sociologists interested in interpersonal dynamics, friendship networks, and social capital. The presented study explores the effect of network structural features and psychological characteristics of SNS users on changes in their friendship networks. The data from the representative and diverse sample of 375 Russian Vkontakte SNS users from Vologda city was used. Two waves of network data collection allow us to estimate changes in the size of the friendship networks. Regression analysis reveals similarities in the factors responsible for the changes in networks for users who attract or reject friends. We discuss possible explanations of this phenomenon, as well as limitations of the study and further research directions.

Keywords: Friendship network · Network dynamics · Propensity to make connections · Ego-network · Social networking sites

1 Introduction

Online social networks (OSN) provide people with a wide range of opportunities to manage social ties: to make new friends, recover lost contacts, establish contact with friends of their friends, or participate in communities of interest. One of the main specific features of OSNs is that they make it possible for a user to articulate their own social circle, formalizing it as a friend list [1]. The way of establishing social ties online raises a lot of questions. For example, how do people form and manage their personal online ego-networks? What factors influence the network dynamics (expansion or reduction)?

As to the network expansion, one of the strategies of forming a social network named by Donath and Boyd [2] at the rise of OSNs as "Friendster whores" refers to the indiscriminate friending activity when the goal is to collect as many friends as possible. In the modern era of OSNs it seems that such a strategy represents not as much as a

A. Antonyuk and N. Basov (Eds.): NetGloW 2020, LNNS 181, pp. 196–208, 2021.
https://doi.org/10.1007/978-3-030-64877-0_13

vanity fair of distinct persons, but an instrument of getting monetization of a personal brand, or becoming an informational influencer [3].

In addition, a well-known mechanism responsible for network growth helps to advance the already gained popularity of a person [4]. The "rich get richer" effect (or the Matthew effect), usually observed in real networks, that received a formal definition in the model of preferential attachment by Barabási and Albert, means that the more friends an individual has, the more likely he or she will have more new friends [5].

Stepping out from the specific phenomenon described above, the extant literature on this topic shows that some kind of selection is generally applied by users in shaping their online environment. Typically people accept not all friend requests [6] and primarily use social media for keeping in touch with those who they have an offline connection with [7]. In addition, the communication activity of users aims at addressing those who users know in person [1, 8, 9].

Another factor that might influence the social network expansion is homophily – strong tendency toward forming relationships selectively with those who possess self-similar demographic parameters, attitudes, personal preferences, appearance, social status, or any other attributes [10]. As Aiello and coauthors show, users with similar interests are more likely to be friends [11]. Thus, various online communities might be treated as a source of homophily-based social ties for individuals on OSNs. Rykov and colleagues demonstrate a significant contribution of being a member of online groups to accruing social capital [12]. In addition, a user's personal profile may play a role of a "social lubricant", which assists people in finding something in common serving as a ground for further relationship development. Lampe et al. found that some specific fields of an OSN profile are positively related to the number of online friends [13].

Despite the aforementioned mechanisms that guide users' online friending behavior, some studies show evidence that personal networks of Facebook users comprise only 25% of those who are claimed by them as an actual friend [11]. Such a granular metric also allows to gain a more accurate estimation of users' bridging social capital. Ellison et al. suggest that the benefits of having a large network are limited and start to decrease after exceeding the threshold of 400–500 friends [14].

Moreover, the so-called "complexity index" [15], which suggests that individuals are able to actively communicate and maintain relationships with a limited number of recipients, found its confirmation in online computer-mediation communication [16, 17].

Thus, if the communication abilities are constrained and the real social capital benefits follow only from a small share of online friends, the natural suggestion would be that people might "clean" their networks and delete irrelevant social contacts. Surprisingly, to the best of our knowledge, there are no studies to date directly addressing the process of network reduction.

Thus, the first gap that we aim to cover is to explore not only how people foster their personal networks but also how they abort their social connections on OSN.

Another research avenue is to place the process of expansion and reduction of social contacts in the psychological context. The role of the individual psychological characteristics that may shape personal networks is considered as understudied [18].

There is evidence that such a personality trait as extraversion explains the difference between people in terms of the size of personal networks and should be considered when predicting network size. At the same time, Kalish and Robins argue that studying the specific behavioral orientation rather than broad personality traits helps to avoid potential diffusion in capturing behavioral outcomes associated with the latter [18].

Following this trend, we adopt the measure of "propensity to make connections with others" (PCO) proposed by Totterdell et al. [19]. It reflects different aspects of networking attitudes and demonstrates higher predictive validity in accounting for the size of a user's friendship network than extraversion.

It should be noted that prior studies devoted to the process of getting friends online mostly focused on the static number of friends in networks rather than on factors driving the dynamic changes of networks such as the growth of a social network. In the current study, we measure the propensity to make connections for predicting the dynamic changes in size of friendship networks of OSN users.

Apart from the personality traits, the respective online behavior aimed at having and maintaining social connections also plays a part in understanding the friending behavior. Ellison and colleagues proposed the so-called "Relationship maintenance behavior" that implies responding to the requests for help of online friends and found that it positively affects the social capital [20–22]. It may happen that users who have already invested a lot of energy in communication with others would be less inclined to get rid of their even irrelevant social contacts.

Finally, while predicting the changes in the network size, it seems reasonable to take into account the structural characteristics of users' ego-networks due to the aforementioned process of preferential attachment [9]. In a similar way to the "rich get richer" effect, when the size of a person's network influences its growth, the network density or number of clusters might follow the same pattern.

To sum up, the aim of the study is to expand understanding of the role of psychological and structural factors in determining the changes in the size of friendship networks of social media users. We explore how the dynamics of growing/reduction of personal networks managed by OSN users themselves are related to their networking attitudes, the respective behavior of investing in the existing social ties and the prior structural ego-network characteristics.

In this paper the terms *networking* and *friending* are used as synonyms and mean establishing social network friendship connections. We regard friends as persons who were marked as friends on a social networking site (SNS) [23] and focus our study on friending behavior, which we define as behavior to form a personal friendship network (expansion or reduction).

2 Data and Methods

2.1 Vologda Project Dataset

In 2017–2018 the Laboratory of Internet studies (LINIS) at National Research University Higher School of Economics conducted an online survey among Vkontakte SNS users. The dataset contained psychological, demographics aspects, and characteristics of ego-networks from the representative and diverse sample of 375 residents of a Russian city Vologda [24]. It is worth mentioning there were two waves of network data collection, which allow investigating the dynamic of personal social networks (changes in number of friends) with a year gap between sampling.

The respondents were recruited by means of Vkontakte targeting advertisement service. The ground truth size and the socio-demographic characteristics of the "active"[1] online population of Vologda was revealed in [12] – it comprises around 196 000 users. The invitation to take part in the survey was transmitted through the groups of interest until the demographic requirements were saturated. The survey app controls the opportunity of survey retake.

Targeting as a recruiting method could be classified as a "river sampling" [25], which demonstrated the response rate comparable to traditional methods [26].

As a measure of efficiency of the recruiting campaign, the Click Through Rate (CTR) and completion rate were used. The CTR comprised 0.022%, and the completion rate was 11.5% – 375 completed questionnaires per 3266 dropped-out. These scores turned to be comparable to those obtained in previous research of this kind [27–29].

Previously, we have analyzed the role of social capital in network expansion on the subsample of the participants who increased their friendship networks [24]. It was found that only two factors have significant influence – propensity to make connections and the actual number of friends [24]. In the current study we continue the analysis of dynamics of the friendship networks considering psychological and structural characteristics.

2.2 Variables

Relationship Maintenance

The Vologda dataset contains answers to three items adopted from the Facebook Relationship Maintenance Behaviors scale (RM) [20]. All items (for example, "When I see a friend or acquaintance sharing important news on Vkontakte, I try to respond") were measured using a Likert-type response scale ranging from "1 = Strongly Disagree" to "5 = Strongly Agree". This scale showed satisfactory reliability (Cronbach's = .71) and was included in the analysis.

Propensity to Make Connections

The concept of "propensity to make connections with others" is used to study the disposition of individuals to foster social ties [19]. The construct consists of three dimensions of individual social activity: the propensity to connect with others, the perception of having social ties at the moment, and the perception of one's own ability to form social ties with others. Furthermore, the "strength" of social ties can be considered by measuring the propensity to form three different types of social connections: making friends, making acquaintances, and joining others [19]. The Vologda dataset contains answers for six items of the PCO scale. All items (for example, "I like to have many friends")

[1] "Active" users are those 1) with non-deactivated Vkontakte accounts and 2) who have visited Vkontakte account no later than half a year prior to the date of data gathering.

were measured using a Likert-type response scale ranging from "1 = Strongly Disagree" to "5 = Strongly Agree". The PCO scale showed good reliability (Cronbach's $\alpha = .8$).

Structural Network Characteristics

The first wave of data collection based on publicly available data from Vkontakte network allowed us to calculate the structural characteristics of ego-networks: the number of friends, density, communities (Girvan-Newman algorithm was applied), and modularity. At the second step we calculated the shift in the number of contacts in the friendship networks (delta of friends). To make the distribution of the number of friends and the shift in the friendship networks close to normal, the log transformation of the variables was made. Running an analysis of the whole data set, we subtracted the minimal value of the shift (as there were participants with a negative shift) and added 1 to avoid zero values. Running an analysis of subgroups, we included participants who do not change the number of friends into a positive change group and added 1 to avoid zero in logarithm calculations. In the reduction group we used the absolute value of the shift in the number of friends to calculate logarithm.

Demographics and Control

As control variables the dataset contains information about age, sex, education, and occupation of the respondents. In addition, there are two questions about self-esteem (positive and negative), frequency of SNS usage, and average time spent using a SNS.

It is worth mentioning that SNS as new media provide a possibility to communicate not only with friends, but also with consumers or for other business purposes. To control this, the dataset contains the question: *"I use Vkontakte for selling goods and services, developing online communities for commercial goals or promoting myself"*. Answers to this question were decoded into a binary variable "*Prof*" – having "0" for response "never" and "1" for all the other answering options. For a detailed description, see [24].

3 Analysis

To achieve the research goal, we ran a series of nested OLS regressions. The nested OLS regressions were used because they allow us to reveal psychological and structural factors determining the changes in the size of the friendship networks of social media users and it is a common approach used in many studies in this area (e.g. [12, 14, 19, 20]).

Regression analysis showed that the main factors responsible for changes in the social networks are the number of communities and modularity (see Appendix, Table 3). Surprisingly, in contrast to the results of previous research, the psychological and behavioral variables and the number of actual friends were insignificant. In addition, the low determinant coefficient was revealed (adj.R2 = 0.038).

For further investigation we decided to divide users into two groups: those who expand their friendship networks in the observed period (expansion group) and those with a negative shift in the number of friends in the networks (reduction group). The delta friends variable was calculated as the logarithm of the absolute value of change in the number of friends.

This decision was driven by the assumption that these groups may differ in their friending strategy which may have led us to the confusing initial results.

The basic correlation check in the expansion and reduction groups showed that density and the number of friends had negative correlation (−0.54 and −0.74 respectively). For the expansion group it was below a threshold for exclusion, so density was included in regression analysis. For the reduction group we decided to remove density from regression analysis.

Regression modeling for the separate groups showed significantly better performance in prediction (for the expansion group adj.R2 = 0.388, for the reduction group adj.R2 = 0.489). The Tables 1–2 present the models' outputs containing only significant variables (full models are presented in Appendix, Tables 4–5).

Table 1. OLS regression predicting shift (logarithm) in ego-network (expansion group)

Predictors	Delta friends (log)		
	Estimates	Std. beta	p
(Intercept)	−0.57	0.28	0.14
Propensity to make connections (PCO)	0.23	0.16	0.01
Number of friends (log)	0.38	0.30	0.00
Number of communities	0.02	0.38	0.00
Density	3.59	0.17	0.02
Observations	274		
R^2/R^2 adjusted	0.442/0.388		

Table 2. OLS regression predicting shift (logarithm) in ego-network (reduction group)

Predictors	Delta friends (log)		
	Estimates	Std. beta	p
(Intercept)	−1.08	0.54	0.13
Propensity to make connections (PCO)	0.26	0.21	0.09
Number of friends (log)	0.76	0.74	0.00
Number of communities	−0.01	−0.38	0.01
Observations	79		
R^2/R^2 adjusted	0.603/0.437		

In the regression model predicting the shift *for the expansion group* the number of friends (log) is a significant factor (std.beta = 0.33, $p < 0.001$), which is in line with the "rich get richer" effect (full model is presented in Appendix, Table 4). Similarly, structural components such as density (std.beta = 0.17, $p < 0.05$) and a number of communities (std.beta = 0.38, $p < 0.001$) significantly contribute to the model. As for the psychological variables, only propensity to make connections reveals its significance (std.beta = 0.16, $p < 0.05$), even in presence of the variable of the number of friends (log).

In the regression model *for the reduction group* we observed almost the same effect of the number of friends variable (std.beta = 1.12, $p < 0.00$) (significant variables are presented in Table 2, full model is presented in Appendix, Table 5).

It turns out that the more friends a user has, the more friends they remove from their network – so the preferential detachment hypothesis could be formulated regarding the model of network reduction.

We found the opposite effect of the variable of the number of communities in comparison with the model of expansion (std.beta = −0.38, $p < 0.00$). The more communities were detected in the friendship network, the lower shift in the friendship network is observed.

The propensity to make connections variable for the reduction group was only borderline significant (std.beta = 0.21, $p = 0.09$). The positive relationship between the intention to have social ties and removing ties from the network seems paradoxical, but it could be assumed that people may be simultaneously eager to have social contacts and delete irrelevant social ties from the network.

4 Discussion

The study aimed at assessing the effect of structural and psychological factors on the shift in the number of contacts in the ego-networks of SNS users. It was found that both of these factors predict the shift in the number of contacts in the network. In line with our expectations, the preferential attachment effect was detected in the group of users who have expanded their networks within the observed period of time. Those who have a greater number of friends are characterized by a larger increase in social ties later. Surprisingly, such effect manifests itself for the group of users who have reduced their ego-networks. The more friends those users have, the more friends they tend to delete.

It should be noted that the initial sample of users was divided into two separate ones. It means that we did not consider a situation when a user both expands and reduces a network. Thus, the effects revealed in models of extension and reduction of a network should be considered separately. In this vein, the previous result on "detachment" of social ties by those who possessed them more in the earlier period seems logical. It may reflect the simple "hygiene" performed by the user in order to reduce the irrelevant information noise.

In support of this, users who have reduced their networks do not report the intention to have more friends (the variable propensity to make connection has borderline effect in the reduction model). As to the expansion group – the propensity to make connections has a positive effect on the growth of the personal network. Users who aim at having more friends demonstrate an increase in the number of social connections in the ego-network in the later period. This result is in line with the earlier work of Totterdell and colleagues [19]. Moreover, it serves as additional support of sufficient predictive power of the psychological construct of "propensity to make connections".

Along with the psychological disposition toward networking activity, the role of the respective behavior was tested. It turns out that investment in relationship maintenance affects neither the growth nor the reduction of social ties in the network. Such metric has revealed its effect on the specific social capital outcomes [14] which are rooted not in the quantity of social ties but in their quality. Thus, the investment into maintenance of existing relationships is related to the resources gained by the agency of fostered social ties but it does not affect the dynamics of network growth.

Finally, the assumption that structural characteristics of users' ego-network might have an independent effect on the dynamics of reduction or extension of personal network found partial support. The number of communities detected in the users' networks is positively related to the number of friends users add to their networks in the later period. And on the opposite – those who have more communities in the network are less inclined to remove friends from the network. In addition, the users with denser networks would be more inclined to add new friends later. We could speculate that the communities which might to some extent represent different social contexts [30] could be the possible sources of new social connections for users. This would also be in line with the classical formula of the "strength of weak ties", the access to which could be gained from the diversity of different groups an individual belongs to [31]. But the limitations of our method do not allow interpreting the results in the causation manner. In other words, from our analysis it is impossible to conclude which ties added in the later period had its origin in the clusters presented before. This question might be considered for further investigation.

Limitations of the Study
The presented study is exploratory by design and the obtained result should be treated through the prism of several limitations. In particular, the reliability and validity of the river-sampling technique is a debatable question in social science [32]. Thus, the results should be generalized with caution to the online population of Russian OSN users. In addition, the study considers the users of one particular social networking site Vkontakte and users of one Russian city, Vologda.

The shift in the network was assessed only in the scope of one-year period. Further research needs a more comprehensive way of assessing the substitution of friends in a social network. The presented procedure did not distinguish between the situations when participants did not establish new social ties or invited and removed the equal number of friends during the observed period.

5 Conclusion

The presented work contributes to understanding the role of psychological and structural factors in the processes of growth and decrease of personal ego-networks.

We found that considering such diverse processes as adding and removing friends on a social network separately is more productive than combining people with different outcomes in one model. Albeit such approach revealed that the same variables are significant in both models (a priori number of friends, the number of communities detected in the network, and the propensity to make social connection), their influence is not homogeneous.

Acknowledgments. This article is an output of a research project implemented as part of the Basic Research Program at the National Research University Higher School of Economics (HSE University).

Appendix

Table 3. Nested regressions for whole sample

	log.d_friends			log.d_friends			log.d_friends			log.d_friends			log.d_friends		
Predictors	*Estimates*	*std. Beta*	*p*	*Estimates*	*std. Beta*	*p*	*Estimates*	*std. Beta*	*p*	*Estimates*	*std. Beta*	*p*	*Estimates*	*std. Beta*	*p*
(Intercept)	5.62	-0.53	0.01	5.59	-0.55	0.01	5.63	-0.52	0.01	5.64	-0.52	0.01	5.88	-0.54	0.01
age	-0.00	-0.05	0.49	-0.00	-0.05	0.52	-0.00	-0.03	0.73	-0.00	-0.03	0.72	-0.00	-0.05	0.51
sex [male]	0.03	0.08	0.51	0.02	0.06	0.58	0.02	0.05	0.65	0.02	0.05	0.67	0.00	0.01	0.96
edu [1]	0.12	0.32	0.22	0.12	0.34	0.20	0.12	0.32	0.23	0.12	0.32	0.23	0.12	0.33	0.20
edu [2]	0.12	0.33	0.21	0.13	0.34	0.19	0.12	0.31	0.23	0.12	0.31	0.23	0.15	0.41	0.11
edu [3]	0.16	0.44	0.10	0.17	0.47	0.08	0.15	0.41	0.13	0.15	0.42	0.13	0.18	0.49	0.07
edu [4]	0.15	0.39	0.14	0.15	0.41	0.12	0.14	0.39	0.14	0.14	0.39	0.14	0.18	0.49	0.07
edu [5]	0.07	0.18	0.51	0.07	0.19	0.50	0.05	0.14	0.61	0.05	0.14	0.61	0.09	0.23	0.41
occup [1]	0.02	0.06	0.80	0.03	0.08	0.77	0.03	0.08	0.76	0.03	0.09	0.74	0.03	0.08	0.77
occup [2]	0.10	0.28	0.20	0.11	0.29	0.19	0.10	0.28	0.20	0.11	0.29	0.20	0.09	0.24	0.28
occup [3]	-0.04	-0.12	0.70	-0.04	-0.11	0.73	-0.04	-0.10	0.75	-0.04	-0.10	0.76	-0.07	-0.19	0.55
occup [4]	0.09	0.24	0.38	0.09	0.25	0.37	0.10	0.26	0.35	0.10	0.27	0.34	0.08	0.22	0.44
occup [5]	0.18	0.49	0.12	0.18	0.49	0.12	0.18	0.49	0.12	0.18	0.49	0.12	0.15	0.40	0.20
occup [6]	0.08	0.22	0.32	0.08	0.22	0.31	0.09	0.23	0.29	0.09	0.24	0.28	0.07	0.19	0.38
Prof	0.02	0.07	0.23	0.01	0.05	0.34	0.02	0.06	0.29	0.02	0.06	0.29	0.01	0.03	0.55
SE_pos	0.03	0.11	0.08	0.03	0.09	0.16	0.03	0.09	0.14	0.03	0.09	0.15	0.03	0.10	0.12
SE_neg	0.03	0.11	0.07	0.03	0.11	0.07	0.03	0.11	0.08	0.03	0.11	0.08	0.02	0.08	0.16
freq	-0.00	-0.01	0.91	-0.00	-0.01	0.89	-0.01	-0.03	0.68	-0.01	-0.03	0.69	-0.01	-0.03	0.64
online	0.01	0.05	0.39	0.02	0.06	0.33	0.02	0.06	0.32	0.02	0.06	0.33	0.02	0.09	0.18
PCO				0.02	0.06	0.36	0.03	0.08	0.21	0.03	0.08	0.21	0.03	0.08	0.25
RM							-0.03	-0.08	0.22	-0.03	-0.08	0.21	-0.03	-0.08	0.22
log.friends										-0.00	-0.01	0.86	-0.03	-0.08	0.41
communities													0.00	0.14	0.04
density													0.20	0.04	0.65
modularity													-0.34	-0.14	0.05
Observations	353			353			353			353			353		
R^2 / R^2 adjusted	0.062 / 0.011			0.064 / 0.011			0.069 / 0.013			0.069 / 0.010			0.104 / 0.038		

Table 4. Nested regressions for expansion group

	log.d_friends			log.d_friends			log.d_friends			log.d_friends		
Predictors	*Estimates*	*std. Beta*	*p*	*Estimates*	*std. Beta*	*p*	*Estimates*	*std. Beta*	*p*	*Estimates*	*std. Beta*	*p*
(Intercept)	2.62	0.23	0.31	1.79	0.09	0.69	-0.56	0.29	0.16	-0.57	0.28	0.14
age	-0.01	-0.08	0.34	-0.01	-0.07	0.36	-0.00	-0.04	0.58	-0.01	-0.07	0.30
sex [male]	-0.19	-0.14	0.27	-0.26	-0.20	0.11	-0.10	-0.07	0.52	-0.11	-0.09	0.43
edu [1]	-0.13	-0.10	0.75	-0.09	-0.07	0.83	-0.21	-0.16	0.57	-0.16	-0.12	0.64
edu [2]	-0.56	-0.42	0.15	-0.45	-0.34	0.23	-0.48	-0.37	0.15	-0.37	-0.28	0.25
edu [3]	-0.60	-0.46	0.13	-0.37	-0.28	0.34	-0.49	-0.37	0.16	-0.43	-0.32	0.20
edu [4]	-0.60	-0.45	0.14	-0.42	-0.32	0.28	-0.56	-0.42	0.11	-0.48	-0.36	0.15
edu [5]	-0.48	-0.37	0.26	-0.43	-0.33	0.30	-0.61	-0.46	0.11	-0.46	-0.35	0.19
occup [1]	0.12	0.09	0.76	0.26	0.20	0.50	-0.05	-0.04	0.88	-0.02	-0.01	0.95
occup [2]	0.37	0.28	0.28	0.51	0.39	0.13	0.13	0.10	0.66	0.02	0.01	0.96
occup [3]	0.07	0.05	0.89	0.28	0.21	0.54	0.07	0.05	0.86	-0.24	-0.18	0.55
occup [4]	0.40	0.30	0.35	0.55	0.42	0.18	0.16	0.12	0.67	0.12	0.09	0.72
occup [5]	0.46	0.35	0.31	0.45	0.34	0.31	0.43	0.33	0.28	0.33	0.25	0.39
occup [6]	0.17	0.13	0.62	0.27	0.20	0.43	0.09	0.07	0.76	0.09	0.07	0.77
Prof	0.16	0.19	0.00	0.09	0.10	0.10	0.06	0.07	0.20	0.02	0.03	0.59
SE_pos	0.11	0.11	0.10	0.02	0.02	0.75	0.04	0.03	0.56	0.04	0.04	0.49
SE_neg	0.01	0.01	0.89	0.02	0.02	0.80	0.01	0.01	0.90	-0.02	-0.02	0.76
freq	0.06	0.06	0.36	0.04	0.05	0.50	0.05	0.05	0.45	0.04	0.05	0.43
online	-0.15	-0.17	0.02	-0.11	-0.13	0.06	-0.06	-0.07	0.25	-0.07	-0.08	0.15
PCO				0.44	0.32	0.00	0.25	0.18	0.01	0.23	0.16	0.01
RM				0.03	0.03	0.66	0.06	0.05	0.38	0.05	0.04	0.43
log.friends							0.54	0.43	0.00	0.38	0.30	0.00
communities										0.02	0.38	0.00
density										3.59	0.17	0.02
modularity										0.57	0.06	0.32
Observations	274			274			274			274		
R^2 / R^2 adjusted	0.124 / 0.063			0.205 / 0.142			0.350 / 0.296			0.442 / 0.388		

Table 5. Nested regressions for reduction group

	log.d_friends			log.d_friends			log.d_friends			log.d_friends		
Predictors	*Estimates*	*std. Beta*	*p*	*Estimates*	*std. Beta*	*p*	*Estimates*	*std. Beta*	*p*	*Estimates*	*std. Beta*	*p*
(Intercept)	3.66	-0.04	0.93	3.17	0.07	0.86	0.25	0.48	0.20	-1.08	0.54	0.13
age	-0.03	-0.36	0.06	-0.03	-0.27	0.18	-0.01	-0.08	0.66	0.00	0.04	0.80
sex [male]	-0.36	-0.30	0.25	-0.50	-0.42	0.11	-0.56	-0.46	0.05	-0.40	-0.33	0.14
edu [1]	-0.02	-0.02	0.96	0.02	0.02	0.97	0.03	0.02	0.96	0.06	0.05	0.89
edu [2]	0.61	0.51	0.35	0.54	0.45	0.40	0.43	0.36	0.45	0.46	0.38	0.40
edu [3]	0.15	0.12	0.82	0.31	0.26	0.64	0.43	0.36	0.47	0.16	0.14	0.77
edu [4]	-0.21	-0.17	0.73	-0.15	-0.12	0.81	-0.16	-0.13	0.77	-0.32	-0.26	0.53
edu [5]	-0.09	-0.07	0.90	-0.20	-0.17	0.76	-0.25	-0.21	0.67	-0.14	-0.12	0.80
occup [1]	0.68	0.56	0.26	0.59	0.49	0.32	0.03	0.02	0.96	-0.27	-0.22	0.62
occup [2]	-0.40	-0.33	0.43	-0.42	-0.35	0.40	-0.79	-0.65	0.09	-0.90	-0.74	0.04
occup [3]	1.17	0.97	0.16	0.87	0.72	0.29	0.10	0.09	0.89	-0.24	-0.20	0.74
occup [4]	1.14	0.94	0.11	0.85	0.70	0.23	-0.38	-0.32	0.58	-0.45	-0.37	0.50
occup [5]	-0.31	-0.26	0.74	-0.20	-0.17	0.83	-0.98	-0.81	0.26	-1.39	-1.15	0.10
occup [6]	-0.23	-0.19	0.62	-0.32	-0.26	0.49	-0.88	-0.73	0.04	-0.86	-0.71	0.04
Prof	-0.04	-0.05	0.68	-0.04	-0.05	0.70	-0.05	-0.06	0.60	0.00	0.00	0.96
SE_pos	0.03	0.03	0.80	-0.05	-0.06	0.69	0.06	0.06	0.63	0.02	0.02	0.86
SE_neg	-0.21	-0.26	0.07	-0.15	-0.18	0.20	0.09	0.11	0.46	0.08	0.10	0.47
freq	0.06	0.06	0.66	0.04	0.04	0.74	0.03	0.03	0.79	-0.01	-0.01	0.95
online	-0.16	-0.18	0.15	-0.13	-0.15	0.22	-0.15	-0.17	0.13	-0.24	-0.27	0.02
PCO				0.40	0.32	0.03	0.25	0.20	0.13	0.26	0.21	0.09
RM				-0.11	-0.10	0.45	-0.09	-0.09	0.47	-0.14	-0.13	0.25
log.friends							0.53	0.51	0.00	0.76	0.74	0.00
communities										-0.01	-0.38	0.01
modularity										0.93	0.13	0.18
Observations	79			79			79			79		
R^2 / R^2 adjusted	0.354 / 0.160			0.407 / 0.202			0.539 / 0.369			0.603 / 0.437		

References

1. Boyd, D.M., Ellison, N.B.: Social network sites: definition, history, and scholarship. J. Comput.-Mediat. Commun. **13**, 210–230 (2007). https://doi.org/10.1111/j.1083-6101.2007.00393.x
2. Donath, J., Boyd, D.: Public displays of connection. BT Technol. J. **22**, 71–82 (2004)

3. Jin, S.V., Muqaddam, A., Ryu, E.: Instafamous and social media influencer marketing. MIP **37**, 567–579 (2019). https://doi.org/10.1108/MIP-09-2018-0375
4. Weiqin, E.L., Campbell, M., Kimpton, M., Wozencroft, K., Orel, A.: Social capital on Facebook: the impact of personality and online communication behaviors. J. Educ. Comput. Res. **54**, 747–786 (2016). https://doi.org/10.1177/0735633116631886
5. Barabási, A.-L., Albert, R.: Emergence of scaling in random networks. Science **286**, 509–512 (1999)
6. Hu, H., Wang, X.: How people make friends in social networking sites—a microscopic perspective. Phys. A: Stat. Mech. Appl. **391**, 1877–1886 (2012). https://doi.org/10.1016/j.physa.2011.10.020
7. Lampe, C., Ellison, N., Steinfield, C.: A face (book) in the crowd: social searching vs. social browsing. In: Proceedings of the 2006 20th Anniversary Conference on Computer Supported Cooperative Work, pp. 167–170 (2006)
8. Mayer, A., Puller, S.L.: The old boy (and girl) network: social network formation on university campuses. J. Public Econ. **92**, 329–347 (2008). https://doi.org/10.1016/j.jpubeco.2007.09.001
9. Subrahmanyam, K., Reich, S.M., Waechter, N., Espinoza, G.: Online and offline social networks: use of social networking sites by emerging adults. J. Appl. Dev. Psychol. **29**, 420–433 (2008). https://doi.org/10.1016/j.appdev.2008.07.003
10. Rivera, M.T., Soderstrom, S.B., Uzzi, B.: Dynamics of dyads in social networks: assortative, relational, and proximity mechanisms. Ann. Rev. Sociol. **36**, 91–115 (2010). https://doi.org/10.1146/annurev.soc.34.040507.134743
11. Aiello, L.M., Barrat, A., Schifanella, R., Cattuto, C., Markines, B., Menczer, F.: Friendship prediction and homophily in social media. ACM Trans. Web **6**, 1–33 (2012). https://doi.org/10.1145/2180861.2180866
12. Rykov, Y., Koltsova, O., Sinyavskaya, Y.: Effects of user behaviors on accumulation of social capital in an online social network. PLoS ONE **15**, e0231837 (2020). https://doi.org/10.1371/journal.pone.0231837
13. Lampe, C.A.C., Ellison, N., Steinfield, C.: A familiar face(book): profile elements as signals in an online social network. In: Proceedings of the SIGCHI Conference on Human Factors in Computing Systems - CHI 2007, pp. 435–444. ACM Press, San Jose (2007). https://doi.org/10.1145/1240624.1240695
14. Ellison, N.B., Steinfield, C., Lampe, C.: Connection strategies: social capital implications of Facebook-enabled communication practices. New Media Soc. **13**, 873–892 (2011). https://doi.org/10.1177/1461444810385389
15. Dunbar, R.I.M.: The anatomy of friendship. Trends Cogn. Sci. **22**, 32–51 (2018). https://doi.org/10.1016/j.tics.2017.10.004
16. Pollet, T.V., Roberts, S.G.B., Dunbar, R.I.M.: Use of social network sites and instant messaging does not lead to increased offline social network size, or to emotionally closer relationships with offline network members. Cyberpsychol. Behav. Soc. Netw. **14**, 253–258 (2011). https://doi.org/10.1089/cyber.2010.0161
17. Gonçalves, B., Perra, N., Vespignani, A.: Modeling users' activity on Twitter networks: validation of Dunbar's number. PLoS ONE **6**, e22656 (2011). https://doi.org/10.1371/journal.pone.0022656
18. Kalish, Y., Robins, G.: Psychological predispositions and network structure: the relationship between individual predispositions, structural holes and network closure. Soc. Netw. **28**, 56–84 (2006). https://doi.org/10.1016/j.socnet.2005.04.004
19. Totterdell, P., Holman, D., Hukin, A.: Social networkers: measuring and examining individual differences in propensity to connect with others. Soc. Netw. **30**, 283–296 (2008). https://doi.org/10.1016/j.socnet.2008.04.003

20. Ellison, N.B., Vitak, J., Gray, R., Lampe, C.: Cultivating social resources on social network sites: Facebook relationship maintenance behaviors and their role in social capital processes. J. Comput.-Mediat. Commun. **19**, 855–870 (2014). https://doi.org/10.1111/jcc4.12078
21. Vanden Abeele, M.M.P., Antheunis, M.L., Pollmann, M.M.H., Schouten, A.P., Liebrecht, C.C., van der Wijst, P.J., van Amelsvoort, M.A.A., Bartels, J., Krahmer, E.J., Maes, F.A.: Does Facebook use predict college students' social capital? A replication of Ellison, Steinfield, and Lampe's (2007) study using the original and more recent measures of Facebook use and social capital. Commun. Stud. **69**, 272–282 (2018). https://doi.org/10.1080/10510974.2018.1464937
22. Ellison, N.B., Vitak, J., Steinfield, C., Gray, R., Lampe, C.: Negotiating privacy concerns and social capital needs in a social media environment. In: Trepte, S., Reinecke, L. (eds.) Privacy Online, pp. 19–32. Springer, Heidelberg (2011). https://doi.org/10.1007/978-3-642-21521-6_3
23. Thelwall, M.: Social networks, gender, and friending: an analysis of MySpace member profiles. J. Am. Soc. Inf. Sci. Technol. **59**, 1321–1330 (2008). https://doi.org/10.1002/asi.20835
24. Sinyavskaya, Y., Porshnev, A.: Propensity to make social connections and structural social capital of SNS users. Ann. Rev. Cybertherapy Telemed. **2019**(17), 33–38 (2019)
25. Lehdonvirta, V., Oksanen, A., Räsänen, P., Blank, G.: Social media, web, and panel surveys: using non-probability samples in social and policy research. Policy Internet. https://doi.org/10.1002/poi3.238
26. Iannelli, L., Giglietto, F., Rossi, L., Zurovac, E.: Facebook digital traces for survey research: assessing the efficiency and effectiveness of a Facebook ad–based procedure for recruiting online survey respondents in niche and difficult-to-reach populations. Soc. Sci. Comput. Rev. **38**, 462–476 (2020). https://doi.org/10.1177/0894439318816638
27. Kapp, J.M., Peters, C., Oliver, D.P.: Research recruitment using Facebook advertising: big potential, big challenges. J. Cancer Educ. **28**, 134–137 (2013). https://doi.org/10.1007/s13187-012-0443-z
28. King, D.B., O'Rourke, N., DeLongis, A.: Social media recruitment and online data collection: a beginner's guide and best practices for accessing low-prevalence and hard-to-reach populations. Can. Psychol./Psychologie canadienne. **55**, 240–249 (2014). https://doi.org/10.1037/a0038087
29. Pötzschke, S., Braun, M.: Migrant sampling using Facebook advertisements: a case study of polish migrants in four European countries. Soc. Sci. Comput. Rev. **35**, 633–653 (2017). https://doi.org/10.1177/0894439316666262
30. Mcauley, J., Leskovec, J.: Discovering social circles in ego networks. ACM Trans. Knowl. Discov. Data **8**, 1–28 (2014). https://doi.org/10.1145/2556612
31. Granovetter, M.S.: The strength of weak ties. Am. J. Sociol. **78**, 1360–1380 (1973)
32. Devyatko, I.F.: Onlayn issledovaniya i metodologiya sotsial'nykh nauk: novye gorizonty, novye (i ne stol' novye) trudnosti [Online studies and the methodology of social sciences: new horizons, new (and not so new) difficulties]. Onlayn issledovaniya v Rossii **2**, 17–30 (2010)

Co-playing with Friends and Social Capital: A Comparative Analysis of Two Dota 2 Communities

Daria Mikhalchuk(✉) and Anna Shirokanova(✉)

National Research University Higher School of Economics, St. Petersburg, Russia
dsmikhalchuk@edu.hse.ru, a.shirokanova@hse.ru

Abstract. Co-playing, or playing video games together, is a social practice that enriches relationships and game experience by providing the players with informational and social support. This study explores how co-playing integrates into friendship in two small (6–7 people), male communities of adolescent and adult friends. Both communities are local and school-based; both focus on co-playing Dota 2. The study focuses on the leadership in these small networks, compares their co-playing patterns, and the ways in which co-playing affects the relationships in both communities, enhancing their bonding social capital. We apply network analysis and personal interviews to compare and contrast how the co-playing communities emerged, are maintained, and evolve along with the friendship. The main conclusion is that such co-playing communities emerge around a single Dota 2 enthusiast in the early secondary school as a common pastime, but co-playing video games increases bonding social capital among the community members. Network analysis demonstrates the differences in leadership in the teen and adult communities. The research shows how video games are embedded in the collective everyday friendship and how co-playing communities function in support of such a relationship. The findings could be further tested against female and mixed co-playing communities.

Keywords: Game studies · Friendship networks · Co-playing · Dota 2 · Social capital · Local communities

1 Introduction

Playing online video games like Dota 2 has become a part of everyday social life for many people – a virtual space where they communicate with friends and strangers. Users inside many online games perform actions typical for everyday real life, and online games can be considered as 'models of the real world' [1]. Moreover, online games can be examined as 'third places' because they have become as important as home and work, i.e., the first and second places. Participation in such virtual 'third places' can provide a variety of social contacts, thus, generating bridging social capital [2].

Online games are often treated as a new environment for individuals that mediates their social relationships [3, 4]. It can be used, however, in traditional offline friendships

A. Antonyuk and N. Basov (Eds.): NetGloW 2020, LNNS 181, pp. 209–221, 2021.
https://doi.org/10.1007/978-3-030-64877-0_14

as part of making those friendships, and become an anchoring activity for a group of friends. The practice of playing video games together with family or friends is described in the literature as ‘collaborative play’ [5], ‘social video game play’ [6], ‘co-use of video games’ [7], ‘group gaming’ [8], or just ‘co-playing’ [9–11].

Co-playing is a process when an individual plays a game with at least one other person in an environment where the players can interact with each other. It refers to simultaneous and active participation of ‘peer’ players [12]. Since massively multiplayer online games (MMOGs) have become everyday routine in many families, many family members play together, which leads to higher family satisfaction and family closeness [6–8, 10]. The same pattern appears in relations between friends. Such peer-to-peer communities not only co-play but also share attitudes, ideology, values, and common goals. Gaming-based shared identities affect offline social relationships and seem to motivate friendship formation among teenagers [13].

We focus on offline game-related ties between video game co-players for two reasons. First, even though online friendship can be of high quality [14] and online game interaction develops different feelings and less social anxiety as compared to the real life [15], offline friendships are more efficient at building the bonding social capital [16]. Offline friendship within the co-playing community typically involves more interdependence, depth, understanding, and commitment than online friendships [14]. Second, offline relationships among players are associated with positive social outcomes such as social support, trust, reciprocity between players, and benefits for psychological well-being [6, 17–19].

In this paper, we explore the effects of co-playing from the perspective of such potential benefits as emotional support, trust, help, and access to scarce information. We take Dota 2, the second most popular (after the ‘League of Legends’) Multiplayer Online Battle Arena (MOBA) game with more than 11 million unique players per month (as of 2019) where interactions between players are necessary and are very broad [20]. Previous research on building social capital during co-play already exposed how parent-child co-playing develops the bonds between girls with their parents [9], how fathers play a major role in initiating video game co-playing with sons [7, 8]. Peer communication gains in importance during the teen years, when offline friendships can develop into co-playing communities. Less is known about the ways in which members of such communities increase their social capital through gaming. This study provides an in-depth look into one adult and one teen community, tracing both individual development of social capital and the evolution of playing communities.

This study demonstrates how co-playing maintains larger offline friends’ communities during high school and early adulthood years. Co-playing at the age of 16 brings primarily social and emotional support, while young adults engaged with work and families co-play less frequently for relaxing, as a habit. The teen co-playing group is a tight network around the leader, while for young adults the leading role is distributed and varies according to activity. We have also learned how playing together helps maintain and develop offline social relationships. Notably, even frequent co-playing (5 times a week) in the teens’ group does not cut off other social connections. In this respect, playing together video games like Dota 2 reinforces the feelings of community among those who are already friends at school. After the end of school, playing together can become an anchoring activity, keeping the community together for years.

The general research question for this paper is how co-playing Dota 2 in a group of friends is related to the bonding social capital of players. We go through the network structure of the group, the members' motivation for play, their co-playing patterns, and their reflection on the gains of co-playing with friends. First, we explore the communities' network structures. Second, we describe and compare the co-playing patterns of two communities. Then, we investigate how co-playing Dota 2 leads to maintaining (or ruining) the existing social capital of community members. Lastly, we reveal which resources gained from this community are valuable to individuals.

2 Data and Methods

All informants are males who were born and who live in the same area of a big city in Russia. The two communities are not connected. Both were selected for the study based on convenience sampling, as they are groups of friends who play Dota 2 when they meet. The members of both communities appear as typical dwellers of a big city in Russia.

There are six members in the teen group. All informants are 15–16 years old; they come from middle-class families and get about $200 of pocket money per month. They all attend the same secondary school and were studying at the moment of the interviews in the last, ninth, year of compulsory school.

The adult community consists of seven male friends, 24–25 years old. All of them attended the same secondary school, some of them were classmates and then enrolled in the same university. All the adults have full-time jobs earning about $920/month. They are also middle-class peers in socio-economic terms.

The two communities differ by the experience and weekly time spent playing Dota 2: 4 years of experience and 8 h/week among the teens, and 11–13 years of experience with 2 h/week among the adults. Informants from both communities have friends who do not play Dota 2.

Co-playing patterns are defined as typical behavior during game practices, which individuals follow while playing MMOGs. It includes such characteristics as relationship to game partners, intensity, and the frequency of game play [21]. In this paper, co-playing patterns are operationalized as a combination of five indicators:

- the average playing session length,
- frequency of game play,
- context of game play,
- relationship type between co-players, and
- their motivations to play.

Social capital is a set of resources obtained and accumulated through social connections [22, 23]. Putnam's [24] distinction between the bridging and bonding social capital can be used for measuring the social effects of video games. Social capital here includes the links between people but also the resources which they gain from these relationships [22], with a focus on help and emotional support. The interviews focused on the following types of resources:

- emotional support (listening, understanding, explaining, giving advice, alleviating the suffering or problem),
- material help (lending money, help with relocation, etc.),
- information and spending time together,
- emotions (joint activities, traditions, celebrating holidays), and
- trust between the members of one community.

2.1 Methods

The first stage of the study was the 'immersion' in the field, i.e. in the communities of Dota 2 players. To understand the behavior of the respondents, we had to spend time playing Dota 2 on our own and together with the members of two groups separately. Participant observation enabled better understanding of the game play, some co-playing patterns, and the ways the members of the communities interact with each other.

In the second stage, we conducted personal interviews. The interviews with adults lasted longer than the interviews with teens because the former were more open to tell about their experience and have a longer history of community. The interviews with teens lasted from 20 to 25 min, whereas interviews with adults lasted from 35 to 40 min. All the interviews were conducted in Russian, the native language of players. Parents' consent was obtained before conducting the interviews with adolescents. All the information was anonymized. The interviews with adults were held at the home of one of the respondents. The interviews with teens were conducted either at their school (3 interviews) or at the residence of one of them (3 interviews).

2.2 Strategy of Analysis

The interview guide consisted of several groups of questions. We asked the respondents about their experience in playing Dota 2 (session length, place, motivations, relationship to partners, etc.), the ways in which they supported their playing community and friendship within it, and how their community was structured. In addition, we collected some socio-demographic information about the informants.

All the interviews finished successfully. The adults were enjoying the process and were willing to give answers. The teens, however, behaved somewhat shier and sometimes were initially hesitant to share personal information but changed their mind as they got used to the interviewer and felt more comfortable. The most difficult question within the whole interview asked about the skills they obtained or lost while playing Dota 2; the question itself was not always clear to informants, sometimes they did not know how to formulate an answer.

The analysis started with the records which were taken by hand during the participant observation ethnography. These records were useful as they contained information about the behavior of informants during the game play and observations on co-playing patterns. After that, the interviews were coded thematically around the co-playing patterns, social capital, and the structure of communities. Then, the network information about friendship ties was analyzed, which was collected after the interview. The informants wrote on a piece of paper the names of the members of their community with whom they had strong friendship ties and who first started the conversation or kept the company. Further prompts were given on the roles of each member.

3 Results

3.1 Leadership in Communities

Figure 1 shows who initiated friendships in each community. In the teens' group, A was the newcomer in the first year of secondary school, the first person who started playing Dota 2 and engaged others. This informant has also the best gaming equipment, making his parents' flat the place where the community plays most frequently. Remarkably, hosting the playing sessions is not what defines A as a group leader for the others. Informant A reported training at the gym of a European champion and getting the highest amount of pocket money. All this makes A the person with the highest status in the group and an example for the others.

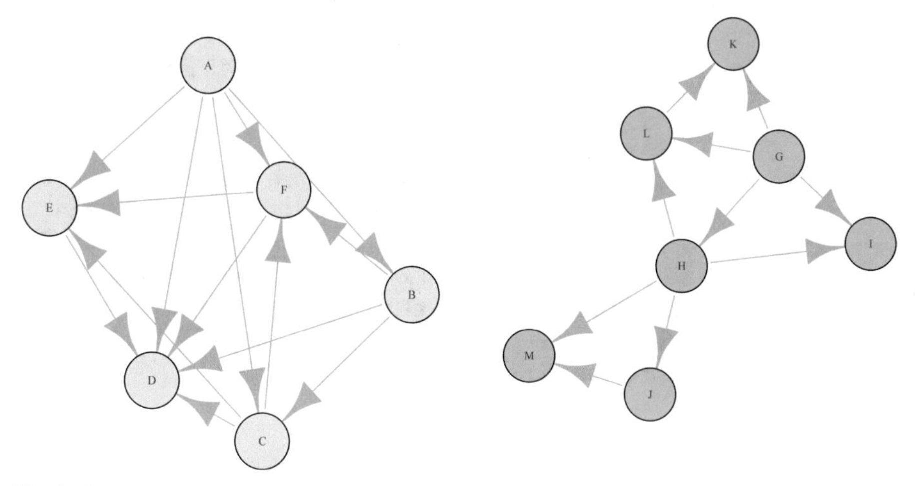

Fig. 1. Network of friendship initiative. **(Left)** Teens: A is the group leader, the play initiator, and host; **(right)** Adults: G initiated the play, H is the play host and former group leader.

Among the adults, G was the first one to play Dota 2 in the school class where respondents H, I, K, and L were studying. After that H, who was friends with M and J from a parallel class, contaminated them. All of them were close friends 8–10 years ago. H used to be the leader of the group as he lived closest to the school and organized joint Dota 2 sessions.

Figure 2 complements the picture, showing the current strength of ties in the communities. The teen network is densely connected. The adult network is sparser because not all the members are good friends now. M and J have had a quarrel with G, I, K, and L, and now communicate rarely. Informant H can be considered a network broker [25] who maintains friendship with all the members of community and hosts the co-playing sessions. The most interesting fact is that neither G nor H are considered the leaders of the adults' community because, the informants said, each of them can initiate some activity (e.g., G can invite everyone to the cinema, while H can call everyone for a quest).

The teens community has a leader – the one who first started playing Dota 2 and has a comfortable place for co-playing. In the adults' community, the leader from ten

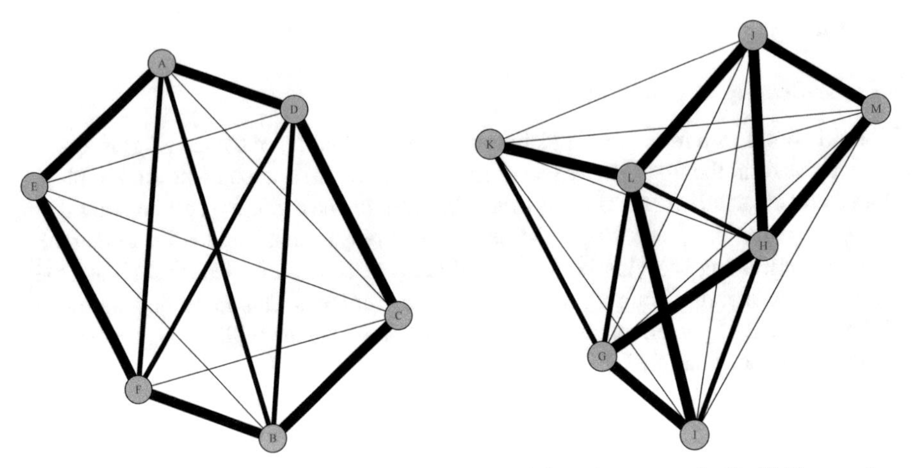

Fig. 2. Network of current friendship strength (the thicker, the stronger). **(Left)** Teens: close friends; **(right)** Adults: H is the peacekeeper, not everyone is a close friend anymore.

years ago was also the one who hosted (but not initiated) the game and contaminated others. However, at present, the informants claim that they do not have a stable group leader as this role is shifting from one person to another depending on the activity. This happens because the adult community has a longer group history than the teens, and the leadership pattern in their group today is rather distributed, as opposed to the past. Thus, leadership is strong in the adolescent group and it used to be similar in the adult group, but it became more distributed with time.

3.2 Co-playing Patterns

The teens play with their real-life friends more often than with randomly assigned teams, whereas the adults report a different story. When the adults were studying at school, they played together 5–6 times a week, while currently they meet to play 3–4 times a month, or less than once a week. They explain this decrease by having grown up and getting more duties connected with love relationships, families, work, and everyday chores such as shopping or renovating the home. Despite this, they always find time to play alone (5–6 times/week in the evening or at night). The teens play together much more often. Both the teens and the adults have hobbies, and Dota 2 is on par with those activities: 'Well, I come home after school, eat fast and we talk with [other group members] and, if everyone is ready, then we play online… Well, on the weekends we go out and, after that, play together at someone's place' (teen, 15 y.o.).

Motivations for play are diverging between the communities. The teens treat playing Dota 2 as a competition with each other (about who will gain more gold or victories, who will be the best hero, or who will level up faster). They also really enjoy the process, but this emotion differs from the adults' attitude. For the adults, Dota 2 is more of a habit that comes from their childhood. They treat the game as an opportunity to relax, to 'switch off' their negative thoughts, and forget about the problems:

> I play for fun; I really like the process of gaming… So, I really relax when playing. While my girlfriend lies in the bathtub resting, I have the same with Dota 2. (adult, 25 y.o.)

> After a hard day at work and traffic jams on the way back home, it is a pleasure to play Dota, to unload the brain. I just need to let emotions and anger off somewhere. (adult, 24 y.o.)

> It's cool to play with friends, it unites us, and so we play together more often. For me, it is like going to the cinema, or watching a TV series. By the way, it is even more interesting to watch streams sometimes. I have given up on the movies. (teen, 15 y.o.)

All in all, the average playing session length, frequency, and context of game play are different between the observed communities due to their 'pace of living' and lifestyle typical of their age. As for the session length, the adults' average session lasts 1–1.5 h. per two or three days (i.e., 20–30 min./day) while for the teens it is about 3–4 h. per day. As for the relationship type between co-players, all of them are friends and they made offline friendship before joining the game.

Both groups have some traditions or shared activities outside playing Dota 2, which help them to maintain the community. Those are, among others: going to the gym, to the sauna, fishing, playing football or tennis, going to the cinema, and eating out. The adults spend all the holidays together, from winter holidays (10–14 days in Russia) to summer vacations when they travel together. The teens celebrate birthdays and some holidays together, but spend more time with their families. Playing together helps the group members to find shared interests. The adults tell that, at the beginning, playing together was an opportunity for meeting and having some conversation. For the teens, Dota 2 is a way of spending leisure time, it is their way of entertainment.

3.3 Social Capital

All the informants value their community because of the existing trust and support between the members. The adults say that key features of their community are reliability, loyalty, and shared interests and topics. The teens value the sense of humor, understanding, and closeness in friendships, the opportunity to ask for help or advice, or to tell a secret. Both communities report 'always helping each other': the adults speak about emotional support on par with material help (lending money/a car, helping with a car/repairs, etc.). The teens do not speak about emotional support, but they say they can always cover up their friends, help with homework, and stand up for them. So, friendship between the adults is more versatile and deliberate as they grew up together and shaped their future hand in hand: 'I have grown up with them, experienced the most difficult moments, they are like brothers to me' (adult, 25 y.o.).

The bonding social capital of the teen community is developing in the following way: all of them have been classmates since age seven. They formed a group of three at first. This small group started to play 'FIFA' and 'Witchcraft'. Then, a new member joined who was the first one to play Dota 2. This player spread the game to the others who joined the group. Nowadays, the teens go out together and claim they do not invite anybody else to their co-playing sessions:

> We spend almost every day together. We have become friends: we come to [A], play, eat, sometimes do homework together… Of course, it is more interesting to spend time with them than with adults because my friends understand me. (teen, 16 y.o.)
>
> It all started with the games; at first, three of us played the World of Tanks together. Then [A] got us hooked on Dota, and the others joined, it happened by itself. (teen, 15 y.o.)

In the adults' community, the bonding social capital also started offline, in a school class, and these connections grew into co-playing. The friendship links appeared in the first year of secondary school. First, they played solo the Counter-Strike or the World of Tanks and shared emotions, but never co-played. Then their ties grew closer with the arrival of a new classmate who played Dota 2. At times, the adults used to co-play at the Internet café where they met other players and had conversations, but those 'strangers', as they call them, never became their friends. The same pattern was observed through the interviews when both the adults and teens talk about their online interactions in Dota 2. They say that sometimes they interact with other players online, but these links do not become the close ones:

> I recall that we went to Internet cafes for a change… We met there a lot of interesting people – some were single; others came in pairs. We were not particularly interested in them as we came together. So, we could talk for a minute and that's all. (adult, 25 y.o.)

Thus, co-playing Dota 2 helps the community members to maintain their existing bonding social capital (but not expand their bridging capital, at least directly). Informants from the adults' community claimed that friendship in the group brings them emotional support and material help, shared interests, goals, and activities. Moreover, they noted that their relations were trustful as those who have links in the network can tell each other the truth, get or give advice, do business together. The teens did not use the words of social support or reciprocity in the interview, but they said that they all understood each other in the best way and spent their free time together. In addition, they often helped each other with homework or protected their friends from parents' anger.

4 Discussion

All informants report having positive experience during the process of game co-playing, such as developing and maintaining friendship and increasing the sense of community. This matches the results of Cole and Griffiths [26] where the authors argue that gaming activity may be the good soil for building strong emotional friendships. This is further supported by Eklund and Roman [13] and De Grove [27] who report that online gaming motivates friendship formation. Eklund and Roman note that if teenagers identify themselves as gamers or players not at the start of schooling but later at school, then it makes starting a friendship 1.5 times more likely. This is the situation observed in the case of these two co-playing communities. The respondents started co-playing Dota 2 after the primary school, approximately at the age of 12.

Another interesting finding was that Dota 2 develops the bonding capital within communities. This result is similar to the conclusions on the relationship between the social capital and the frequency of the game play with friends [6, 28]. Meng et al. [28] found out that co-playing with offline friends was associated with bonding social capital while playing with offline and online friends first met in the game was connected with bridging social capital in the League of Legends, another MOBA game like Dota 2.

In our study, some respondents from the teen community were friends with teenagers from different countries whom they met online in Dota 2, but they did not play or talk with each other frequently. For them, offline friendship already brings trust, understanding, support, and emotional help, as the informants explained. However, Shen and Chen [29] discovered that gamers who often play with friends first met online, in an MMOG, have strong ties with people which are included in their online networks. This dissimilarity can be explained by the genre of an MMOG game where interactions between players are more social and the game universe is a constant world, unlike Dota 2 where each session generates a new combination of players on the map. Another study [30] reports that some players use the game to spend some time with their offline friends but also to make new acquaintances, so that they switch between two playing spheres (offline-online), and this skill might increase their social capital.

We can conclude that co-playing video games with offline friends is related to the increase in bonding social capital. Players are of the same gender, come from homogeneous social backgrounds, live in the same neighborhoods and attend the same school. Both communities have a game leader who would lead the group in general in school days. In the adult community, the leadership is distributed, and the group itself is less densely connected. Perhaps, co-playing with friends for them was an efficient way to keep together when becoming adults, and has transformed into more of a habit later.

The adults claim it is easy for them to give up playing (e.g., if the game starts to bring them negative feelings). By contrast, the teen community is to a greater extent spinning around the process of playing and they do not see themselves seizing to play Dota 2 in the near future as they have already had some achievements in the game. For the teens, these achievements are more meaningful than for the adults.

This community-binding function is important as groups of friends are prominent in the socialization of teenagers. Friendship provides them with emotional and individual growth, social support, trust, and social identity [31]. Additionally, several studies have shown that the quality of adolescent friendships is related to happiness and well-being [32–35]. Furthermore, trust, intimacy, and social support, which strong friendship ties provide, are related to improved health and better psychological state in later life [36, 37]. Unsurprisingly, every fifth video game player 50 years old or over plays together with friends [38]. To conclude, co-playing unites the teens as friends during their school time. Adults also claim that the game has helped them to maintain friendship, and now they continue doing a lot of other activities together.

To what extent are these results representative of the friends' co-playing groups of other gender makeups and from other countries? Both communities were selected for the study based on convenience sampling. The gender composition was not controlled at the selection. Also, both groups are male only, which could be representative of a typical Dota 2 co-playing community, but it by no means covers the variety of existing co-playing groups. Literature on co-playing says that parents more often play video games with sons than with daughters [8]. We also know that boys are described as more game-oriented in friendship, while girls as more active on social media [39]. From a historical perspective, social play is a recognized part of socialization, and single-player video games are much more of a novelty than a social video game like Dota 2. In this regard, the all-male gender makeup of both studied communities can be interpreted as rather typical for such friends' gaming communities. However, there is a discussion that cross-gender online friendship is of higher quality than offline friendships [14]. Therefore, it is possible that online gaming communities are more likely to involve mixed co-playing communities. For further exploration, we would recommend recruiting all-girls co-playing communities or a mixed gender community of co-playing friends [9, 11] for comparison.

This study's conclusions are limited by the small size of both communities and the homogeneity of their composition. Small-n studies are not generalizable to larger groups. However, maintaining a close friendship with 5–6 people can be considered neither too small nor too big for a co-playing community, as co-playing is much more likely to occur in dyads. Moreover, homogeneous social background contributes to making stronger ties between community members, both offline and online.

All the informants in this study are local residents of a big city in Russia. Whether spatial localization or town size affects social capital formation in co-playing groups or it is similar to that in towns or in the countryside is another question that goes beyond the scope of this paper. Belonging to Russian gamers as such can also be treated as a segment of players with particular features. According to the study of the EVE Online MMOG by players' behavior such as making alliances and trading, Russian players belong to the Eastern European cluster who demonstrate similar game behavior [15]. We would suggest that this study's results are relevant for this region as well.

To overcome memory faults when recalling earlier times among the adults, additional methods of data collection can be applied, e.g., a longitudinal survey. There is a difficult trade-off, however, between the length of such a study and the uncertainty of delivering additional knowledge from such research. These conclusions are also relevant for moderate Dota 2 players as the informants in this study treat the game as a hobby on par with fishing or doing sports.

5 Conclusions

All things considered, the following conclusions about the social role of Dota 2 co-playing communities of teens and young adults were obtained. First of all, members of both groups treat Dota 2 as a space for leisure time that helps them relax from work or studying and can last as a community for years. Secondly, co-playing Dota 2 helps the players to develop their bonding social capital as their joint activity, which unites them and creates opportunities for regular meetings. Moreover, members of both communities reap additional gains from co-playing, such as material help, emotional support, trust, and shared social activities.

The results bear implications and can be used by game designers, psychologists, and educators exploring the potential of online gaming as an everyday-life activity of adolescents and young adults for generating the bonding social capital within the players.

For network researchers, this study outlines the basic network dynamics in small groups of school friends.

References

1. Castronova, E., Bell, M.W., Cornell, R., Cummings, J.J., Emigh, W., Falk, M., Fatten, M., LaFourest, P., Mishler, N.: A test of the law of demand in a virtual world: exploring the Petri dish approach to social science. Int. J. Gaming Comput.-Mediat. Simul. **1**(2), 1–16 (2009). https://doi.org/10.4018/jgcms.2009040101
2. Steinkuehler, C.A., Williams, D.: Where everybody knows your (screen) name: online games as 'third places'. J. Comput.-Mediat. Commun. **11**, 885–909 (2006). https://doi.org/10.1111/j.1083-6101.2006.00300.x
3. Nardi, B., Harris, J.: Strangers and friends: collaborative play in World of Warcraft. In: Proceedings of the 2006 20th Anniversary Conference on Computer Supported Cooperative Work, pp. 149–158. Association for Computing Machinery (2006). https://doi.org/10.1145/1180875.1180898
4. Voida, A., Greenberg, S.: Wii all play: the console game as a computational meeting place. In: Proceedings of the SIGCHI Conference on Human Factors in Computing Systems, pp. 1559–1568. Association for Computing Machinery (2009). https://doi.org/10.1145/1518701.1518940
5. Schott, G., Kambouri, M.: Moving between the spectral and material plane: interactivity in social play with computer games. Convergence **9**, 41–55 (2003). https://doi.org/10.1177/135485650300900304
6. Perry, R., Drachen, A., Kearney, A., Kriglstein, S., Nacke, L.E., Sifa, R., Wallner, G., Johnson, D.: Online-only friends, real-life friends or strangers? Differential associations with passion and social capital in video game play. Comput. Hum. Behav. **79**, 202–210 (2018). https://doi.org/10.1016/j.chb.2017.10.032
7. Marchenko, F.O., Matweeva, L.V., Makhovskaya, O.I.: Fathers' role in co-use of video games in Russia. Procedia – Soc. Behav. Sci. **233**, 455–458 (2016). https://doi.org/10.1016/j.sbspro.2016.10.184
8. Chambers, D.: Changing Media, Homes and Households: Cultures, Technologies and Meanings. Routledge, Abingdon-on-Thames (2016)
9. Coyne, S.M., Padilla-Walker, L.M., Stockdale, L., Day, R.D.: Game on… girls: associations between co-playing video games and adolescent behavioral and family outcomes. J. Adolesc. Health **49**, 160–165 (2011). https://doi.org/10.1016/j.jadohealth.2010.11.249
10. Coyne, S.M., Jensen, A.C., Smith, N.J., Erickson, D.H.: Super Mario brothers and sisters: associations between coplaying video games and sibling conflict and affection. J. Adolesc. **47**, 48–59 (2016). https://doi.org/10.1016/j.adolescence.2015.12.001
11. Lopez-Fernandez, O., Williams, A.J., Griffiths, M.D., Kuss, D.J.: Female gaming, gaming addiction, and the role of women within gaming culture: a narrative literature review. Front. Psychiatry **10**, 454 (2019). https://doi.org/10.3389/fpsyt.2019.00454
12. Lim, S., Lee, J.R.: Effects of coplaying on arousal and emotional responses in videogame play. In: International Communication Association Conference, San Francisco, CA, USA (2007)
13. Eklund, L., Roman, S.: Do adolescent gamers make friends offline? Identity and friendship formation in school. Comput. Hum. Behav. **73**, 284–289 (2017). https://doi.org/10.1016/j.chb.2017.03.035
14. Chan, D.K.-S., Cheng, G.H.-L.: A comparison of offline and online friendship qualities at different stages of relationship development. J. Soc. Pers. Relationsh. **21**, 305–320 (2004). https://doi.org/10.1177/0265407504042834

15. Belaza, A.M., Ryckebusch, J., Schoors, K., Rocha, L.E.C.: On the connection between real-world circumstances and online player behaviour: the case of EVE Online. PLoS ONE **14**, e0240196 (2020)
16. Williams, D.: On and off the'Net: scales for social capital in an online era. J. Comput.-Mediat. Commun. **11**, 593–628 (2006). https://doi.org/10.1111/j.1083-6101.2006.00029.x
17. Ratan, R.A., Chung, J.E., Shen, C., Williams, D., Poole, M.S.: Schmoozing and smiting: trust, social institutions, and communication patterns in an MMOG. J. Comput.-Mediat. Commun. **16**, 93–114 (2010). https://doi.org/10.1111/j.1083-6101.2010.01534.x
18. Trepte, S., Reinecke, L., Juechems, K.: The social side of gaming: how playing online computer games creates online and offline social support. Comput. Hum. Behav. **28**, 832–839 (2012). https://doi.org/10.1016/j.chb.2011.12.003
19. Shen, C., Chen, W.: Social capital, coplaying patterns, and health disruptions: a survey of massively multiplayer online game participants in China. Comput. Hum. Behav. **52**, 243–249 (2015). https://doi.org/10.1016/j.chb.2015.05.053
20. Mora-Cantallops, M., Sicilia, M.Á.: Exploring player experience in ranked League of Legends. Behav. Inf. Technol. **37**, 1224–1236 (2018). https://doi.org/10.1080/0144929X.2018.1492631
21. Chen, W., Shen, C., Huang, G.: In game we trust? Coplay and generalized trust in and beyond a Chinese MMOG world. Inf. Commun. Soc. **19**, 639–654 (2016). https://doi.org/10.1080/1369118X.2016.1139612
22. Coleman, J.S.: Social capital in the creation of human capital. Am. J. Sociol. **94**, S95–S120 (1988). https://doi.org/10.1086/228943
23. Lin, N., Fu, Y., Hsung, R.-M.: Measurement techniques for investigations of social capital. In: Social Capital: Theory and Research, pp. 57–81. Aldine de Gruyter, New York (2001)
24. Putnam, R.D.: Bowling Alone: The Collapse and Revival of American Community. Simon and Schuster, New York (2000)
25. Burt, R.S.: Structural holes and good ideas. Am. J. Sociol. **110**, 349–399 (2004). https://doi.org/10.1086/421787
26. Cole, H., Griffiths, M.D.: Social interactions in massively multiplayer online role-playing gamers. Cyberpsychol. Behav. **10**, 575–583 (2007). https://doi.org/10.1089/cpb.2007.9988
27. De Grove, F.: Youth, friendship, and gaming: a network perspective. Cyberpsychol. Behav. Soc. Netw. **17**, 603–608 (2014). https://doi.org/10.1089/cyber.2014.0088
28. Meng, J., Williams, D., Shen, C.: Channels matter: multimodal connectedness, types of co-players and social capital for multiplayer online battle arena gamers. Comput. Hum. Behav. **52**, 190–199 (2015). https://doi.org/10.1016/j.chb.2015.06.007
29. Shen, C., Chen, W.: Gamers' confidants: massively multiplayer online game participation and core networks in China. Soc. Netw. **40**, 207–214 (2015). https://doi.org/10.1016/j.socnet.2014.11.001
30. Domahidi, E., Festl, R., Quandt, T.: To dwell among gamers: investigating the relationship between social online game use and gaming-related friendships. Comput. Hum. Behav. **35**, 107–115 (2014). https://doi.org/10.1016/j.chb.2014.02.023
31. Bukowski, W.M., Newcomb, A.F., Hartup, W.W.: The Company they Keep: Friendships in Childhood and Adolescence. Cambridge University Press, Cambridge (1998)
32. Parker, J.G., Asher, S.R.: Friendship and friendship quality in middle childhood: links with peer group acceptance and feelings of loneliness and social dissatisfaction. Dev. Psychol. **29**, 611 (1993). https://doi.org/10.1037/0012-1649.29.4.611
33. Gauze, C., Bukowski, W.M., Aquan-Assee, J., Sippola, L.K.: Interactions between family environment and friendship and associations with self-perceived well-being during early adolescence. Child Dev. **67**, 2201–2216 (1996). https://doi.org/10.1111/j.1467-8624.1996.tb01852.x

34. Hartup, W.W., Stevens, N.: Friendships and adaptation across the life span. Curr. Dir. Psychol. Sci. **8**, 76–79 (1999). https://doi.org/10.1111/1467-8721.00018
35. Greca, A.M.L., Harrison, H.M.: Adolescent peer relations, friendships, and romantic relationships: do they predict social anxiety and depression? J. Clin. Child Adolesc. Psychol. **34**, 49–61 (2005). https://doi.org/10.1207/s15374424jccp3401_5
36. Bagwell, C.L., Newcomb, A.F., Bukowski, W.M.: Preadolescent friendship and peer rejection as predictors of adult adjustment. Child Dev. **69**, 140–153 (1998). https://doi.org/10.1111/j.1467-8624.1998.tb06139.x
37. Chow, C.M., Ruhl, H., Buhrmester, D.: The mediating role of interpersonal competence between adolescents' empathy and friendship quality: a dyadic approach. J. Adolesc. **36**, 191–200 (2013). https://doi.org/10.1016/j.adolescence.2012.10.004
38. Marston, H.R., del Carmen Miranda Duro, M.: Revisiting the twentieth century through the lens of generation X and digital games: a scoping review. Comput. Game J. **9**, 127–161 (2020). https://doi.org/10.1007/s40869-020-00099-0
39. Lorenz, T., Kapella, O.: Children's ICT use and its impact on family life (DigiGen - working paper series No. 1) (2020). https://doi.org/10.6084/m9.figshare.12587975.v1

Political and Social Movement Networks

The International Reach of the Koch Brothers Network

Patrick Doreian[1,2](✉) and Andrej Mrvar[2]

[1] University of Pittsburgh, Pittsburgh, USA
pitpat@pitt.edu
[2] University of Ljubljana, Ljubljana, Slovenia
andrej.mrvar@fdv.uni-lj.si

Abstract. We provide results of the analyses of two important and integrally linked topics. One is the extent of the international reach of the Koch Brothers. The other is their very prominent role in promoting climate change denial. Using a variety of social network tools, we display both the international reach of the Koch Brothers and their focus and motivating force driving the formation of units promoting climate change denial.

Keywords: Koch Brothers · International reach · Climate change denial

1 Motivation

We have two primary, but interlocked, goals for this paper. One is to assess and document the international reach of the Koch Brothers (KB) by examining the network of their allies which they formed to create a powerful social movement. The other is to couple this to climate change denial, the motivating force, for them in creating, funding and supporting the allies in this network.

We provide a brief introduction to earlier relevant work in Sect. 2. We needed to identify units in this network that are not in the US. We describe our procedures for doing this in Sect. 3. Section 4 provides a closer look at some of these units. The results are shown in a visual fashion as networks. Throughout, we use black circles for the non-US units and white circles for the units located in the US. In Sect. 5, we turn to identifying the active climate change deniers in the KB network of allies. A partial summary, along with an assessment of some limitations in our study and a brief statement of future work, is presented in Sect. 6. All of the analyses leading to our results were done in Pajek [1].

2 Introduction

The Koch Brothers have constructed a large network of allies to support and promote their libertarian agenda. An initial documentation of this was provided by MacLean [2] and Mayer [3]. It is important to document the true range of this network, as it is

A. Antonyuk and N. Basov (Eds.): NetGloW 2020, LNNS 181, pp. 225–235, 2021.
https://doi.org/10.1007/978-3-030-64877-0_15

more extensive that these authors claimed. In a recent paper and also an invited chapter, Doreian and Mrvar [4, 5], provided some preliminary results based on their studies of the Koch Brothers (KB) network of allies. Their methods are described in these documents, but we add some details here as well. They involved using VOSON (a web crawler that is designed to determine and record the directed links between websites). We started with a known list of KB allies as revealed by MacLean [2] and Mayer [3]. We established the list of URLs for these outfits and fed them to VOSON for its search. The result was a large directed network. Doreian and Mrvar [4, 5] documented its large size and discussed some of the implications of their results regarding the economic and political consequences of the actions of this network for the US. Their arguments need not be repeated here, even though some of them are rather dire regarding the consequences if the KB allies are successful in imposing their will on the US and the world. However, one feature of the behavior of the KBs and their allies must be mentioned: their complete secretiveness and deception.

This secretiveness is not surprising – see [6–9] for a more general discussion about corporate deception and their attempts to minimize the role of science. Put more forcefully, these attempts were made to undermine the credibility of science because sound empirically based scientific knowledge poses many threats to the bottom line of large corporations, especially when they involve environmental stewardship. This has been practiced by the KBs in an extremist fashion, as documented by Leonard [10, 11].

The studies of Doreian and Mrvar were inspired by the books of MacLean [2] and Mayer [3]. The nature (methodologies) of how they built their work on these inspirational books to obtain networks linking KB allies has been described in the above references. But while units in the KB network are secretive in the extreme, they do provide links to the URLs of other units. Our network data regarding the links between all of the KB allies rests on the use of the VOSON web crawler [12] to create links between websites. In our view, the KB secretiveness has been undone – at least, partially – by their willingness to link with other units sharing their world view.

The activities of the KB allies are becoming more well known. They include the great funding of right-wing outfits to promote their agenda. This includes limiting the role of governments, both national and local, so that they do not interfere with organizations seeking to not be regulated. This is particularly acute with regard to environmental pollution and running production facilities that are unsafe for many workers. At face value, given that most of the KB allies are located in the US, this seems a uniquely US problem. However, many of the identified KB allies exist in other countries both as sources for intellectual ideas and with production facilities of their own. It is important to establish the identities of these units as they form part of a coordinated international movement seeking to avoid governmental regulation. While the efforts of Doreian and Mrvar were focused on the American KB allies, this presentation documents the international reach of the Koch Brothers. This network reaches outside the US.

3 Identifying the Non-US Units

It was straightforward to use the many URLs for the known KB allies, as listed by MacLean and Mayer, to identify the countries where units are located. This is important given our focus on identifying the international reach of the KBs. Here, we list the countries involved, the number of these units, their nature and the roles they play in supporting the goals of the KBs. All are important in examining the global reach of the KBs.

The network that was initially identified had well over 17,000 units. But links between websites involve information streams with widely varying content. We had to 'clean' these data to focus only on those units in the network of KB allies. The result was a network of 1081 units known as KB allies. These units are varied and include organizations, think tanks, institutes, foundations and blogs. The number of directed links in this network is 5629. This is not a small network.

We identified the countries having units in the KB network that are not in the US. At face value, this number of countries is not very large. They are Australia, Austria, Belgium, Brazil, Bulgaria, Canada, Chile, China, France, Germany, India, Ireland, New Zealand, Portugal, Russia, Spain, Switzerland, the UK and Vietnam. In addition, we used 'International' as a code for two huge oil companies – Shell and ExxonMobile – operating in many countries across the globe and 'Europe' as another generic category.

All told, the total number of these units is 44 which also seems rather small. But some of these organizational units and countries where they are located have considerable influence in the world system of business interests and countries across the world.

We opted to include every non-US outfit that we identified. We provide our rationale for doing this. The crucial issue for doing this is to consider what these units do as they play two roles in the network. Some of these units are clear KB allies and must be included as they are active in promoting and supporting the libertarian aims of the KBs. Some of them are considered in more detail in Sect. 4. The other role is played by news media outlets.

While some conservative news organizations may be sympathetic to the aims of some of the KB allies, for example, limited government, promoting free markets and lowering taxes, and may support what some of the KB allies do, most likely, they are reporting on these organizations. Clearly, when doing so, they are not acting as KB allies. However, by reporting about such units, their behavior and the positions they take, these venues give attention and, therefore, greater visibility to the KB units about which they report. We have noticed, as reported below, some of these KB units send reciprocated ties to some news outlets that wrote about them. At a minimum, this suggests that they are paying close attention to what is being written about them. This implies that if they find this coverage to be useful to them, they will use it to further their aims.

4 A Closer Look at Some of the Non-US Units

The units in the KB network of allies vary in both size and importance. The units considered here were identified in two ways. One was whether they were important authorities as defined by Kleinberg [13]. Authorities are sources of critical importance for distributing ideas that are valuable for the members of the network to which they belong. Doreian and Mrvar [5] considered the primary authorities in the US. The second was to extend this to units located outside the US.

The first unit we consider is the Wirth Institute based at the University of Alberta in Canada. Its website states that its primary focus is on Austrian and Central European

Studies. It was established through the initiative of three Austrian agencies. This institute reports receiving funds from some governments in Eastern Europe including Croatia, the Czech Republic, Hungary, Poland, Slovakia and Slovenia. Given its support for positions taken by the KBs, this also helps extend their international reach. They emphasize the work of Hajek, an Austrian-British economist and philosopher who defended classical liberalism as well as the ideas promoted by the Mises Institute (based in Alabama). Figure 1 shows the ego-network of the Wirth Institute.

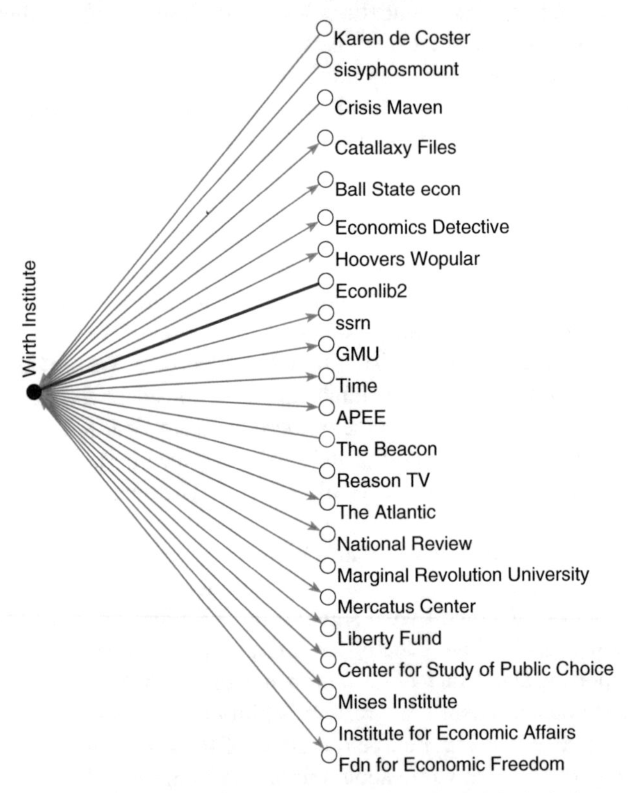

Fig. 1. The ego-network of the Wirth Institute. Notes: henceforth, black circles are non-US units, and white circles are US units. The solid line is a reciprocated link.

There are well-known KB allies shown in Fig. 1. They include the George Mason University (GMU) which has received many millions of dollars from the Koch Brothers directly and from other foundations, the Mercatus Center and the Marginal Revolution University (both of which are programs at GMU), the Foundation for Economic Freedom, Reason TV, a part of the Reason Foundation that was created by the KBs and funded heavily by them, the Beacon, Econlib2 (an online library promoting libertarian economic ideas), the Economics Department at Ball State University, the Association of Private Enterprise Education (APEE) and the Mises Institute. In this Canadian context, we note the work of Brownlee [14] and Carroll et al. [15] who discussed the role of

Canadian organizations obstructing efforts devoted to environmental reform, environmental protection and climate change denial, a topic we tackle in Sect. 5. There are blogs also including one by Karen de Coster and sisyphosmount. The unit ssrn is the Social Science Research Network.

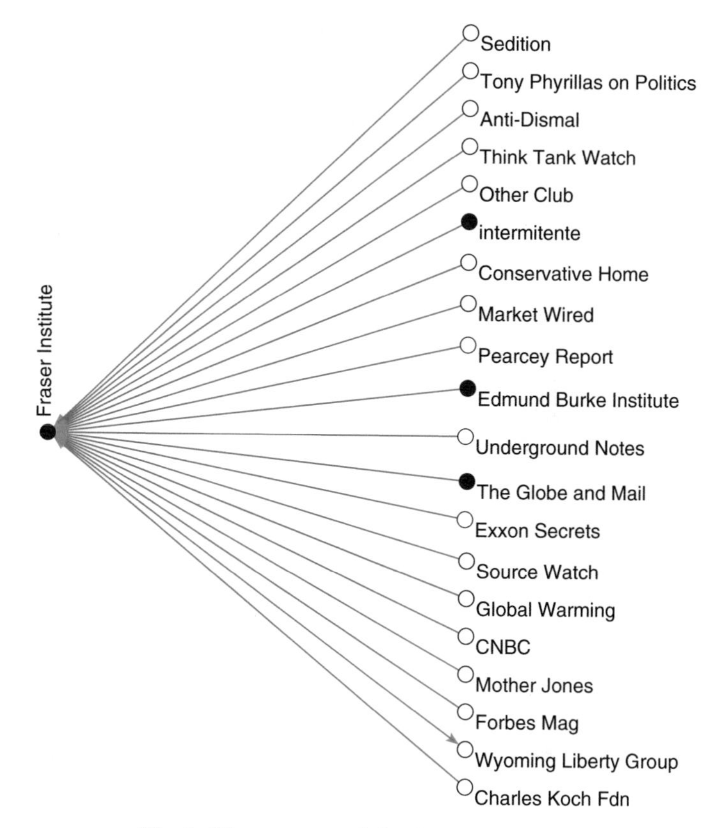

Fig. 2. The ego-network for the Fraser Institute

The second unit we consider here is the Fraser Institute. It is another Canadian public policy think tank founded in 1974 and based in Vancouver, British Columbia. It is libertarian and politically conservative. Its initial funding came from the Tobacco Industry (see [6]) which attacked scientific efforts to link smoking with lung cancer. It has received considerable funding from the KBs (also to attack scientific evidence) and from ExxonMobile (for the same reason) who are both opposed to tackling environmental problems. Indeed, the Fraser Institute is one of the leading climate change deniers in Canada and the world. The Fraser Institute's ego-network is shown in Fig. 2.

As was the case with the Wirth Institute, there are known KB allies in its ego-network. They include the Charles Koch Foundation, the Wyoming Liberty Group, Global Warming (a climate change denying unit), Conservative home and the Edmund Burke Institute (based in Ireland). However, there are also units that are severe critics of the Fraser

Institute. One of them is Exxon Secrets[1], a compilation from Greenpeace showing the extensive efforts of ExxonMobile to fund the activities of many of the KB allies [16]. Mother Jones, a left-leaning outfit pursuing investigative journalism, is likely also to be highly critical of the Fraser Institute. The ego-network of the Edmund Burke Institute is tiny, so we do not show it to save space. It has links to a mere three units – the Fraser Institute, the Institute for Economic Affairs and The Independent Institute. All are strong KB allies.

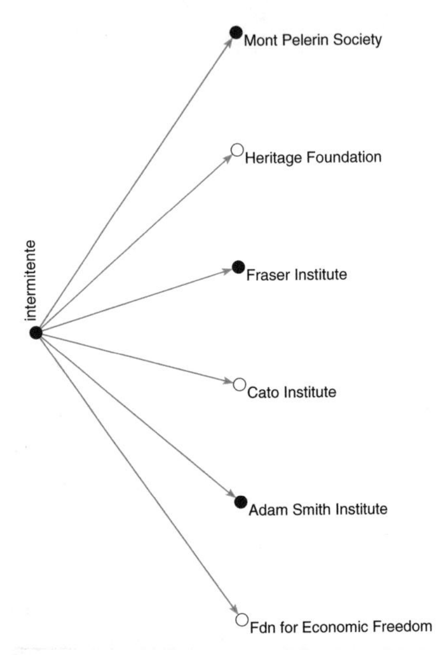

Fig. 3. The ego-network for Intermitente

The final ego-network we show is for the Brazilian think tank, Intermitente (see Fig. 3). All six of the units to which it points to are KB allies, including the Fraser Institute (again!). Another one is the Adam Smith Institute based in the UK. Even with using only these four ego-networks, the counties involved are Austria, Brazil, Canada, the Czech Republic, Ireland, Switzerland, the UK and the US.

While suggestive, these ego-networks provide only hints regarding the international reach of the KBs. Clearly, we need more compelling evidence. To do this, we constructed a network by starting with these non-US units. We then identified the other units in our large network that can be reached through one step out from them and one step into them. These identified units were joined with the starting set of non-US units. The result is a much clearer image of this reach. It is shown in Fig. 4. All of the non-US units, including both the known KB allies and the international media outlets, are placed in one oval with the 44 non-US units in the center of this image. There are 192 units in this network. The remaining 148 units are placed in the outer oval. As shown in Fig. 4,

[1] https://exxonsecrets.org/html/index.php.

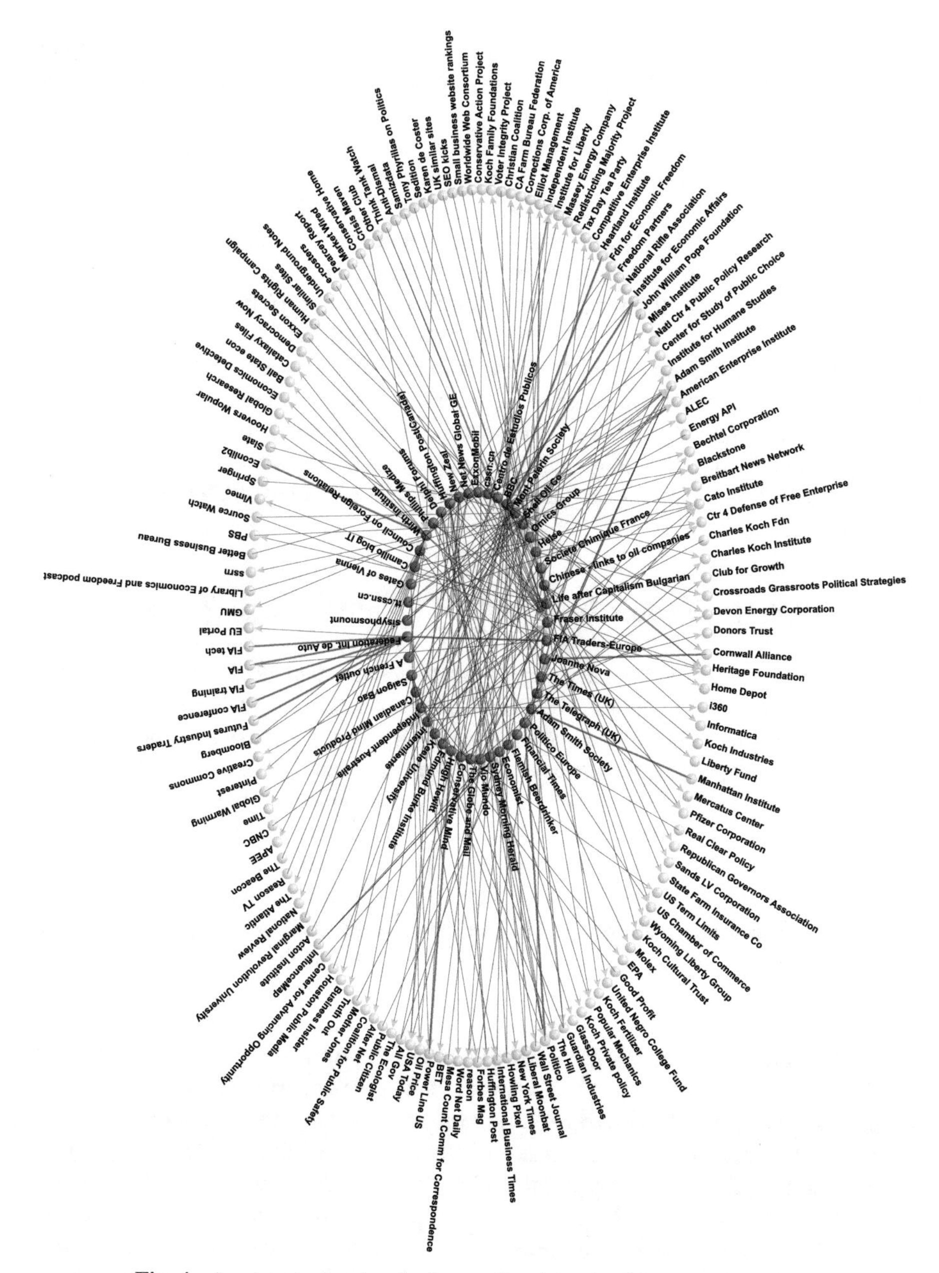

Fig. 4. A network showing the international reach of the Koch Brothers

the international reach of the KBs is considerable and, without doubt, is likely to get larger in the future as the KB allies are active and very aggressive in pursuing their goals, especially limiting the regulatory action of governments. We note that the inner circle of units is composed of units that are all located outside the US and the outer circle has

US units. Clearly, the non-US units are well connected to units in the US. This figure cements the idea that this is an international network. The solid lines are reciprocated ties.

While we have documented this international reach of the KBs, doing so is not sufficient because an integral, and perhaps more important, reason is their promotion of climate change denial. This has mobilized many of its allies seeking to obstruct all efforts to tackle one of the greatest threats to the future of the planet. Many scholars [7, 17–20] have written on the threats to the environment, including climate change, and organizations opposing all efforts to deal with them. Their work provides a continuous stream of cogent arguments regarding environmental assaults and denial of climate change arguments. Doreian and Mrvar [4] provided an initial set of results regarding this movement. In retrospect, the result, while informative, was limited. Here, we go much further by using a very different approach. There are many methods for analyzing networks. We are sure, that many of them need to be used in order to provide a fuller and more complete understanding of the operation of this extremely potent and politically destructive network.

5 Identifying the Units Involved in Climate Change Denial

To document the size of the network of climate change deniers, as part of a separate project, we established a set of keywords and phrases describing the basic concerns and interests of the units in the network of KB allies. The result was a two-mode network with KB allies in one mode and the keywords in the other mode. One keyword was 'climate change denial'. We identified all of the KB allies that explicitly embrace, support and act upon pursuing climate change denial. This comes from their own statements as declared on their websites. Their prose is abundantly clear. We then extracted, from the large overall network obtained from using the VOSON web crawler, the network having only these active denial units. There are 212 of them. They form a potent political force as they are supported by the enormous financial resources of the KBs (see below). The visual result is shown in Fig. 5.

In this image, as for all of the figures we present, the non-US units are represented by black circles with the US units being represented by white circles. We comment briefly on some of these units in this network having the largest number of ties incident to them or from them. Koch Industries is one, without surprise. Its importance is due to this set of companies being the source of the profound wealth supporting the climate change denial social movement promoted by the KBs. According to the Forbes annual rankings, Koch Industries is the second-largest privately held company in the US with an annual revenue of about $110 billion. The Koch family is the second richest family in the world, being worth $99 billion. They are using their enormous wealth to attempt to bend the US government and other national governments to their will based on their extreme libertarian view that their interests, and only those interests, must be promoted.

Also prominent is the Reason Foundation and the Cato Institute. Both have been funded heavily by the Koch Brothers. Doreian and Mrvar [5] in their analysis of hubs and authorities [13] showed how these two units are especially influential in the KB network, especially the Reason Foundation, as it is both a prominent hub and a prominent

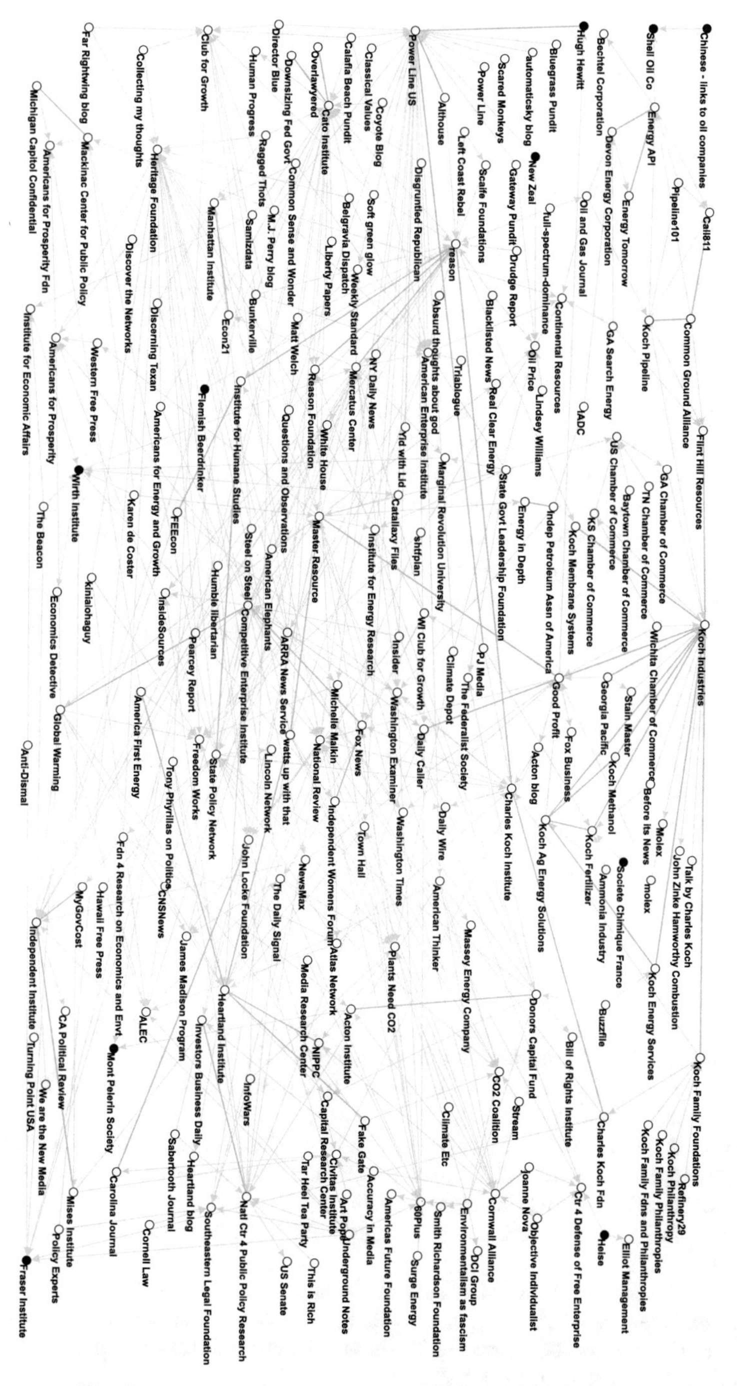

Fig. 5. The climate change denying units in the KB network

authority. The Heritage Foundation is another important authority. As is the Charles Koch Foundation through which many resources flow. Another prominent unit is the Heartland Institute, another very strident climate change denier.

6 Summary, Limitations and Further Work

We have provided some results regarding the extensive international reach of the Koch Brothers, given their basic motivating aim to destroy all efforts intended to deal with the reality of climate change as they wish to protect their massive profits. Without surprise, these two themes are integrated closely. But, in addressing these issues, we have to acknowledge some limitations in the work we have done. One is that our results are based on a single cross-sectional study. But another web crawl search has been completed by Robert Ackland, whose work in doing this we appreciate greatly. We would not, in any way, be surprised if both the international reach of the KBs has been extended further as they are very active, as we have noted, when we study the results of this more recent web crawling search. Also, the ideal data would be to see the interactive links between the Koch Industries. But their secretiveness, as noted above, precludes this. We are able only to deal with the links between all of the units in the KB network of allies as they provide them. We think that this is the best any researchers can do, given the incredible secretiveness of the Koch Brothers and their deeper intentions.

Our future research plans include making a deeper study of the keywords capturing the goals and actions of the KB allies. This will be extended to analyzing the two one-mode projection networks from the two-mode network described above. We will provide the same sorts of treatments to the second network generated by the VOSON web crawler and make temporal comparisons.

References

1. Batagelj, V., Mrvar, A.: Pajek - a program for large network analysis. Connections **21**(2), 47–57 (1998)
2. MacLean, N.: Democracy in Chains: The Deep History of the Radical Right's Stealth Plan for America. Penguin Random House, New York (2017)
3. Mayer, J.: Dark Money: The Hidden History of the Billionaires Behind the Radical Right. Anchor Books, New York (2017)
4. Doreian, P., Mrvar, A.: The Koch Brothers and their climate change denial social movement. In: Tindall, D., Stoddart, M.C.J., Dunlap, R.E. (eds.) Handbook of Anti-Environmentalism. Edward Elgar Publishing, Cheltenham (2020)
5. Doreian, P., Mrvar, A.: Hubs and authorities in the Koch Brothers network. Soc. Netw. **64**, 148–157 (2021)
6. Oreskes, N., Conway, E.M.: Merchants of Doubt: How a Handful of scientists Obscured the Truth on Issues from Tobacco Smoke to Global Warming. Bloomsbury, New York (2010)
7. Supran, G., Oreskes, N.: Assessing ExxonMobil's climate change communications 1977–2014. Environ. Res. Lett. **12**, 1–8 (2017)
8. Bardon, A.: The Truth About Denial: Bias and Self-Deception in Science, Politics, and Religion. Oxford University Press, Oxford (2019)
9. Begley, S.: The truth about denial. Newsweek **150**, 20–29 (2007)

10. Leonard, Ch.: David Koch was the ultimate climate change denier. New York Times, August 13, 2019 (2019)
11. Leonard, Ch.: Kochland: The Secret History of Koch Industries and Corporate Power in America. Simon & Schuster, New York (2019)
12. Ackland, R.: WWW hyperlink networks. In: Hansen, D., Shneiderman, B., Smith, M. (eds.), Analyzing Social Media Networks with NodeXL: Insights from a Connected World. Morgan-Kaufmann (2010)
13. Kleinberg, J.: Authoritative sources in a hyperlinked environment. J. Assoc. Comput. Mach. **46**(5), 604–632 (1999)
14. Brownlee, J.: Ruling Canada: Corporate Cohesion and Democracy. Fernwood, Halifax (2005)
15. Carroll, W., Graham, N., Lang, M.K., Yunker, Z.: The corporate elite and the architecture of climate change denial: a network analysis of carbon capital's reach into civil society. Can. Rev. Sociol. **55**(3), 425–450 (2018)
16. Greenpeace: Koch pollution on campus: academic freedom under assault from Charles Koch's $50 million campaign to infiltrate higher education (2014). https://www.greenpeace.org/usa/global-warming/climate-deniers/koch-pollution-on-campus/
17. McCright, A.M., Dunlap, R.E.: Defeating Kyoto: the conservative movement's impact on U.S. climate change policy. Soc. Probl. **50**(3), 348–373 (2003)
18. Elsasser, S.W., Dunlap, R.E.: Leading voices in the denier choir: conservative columnist's dismissal of global warming and denigration of climate science. Am. Behav. Sci. **57**(6), 754–776 (2013)
19. Dunlap, R.E., McCright, A.M., Yorosh, J.H.: The political divide on climate change: partisan polarization widens in the US. Environment **58**(5), 4–23 (2016)
20. Brulle, R.J., Aronczyk, M., Carmichael, J.: Corporate promotion and climate change: an analysis of key variables affecting advertising spending by major oil companies, 1986–2015. Clim. Change **159**, 87–101 (2016)

'Illegal Community'? Social Networks of Revolutionary Populism in Russia in the 1870s

Aleksei Rubtcov[1,2](✉)

[1] St. Petersburg Institute of History of Russian Academy of Science, St. Petersburg, Russia
rubtsov334@gmail.com

[2] St. Petersburg Electrotechnical University "LETI", St. Petersburg, Russia

Abstract. In the Russian Empire in the second half of the XIX century, the judicial investigative authorities were not independent neither in the interpretation of the revolutionary movement with which they fought nor in the practice of conducting inquiries. Meanwhile, it was the political police that created most of the materials and the main explanations of the Narodnik movement, which became the basis for all subsequent historiography. The article discusses the creation of the concept of "illegal community" as a legal basis for the fight against the revolutionary movement. Then, using the example of the Moscow inquiry of 1873–1874, it analyzes how, under the influence of legislation, the investigation was forced to interpret the crimes. Using the analysis of social networks as the primary tool allows us to identify the gap between the data collected by the police and their final interpretation.

Keywords: Revolutionary populism · Social networks · Social movement · Russian Empire

1 Introduction

The reign of Alexander II (1855–1881) ushered in a decade of transformation in all spheres of life. One of the features of the epoch was an emergence of a mass radical movement which became a new social phenomenon in a situation which made political attitudes of its participants especially dangerous. The movement was loosely united around such political principles as democracy, socialism in its broadest sense, and women's rights. The ethos rather than a specific ideological platform united movement participants into a group. By the 1870s a real radical community existed, defined internally by institutional and ideological bonds and externally by a strong sense of social disaffection. The extreme radicalization of the political struggle was comprehended by all involved groups of experts (lawyers, investigators, psychiatrists, politicians, etc.) in Russia [1].

Changes in the state's criminal policy had tangible progress early in the decade. It was a reaction to the growth of the revolutionary movement. In the summer of 1874, one of the significant political events began. It was a diffuse movement known as "Going to The

A. Antonyuk and N. Basov (Eds.): NetGloW 2020, LNNS 181, pp. 236–246, 2021.
https://doi.org/10.1007/978-3-030-64877-0_16

People" in the course of which hundreds of young people canvassed rural regions. They attempted to enter into direct relations with peasants as teachers, agitators, propagandists, or only as students of folkways and incited the peasantry to rise against the autocracy. There was no organization of any kind to control it, and the different groups of students and young intellectuals were extremely autonomous in their actions. If the participants concentrated in some regions of Empire rather than others, it was because they were responding to the revolutionary literature of the time rather than obeying instructions. But despite all this, the majority of young people set out individually or in small groups. They dressed as peasants (*muzhiks*) and wished to show them (and show themselves too) that they were able to earn a living on their own. But more importantly, they wanted to tell the peasants "the truth," and so the members of the movement tried to propagate revolutionary and socialist ideas among peasants and workers [2, 3].

The spread of the movement "Going to The People" had assumed proportions that seriously worried the government. It led to police persecution resulting from which four thousand people were imprisoned, questioned, or at least harassed by the police. The investigation lasted for three years. Thirty provinces (*guberniya*) had, in varying degrees, been affected by propaganda. Nowhere had the Populists been able to arouse a revolt or upheaval. Everywhere the peasants had listened to these strange pilgrims with amazement, surprise, and sometimes suspicion. But the government understood that in this event, a new form of revolutionary movement had been born.

2 State of Research in the Field

In Russia, the "cultural turn" (and others "turns" too) in humanities and social science has significantly affected historical research on pre-revolutionary Russian radical movement in recent years. This shift towards a cultural history of Russian political protest reflects not only a new fashion in historiography but also the fact that previous historical narratives, which had focused on the genesis of revolutionary ideologies, on strategies of legitimization of political violence or the social basis of radical movements [4, 5], had failed to answer crucial questions. One of the items, how were the extreme political acts perceived by its victims and authorities, and how did the experience of modern political violence and struggle alter contemporaries' concepts of public security? It is crucially important to develop research in this direction. Without a deep study of practices of authorities to describe and control "unreliable subjects" it is impossible to clarify the phenomenon of "revolutionary populism" as an unusual political movement that acted in the specific conditions of the Russian Empire. Some studies have attempted to use new approaches to describe the practices of investigative authorities for inquiries of political cases [6]. Historians have been able to establish main trajectories along which the investigative cases about "revolutionaries' conspiracies" were developed.

The government was not only one of the sides of the political conflict, how often researchers write about it, but yet the influencing power. It was creating the schemes of the description of the revolutionary movement. But despite the attention to different aspects of the revolutionary history of the 1870s, the historiography is characterized by extremely schematic explanations of the law enforcement and decision-making procedures in the field of "public order" protection. Researchers of the populism movement have not yet raised questions regarding the logic of the investigation, principles of information collection, record keeping of police activities, etc. However, studying these issues can explain a lot in the formation of the contemporary state institutions in the Russian Empire. It evolved in the context of the establishment of conditions of a modern state and advanced legal system in the Russian Empire [7, 8]. Legal codification, or at least debates over the problem, became an almost global phenomenon in the nineteenth century as state power was centralized and made uniform. Criminal law rationalization changed not just the concept of crime, but also the idea of evil, victim, and perpetrator [6, 9]. These processes could also be observed in the evolution of the perception of political crime by Russian officials. The revolutionary community considered being "criminal" by the government should have been more rational and understandable, so it had to be inscribed in the existing laws and ideas adopted by the state and society. Laws relating to political crimes were taken under the influence of law enforcement practice, and it was part of a broader legal system that included various actors. Accordingly, it could be affirmed that one of the important actors, who determined the picture of political protest in the Russian Empire, was the judicial investigative bodies.

At the same time, the movement consisted of a network of relationships between different actors. These relationships were uncovered and mapped by the authorities, influencing their perception of the community as a whole and its construction as an "illegal community". Thus, through network analysis, it is possible to investigate the logic of constructing the image of revolutionary movements by authorities. In turn, this makes it possible to understand which political threats the state identified and how its policies contributed to the consolidation of radical communities.

3 Sources and Methods

Primarily sources for the research are materials of investigation cases, which were then united in a big political case named "About propaganda in the Empire". The investigation, in this case, took place in 1873–1877. "The Trial of the 193" enhanced the result of four years of the investigation in the case of "propaganda". The political trial was a reaction of authorities to mass protests movement "Going to the People".

Therefore, it can be said that numerous gendarmes, prosecutors, judges, and their assistants acted as "sociologists" of the Empire. For several years, the Empire's investigative authorities conducted an extensive survey of protest moods in the Empire, conducting investigations in various provinces of the country. These numerous inquiries fit into 11 volumes and 34 cases [10, 11] separately for each region where surveys were conducted. These sources were materials of preliminary investigations and synthesis reports made by prosecutors. To solve the research problems described above, I refer to the approaches of social science. To study mechanisms of investigation of political crimes, I follow the logic of two approaches that I try to combine in my project. Firstly, I have followed the logic of sociology of law [12] for studying the features of investigative process in the Russian Empire. The approach allows to pay attention to how different departments organized their work, what conditions influenced different actors which participated in the investigation, and how did it all affect the final result.

Secondly, I apply methods of social network analysis since some concepts of the approach can facilitate understanding of the structure of a "criminal" revolutionary society

[13, 14]. In this case, the choice of this approach is justified by the following consideration. Applied approaches used by most historians may not help with a study of some complex and mass historical sources. Consider the case where researchers are confronted not just with one event narrative, but with hundreds of such narratives. Clearly, the scale of those numbers makes it imperative for researchers to find more systematic ways of cataloging and retrieving the information contained in the narratives. It is out of that work that will come a shift of focus from the event to the conjuncture, from the single event to patterns of events [15]. So, I use this approach to explore the complex reality of revolutionary society and demonstrate how Russian officials perceived network structure of radical community and designated it as "criminal". Therefore, the research attempts to reconstruct a cognitive social network map of the investigative authorities which produced several important concepts through the prism of this radical community perceived by different political actors. Accordingly, I rely on an approach in which sociologists focus on social ties as they are represented in the cognitions of individuals. Structure, as it exists in the minds of individuals, may be more predictive of important outcomes than has been recognized [16]. It is especially relevant for studying the explored radical movement because it should be constructed in investigative materials by certain rules in order to become a "criminal" society.

4 Constructing Political Threats: The Mechanisms and Tendencies

4.1 Legal Framework for Political Threats in the Russian Empire

Before proceeding to the analysis of the data collected by the investigation, it is necessary to outline the conceptual framework that prosecutors used to structure the investigation cases.

For a long time, Russian legislation on state crimes did not imply "gradualness". It meant the division of punishment of the defendants depending on the degree of their participation in the "crime". Only the monarch's intervention could change the fate of the convicts and introduce the principle of "gradualism" in their punishment. The Criminal Code, introduced in 1845, retained equal penalty for those who participated in conspiracies and riots against the power. However, a similar scheme could not previously be applied to the new conditions for the development of the radical mass movement. Therefore, the practice of investigating such crimes required the production of acceptable approaches to determining the role played by a defendant in the "criminal" community.

The criminal code and its categories presupposed a certain network structure of "criminal communities". In turn, this guided investigation, including the construction of the structure of the Narodniki movement. The "instigators," "accomplices," and "concealers" stood out in the Code. Legislators have identified two degrees of intentional crime, including conspiracy and "criminal" anti-government societies. It was assumed that these crimes were committed "by several persons by prior agreement" [17]. They "always contain the highest degree of intent". The accomplices in such criminal acts are recognized as those "who agreed with the instigators or with other perpetrators to commit, by joint forces or actions, a premeditated crime." Then, "instigators" were considered, according to the law, as those persons "who, having intentionally committed the crime, agreed to the others, and who controlled the actions". The criminal community participants had to be all identified and divided according to the degree of their guilt, depending on the degree of their involvement in the crimes committed. The disclosed "criminal" organization, in this case, should have been strictly hierarchical and have a clearly defined circle of leaders who were the engines of the entire "criminal" process. The adopted laws imposed a particular framework on law enforcement practice, and for understanding this process, we need to study how data that were collected by investigative bodies were used and interpreted. The report placed in Volume I of the "propaganda in the Empire" case [18] was used as a source for the reconstruction. This document proved to be one of the most structured among the available descriptions and investigation reports on the investigation progress. It was made by the prosecutor of the Saratov Chamber of Judges S. S. Zhikharev, who oversaw the inquiries. He used the materials of all previous cases. Thereby Zhikharev created a general picture of the spread of "revolutionary propaganda in the Empire" in the first half of the 1870s. An excerpt from this document, which described the course of the inquiry in the Moscow province, is taken as a basis for subsequent constructions. A preliminary investigation of the activities of the "revolutionary community" in Moscow covered the period of 1873–1874. This inquiry turned out to be one of the largest and most important for the investigation, in particular, because the printing house, in which illegal literature was printed, was located in Moscow and the investigators paid special attention to it and operated there.

Before proceeding directly to the analysis of the material, it is worthwhile to explain the method used in this article. Net Draw [19] was used for visualizing social networks, and Ucinet [20] was used for the obtained data analysis. In my research, I worked with different types of groups formed in social networks. Among them, there were cliques: in such groups, each element of a network is interconnected with every other element, i.e., there are all possible ties between its participants. Clustering is a more complex concept related to finding groups in a network. Clustering characterizes the degree of interaction between the nearest neighbors of the said actor. To analyze the system as a whole, we use the concept of network density, i.e., the ratio of the number of existing links to the maximum possible.

My approach to mapping the networks was to identify all the links mentioned in the investigation case. Mapped ties constitute any contact between political activists connected with joint protest activities (for example, transfer of illegal books, money, or other recourses, participation in the underground meetings). Police agents regularly monitored any contact between the revolutionaries, and any joint action was regarded as part of political activity.

When in investigative cases, it was about meetings where more than two people took part, I map the network links based on common attendance of such events. If A, B, and C attended event Y, I consider A, B, and C all tied to each other. The significance of an actor in the investigation network depends on the number of meetings and events they participated in. The more often a person is mentioned as a participant in any meetings or events, the more different contacts they will have in the network represented. For example, a person who visited two clubs of 8 people would be more important in the network than a person who was a member of only one circle of 10 people. But if people were in the same club, but there was no mention of their contacts with other groups, they will have connections only between each other. On the social network graph, they will be placed side by side, as they will be similar to each other in terms of the contacts present.

4.2 Social Network of Revolutionary Populists

The network represented on Fig. 1 demonstrates the social activity of actors and reflects the general structure of the interactions of participants within the Moscow protest community in the winter – spring of 1873–1874, according to the materials collected and processed by the investigation. First of all, the graph presented shows the connected community. All points in the connected graph can reach one another through one or more paths, i.e., each of the participants was achievable for any other on the network. It perfectly reflected the nature of identifying members of the "criminal community" in the course of the inquiry. All those who came under suspicion were under investigation primarily because they were somehow connected with each other, either directly or through mediation. Therefore, it was impossible to exist isolated from all other groups in such a network.

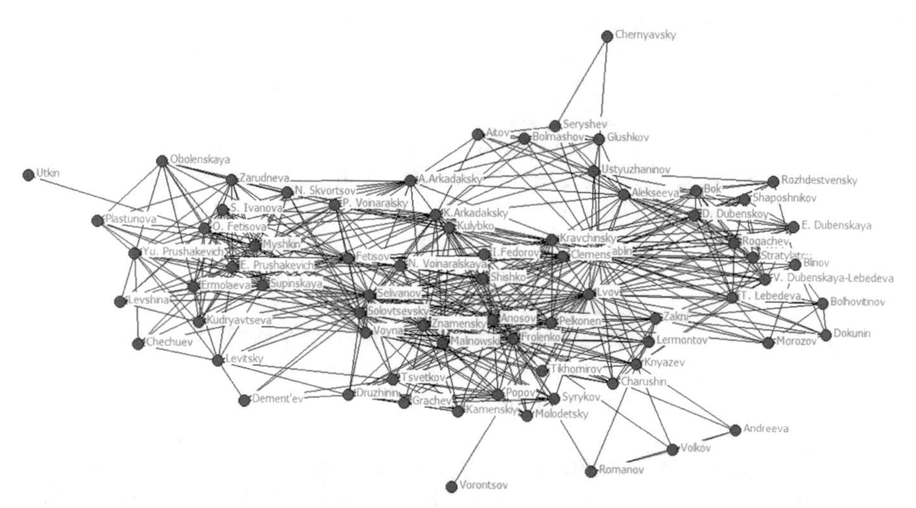

Fig. 1. The network demonstrates the social activity of actors. Nodes represent members of protest communities from Moscow. Ties constitute any contact between political activists that are connected with joint protest activities (for example, transfer of illegal books, money, or other recourses, participation in the underground meetings).

The size of the social network represented is relatively small. The average path from one participant to another is 2.034, and the longest path between a pair of actors in the network is 4. For example, it means that in such a community, information from one network member to another could pass through approximately two contacts. Each actor in this network was achievable, and the average distance between them was small, given that the network consisted of 73 people.

However, the sufficient level of interconnection of individual elements and its small size does not yet indicate the cohesion of this social network as a whole. The index of its density is low and equals 0.213. The maximum density is 1, i.e., the number of existing connections is much less than could be between all community members. This social network also includes 45 primary cliques, which for a network of this size (73 participants) is a low indicator. Therefore, these numbers indicate a low degree of cohesion in the network.

Based on the data presented in the investigation materials, it can be concluded that, despite the potential accessibility of each participant of the network by its other members, its segments were not sufficiently tightly connected and did not represent a tight-knit community. In this regard, this network was not a group that was a stable community with close contacts and a communication network.

4.3 The Role of "Ringleaders" in Construction of Revolutionary Community

In their interpretations, the investigation insisted that the presented data prove the existence of a vast "revolutionary community" that covered dozens of provinces of the Empire with its network. How, then, did the network retain its apparent integrity and give the impression of a serious political movement? Further analysis of the network shows that this graph has a high clustering factor of 0.78. It means that the connections between all segments are distributed unevenly, and several actors accumulate most of them. They constituted a structural center (17 people), which was the core of this network. In this case, the center was a tightly connected segment, and on the periphery, there remained actors with fewer contacts. Therefore, the identified structural center members played an essential part in producing a general picture of the activities of the "revolutionary community," as they pulled together the groups included in this network into a single structure.

In his report, the prosecutor S. S. Zhikharev, who supervised the conduct of investigative actions, highlighted the most important figures of "propaganda". Such people, according to "The Penal Code" were "instigators" of committing state crimes and should have suffered the most significant punishment [17]. In S. S. Zhikharev's report, the part about the Moscow inquiry began as follows: "Under the influence, in particular, of the student Charushin who came from Petersburg, the people's teacher Lermontov, Lev Tikhomirov […] and their accomplices and comrades, clubs began to acquire a political connotation" [18]. All of the listed individuals were engaged in propaganda among students; the investigation particularly distinguished some of them who "turned out to be strong, vivid, and acquired the character of political unreliability. Among them are the following: students of the Moscow University: Lvov, Sablin, and Zakni, and students of the Petrovsky Academy: Frolenko, Selivanov, Anosov, Voina, and Knyazev" [18].

This part of the report was an example of a model according to which, from the investigation, the protest movement developed. One of the defining judgments about the development of the "revolutionary community" is that propaganda is the main driving force. Indeed, how it spread determined the structure of protest communities. Therefore, the resources that allowed the distribution of propaganda were for the investigation the most significant for the protest movement's development. One such resource, in particular, was people who had a large number of acquaintances. Revolutionary "ringleaders" could rely on them and take advantage of these people's positions in local communities. Having access to such people and gaining influence on them was for the "leaders" of this movement a necessary condition for further development. What simplified this access was, among other things, the existence of clubs of "self-education" that had already organized their members in certain groups, which facilitated the spread of propaganda in pre-formed communities. In general, gendarmes and prosecutors used precisely such schemes to describe the formation and development of the "revolutionary community".

Thus, the influence was considered as the ability to propagandize in direct contact with others. By this principle, for example, Charushin, Lermontov, and Tikhomirov were particularly marked. In the investigation's materials, these young people were seen as visiting "propagandists" who gave impetus to the development of "revolutionary" groups in Moscow. Although subsequent events developed without their participation, the propaganda activities they were engaged in seemed to the investigators to be decisive in involving other people in the revolutionary movement. Their "propaganda" work was, as it were, the impetus that determined the subsequent growth of the "revolutionary community".

Simultaneously, the analysis of the amount of connections (degrees) of actors demonstrates that the position of these people in the network of interactions might not be too central. According to the collected data, they had only 10 contacts each. Considering that the average degree of an actor was around 15, this is less than the average in this network, which indicates that they could not directly "reach out" to the majority of participants.

Often the value of a particular position in a social network depends not on how many contacts an actor has, but on how useful they are. Therefore, the "revolutionary" network constructed by the investigation could look successful only if the influence of the "ringleaders" mentioned in the report was distributed through intermediaries to all other community members. That is if the influence of the most significant actors spread throughout the rest of the network. Thus this "criminal" propaganda infected all of its segments, to which the influence of "propagandists" could reach through social connections. It is an important point in the design of the development of revolutionary or any other kind of political communities represented in political cases. Access to the activists listed in the report, who came into contact with Tikhomirov, Lermontov, and Charushin, guarantees the latter access to the entire network. The three "propagandists" listed here were the source of propaganda and the core around which the network of the protest movement in Moscow was formed in the interpretation of the investigation. The people grouped around these influential agents were the transmitters of their propaganda to the rest of the network.

With such a view of the network, the importance of the interconnections marked out by the investigators increases even more. In this social network, the structural center consisted of 17 people (in Fig. 2, they are marked as larger nodes), with the largest number of contacts in the network (16 and above). Among the people with whom Tikhomirov, Charushin, and Lermontov were directly in contact, five out of seven (Anosov, Lvov, Sablin, Selivanov, Frolenko) belonged to this center of the network. At the same time, the two actors with the largest number of contacts in the entire network – Sablin and Lvov – belonged to the group directly influenced by the "visiting propagandists". They had so many connections because they were mentioned in the report more often than others, being members of several different clubs and circles. According to the received results, they occupied the central position in the network and played an essential role in spreading propaganda. Without them, important mediation contacts would have been lost, which would have affected the functioning of protest communities, in particular, the involvement of new members. In this situation, the five people possessed a significant resource – a lot of connections.

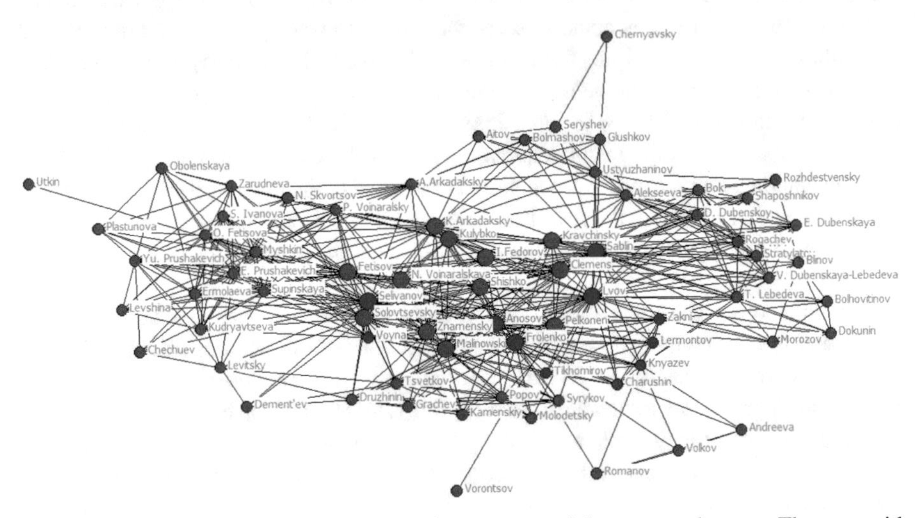

Fig. 2. In this social network, the larger nodes are parts of the structural center. They are with the most significant number of contacts in the network (16 and above). Those actors who were directly in contact with the "influential" propagandists are marked in the blue.

Each of the people directly in contact with the "influential" propagandists brings a specific set of connections to this segment of the network. If we single out the initial network of interactions between Charushin, Lermontov, and Tikhomirov, the investigation indicates, it will consist of 11 people (in blue in Fig. 2). For example, further indications of Lvov's and Sablin's contacts demonstrate that they proved to be essential intermediaries between a group of "visiting propagandists" and members of the "revolutionary community" in Moscow, since they included 39 more people in the circle of influence of the "propagandists".

The described community of Moscow was built into a wider network of interactions in other regions of the country. In the report, prosecutor S. S. Zhikharev mentioned:

> In the note filed on July 29 to the Konotop (Chernigov province) police officer, a village school teacher from the village of Komary, a nobleman Trudnitsky explained that Zakni and Frolenko belong to that one party of 'Saint-Zhebunists', which was formed in 1873 in Zurich and consisted mainly of the brothers Zhebunevs, after whose name the party itself was called 'Saint-Zhebunists' […] the circle to which Frolenko and Zakni belong decided that a revolution is what is now necessary for Russia. [18]

And then the following conclusion was made: "The representatives of the Saint-Zhebunist party in Moscow were, according to Trudnitsky, exactly those Zakni and Frolenko" [18].

A connection between the two inquiries was created, and the Moscow network was included in the broader field of relationships. This example confirms the importance of investigating such intersections between local networks, as it allowed building a wider social network. First of all, such points of intersection formed the structure at all levels of the constructed revolutionary movement. Participants of protest communities served in this capacity: they established links between various groups, often isolated from each other, through their ongoing work in various clubs and circles.

It shows the peculiarity of assembling the "revolutionary community" in the materials of the inquiry. Thanks to the activity of some members of the revolutionary movement, the investigators could put together scattered individuals or groups of people and enclose them within the framework of the protest movement. In this manner, for the judicial and investigative bodies of the Empire, protest is, in many respects, not an explicit "revolutionary activity," but existing connections with people who can carry out this activity.

In this case, active members of the revolutionary movement were subjected to the fact that they were inscribed in certainly designed schemes made by authorities. In the network, an individual was given a specific role and meaning. In the perception of political protest space, the actor's mediating position was the most important. At all levels of the construction of the political case, this tendency to identify figures necessary for the network could be traced. For the investigation, some were important as influential propagandists who organized clubs and circles. Others did not have top priority, but turned out to be essential links for the purposes of these investigations.

5 Conclusion

The view presented in this article makes it possible to understand how the work of the investigative bodies influenced the formation of the image of the revolutionary movement in the Russian Empire. Previously, historians practically did not pay close attention to the specifics of the narrative of investigative documents. The investigation materials were often used to study the "immediate" history of the populist movement. However, it is crucial to understand how the investigators' data about the opposition were used. This, in the future, in itself influenced the formation of various kinds of narratives about the Russian revolutionary movement. Thus, network structures of the communities were an important part of the image of the movement and its narrative construction.

The protest movement of the 1870s was a complex interweaving of relationships, ideologies, political orientations, rather than a world of united revolutionary groups. In this form, it was extremely inconvenient for the forensic investigation authorities' perception because it still did not comply with the law. Mutual misunderstanding of lawyers, who relied on outdated ideas about forms of political protest, and political police who faced new ways of revolutionary life in early 1870 predetermined the features of the inquiry about propaganda in the Empire and the Trial of the 193. Political police and prosecutors could look at youth protest only through the prism of an "illegal community" and a centralized organization created by legislators. More amorphous forms, youth gatherings, and student circles were inaccessible to them legally. In practice, this led to the fact that the protest movement of the 1870s in the eyes of the authorities looked as a united and well-coordinated structure that stretched its nets across the entire Empire, which was not in line with the real picture. Therefore, the investigative bodies sought to unite this multidimensional political space in one way or another into a single whole and make it accessible to interpretation and further litigation. It is not surprising that accusations against the participants of the movement quickly disintegrated during the process. The evidence of the existence of a single agitation center that was carrying out a general “program” was ephemeral, as evidenced by the disgraceful failure of the imperial justice during the Trial of the 193.

References

1. Borisova, T.: Public meaning of the Zasulich trial 1878: Law, Politics Gender. Russ Hist. **43**, 221–244 (2016) https://doi.org/10.1163/18763316-04304002
2. Field, D.: Peasants and propagandists in the russian movement to the people of 1874. J. Modern History **59**, 416–438 (1987) https://doi.org/10.1086/243223
3. Kinsey, P.S.M.: From city workers to peasantry: the beginning of the russian movement. Slavic Rev. **38**, 629–649 (1979) https://doi.org/10.2307/2496567
4. Morrissey, S.K.: Heralds of Revolution. Oxford University Press (1998)
5. Morrissey, S.K.: The “apparel of innocence”: toward a moral economy of terrorism in late imperial russia. J. Modern History. **84**, 607–642 (2012) https://doi.org/10.1086/666051
6. Daly, J.: Crime and Punishment in Russia. Bloomsbury Publishing (2018)
7. Antonov, S.: Russian capitalism on trial: the case of the Jacks of Hearts. Law Hist. Rev. **36**, 35–76 (2018) https://doi.org/10.1017/s0738248017000517
8. Dahlke, S., Templer, B.: Old Russia in the dock. monderusse. **53**, 95–120 (2012) https://doi.org/10.4000/monderusse.9368
9. Farmer, L.: Making the Modern Criminal Law. Oxford University Press (2016)
10. Rossijskii Gosudarstvennyi Istoricheskii Arkhiv (RGIA) [Russian State Historical Archive]. f. 1405, op. 72, d. pp. 7173 – 7180
11. Gosudarstvennyj Arhiv Rossijskoj Federacii (GARF) [State Archive of the Russian Federation]. f. 112, op. 1, d. 241, 246, 247, 253, 256, 356, 357, 359
12. Black, D.: The behaviour of Law. Emerald Group Publishing, New York (1980)
13. Anheier, H.: Movement development and organizational networks: The role of ‘single members’ in the German Nazi party, 1925–1930. In: Diani, M., (ed.) Social Movements and Networks: Relational Approaches to Collective Action. pp. 49–74. Oxford University Press (2003)
14. Gould, R.V.: Multiple networks and mobilization in the Paris commune 1871. Am. Sociol. Rev. **56**, 716 (1991) https://doi.org/10.2307/2096251
15. Franzosi, R.: Quantitative Narrative Analysis. SAGE, Thousand Oaks (2010)
16. Kilduff, M.: Interpersonal Networks in Organizations. Cambridge University Press (2008)
17. Ulozhenie o nakazaniyah ugolovnyh i ispravitel’nyh [Penal Code], (1876) https://www.prlib.ru/item/459768
18. Rossijskii Gosudarstvennyi Istoricheskii Arkhiv (RGIA) f. 1405, op. 72, d. 7174
19. Borgatti, S.P.: NetDraw Software for Network Visualization. Analytic Technologies, Lexington, KY (2002)
20. Borgatti, S.P., Everett, M.G., Freeman, L.C.: Ucinet for Windows: Software for Social Network Analysis. Analytic Technologies, Harvard, MA (2002)

From Formal Political Contacts to Cross-Border Collaborative Networks: A Network Analysis of Partnership Operations in the Spree-Neiße-Bober Euroregion

Joanna Frątczak-Müller(✉) and Anna Mielczarek-Żejmo

University of Zielona Góra, Zielona Góra, Poland
{j.fratczak-muller,a.mielczarek-zejmo}@wns.uz.zgora.pl

Abstract. This article discusses the cooperation network emerging as a result of the actions of a cross-border organization, whose scope of responsibility is the Spree-Neiße-Bober Euroregion. The task of Euroregions is to support local communities in cross-border regions in dealing with the complexity associated with the problems of the modern public sphere. Partnerships that emerge as part of Euroregions, form governance networks. A distinctive feature of their construction on the border is the inter-sectoral as well as the international nature of the actors. The basic questions formulated by this presentation refer to three areas: network construction, network relations, and the network's contribution to policy-making. The main research area is in analyzing the development of relationships between participants of the Euroregion (the inhabitants and organizations), and its role in policymaking and meeting the needs of local communities in the cross-border region. The theoretical framework has been designed around the three main explicatory perspectives: regionalization and soft spaces, network theory, and governance networks. The presented conclusions are the result of the SNA (Social Network Analysis) of participants in projects implemented in the Euroregion in 2019 under the European Territorial Cooperation Fund, Interreg IVA, and qualitative interviews (the leaders of 34 projects).

Keywords: Euroregion · Governance network · Cross-border partnerships · Network analysis

1 Introduction

The ability of social networks to function within the complexity of the problems in the modern public sphere is conducive to entities joining together in relationships based on trust and reciprocity, despite the fact they are independent of each other, dispersed, and often competitive. The network participants involved in implementing public policies generally have different views on the sources of the problems, their nature, and the appropriate solutions. In the border regions, the diversity of views concerning everyday matters is reinforced by cultural, systemic, formal, and legal differences; and often also

A. Antonyuk and N. Basov (Eds.): NetGloW 2020, LNNS 181, pp. 247–265, 2021.
https://doi.org/10.1007/978-3-030-64877-0_17

by the uneven amount of resources at the disposal of cooperating entities. The European Union has recognized both the importance of cooperation and its conditions by establishing an appropriate legal framework[1] and creating a fund for the activities of Euroregions sanctioned in this way; these regions have become one of the elements of multi-level governance (MLG) – a form of managing the public sphere in the EU. The aim of this article is, therefore, to establish the characteristics of Euroregions as participants in policymaking processes. *The basic questions formulated by the article refer to three areas: network construction, network relations, and the network's contribution to policymaking.* An example is the Spree-Neiße-Bober Euroregion (SNBE) in the Polish-German border region.

The discussion is based on three theoretical perspectives. These include, first of all, regionalization, with particular emphasis on the processes occurring in the border regions; second, social network theories; and third, governance networks. The theoretical context of our analyses is presented in the first part of the article. The second part describes the research procedures used. The third part concerns the development of the SNBE. The presentation of the results and a discussion are included in part four, and in this section, we also attempt to answer the question about the structure of the network and relationships between actors. This part also contains an analysis of the Euroregion as a participant in the policymaking processes.

2 Theoretical Background

According to David Knoke [1, 2], theories of social networks formulate three underlying assumptions about the interaction between the networks and their actors. The first assumption indicates the existence of constant patterns of repeated interactions in networks, which connect actors belonging to structures in complex social systems. According to the second, the subjects of network analysis are not the features of their participants, but, instead, the relationships between them. The third relates to the impact of networks – in which actors are embedded in their perception, attitudes, and actions, and vice versa – on the changes that actors cause in the network.

One of the factors contributing to the formation of bonds between members of the social network is homophily. As McPherson et al. state:

> Homophily implies that distance in terms of social characteristics translates into network distance, the number of relationships through which a piece of information must travel to connect two individuals. It also implies that any social entity that depends to a substantial degree on networks for its transmission will tend to be localized in social space and will obey certain fundamental dynamics as it interacts with other social entities in an ecology of social form. [3]

The features that bring organizations closer together are, primarily, their type, which is closely related to the legal and organizational framework that determines the ways and scope of their activities. An appropriate level of homophily in the network is one of the

---

[1] The European Outline Convention on Transfrontier Co-operation between Territorial Communities or Authorities (May 21, 1980). The European Charter of Local Self-government (October 15, 1985).

conditions of the activity undertaken by its actors. While being similar is important from the perspective of establishing relationships, it does limit innovation and the inflow of new information and also, through the similarity of the actors' experience, its possible interpretations [4]. In addition, a low level of diversity among the actors within the network can limit the network's adaptive capacity [5]; this is an important attribute of governance networks and is one of the factors of their effectiveness in implementing public policies.

Understood in this way, the participation of social networks in the creation of public policies has been the subject of many authors' considerations [1, 6–8]. Their findings relate to the composition of policy networks, the flows between actors, and the nature of the relationship between them. Political networks include

> a relatively stable set of mainly public and private corporate actors. The linkages between the actors serve as channels for communications and the exchange of information, expertise, trust and other policy resources. The boundary of [a] given policy network results from processes of mutual recognition dependent on functional relevance and structural embeddedness. [7: 41]

These features are represented by governance networks. They are

> more or less stable patterns of social relations between mutually dependent actors, which cluster around a policy problem, a policy program, and/or a set of resources[,] and which emerge, are sustained, and are changed through a series of interactions. [9]

Governance networks are created by the representatives of authorities from various levels of social life (local, regional, national, and international) as well as representatives from other sectors (private and civic). This type of composition makes it possible to improve the coordination and quality of co-created public policies and public services. A network management process of this type consists of activities conducive to the maintenance of the interaction between actors, such as their activation and diagnosis, and the collection of information of which they are the source. The objectives of the action are negotiated during the interaction without clearly distinguishing the various stages of the political process, for example, the formulation and implementation of public policies and the provision of services. In this way, actors are mutually dependent, on the one hand, but also contribute to the flexibility and efficiency of the actions taken, on the other [9–11]. These are the conditions that are implemented in the multi-level governance system in the EU.

MLG includes procedures for making decisions and regulating the political negotiation process. What is characteristic of this model is the extension of the hierarchical course of the decision-making process by non-hierarchical elements as a result of including entities representing all sectors: public, non-governmental, and private [12]. In this way, political processes are decentralized, with regions becoming important actors. Decentralization results in the regions expanding their competences in the area of creating and implementing public policies [5]. Therefore, the regions are at the center of the EU's territorial policy, and their appreciation is a manifestation of the regionalization process taking place within it. 'The goal of regionalization is to intensify the subjectivity of regions and the role of their bodies in the decision-making procedure not only at the local level, but also at the national, European and international level' [13]. In the literature on the subject, regions are presented as being the result of grass-roots identification and cultural separateness, or institutional design and formation. They are defined as territories established by familiarity and regional identity, or as constructs formulated by the elites [14]. Regions are historically shaped structures whose institutionalization is based on their territorial, symbolic, institutional, and identity dimensions [15].

Under the concept of regionalization, regions are recognized as critical entities in meeting the needs of local communities [16]. Although they are characterized by smaller or greater degrees of autonomy [17], the sources of regionalization can be seen in the intentional actions of central authorities [13]. An example of such an impact are instruments (administrative, organizational, financial), which are created within the EU to support regions. Today, the functioning of cross-border regions in Europe is closely related to the process of European integration and EU regional policy [18]. The main tool for supporting links between local and regional actors along EU borders is the Interreg program, created in 1990, and its component A, which is dedicated to cross-border cooperation. The issue of cross-border multi-level governance is key to understanding both the driving forces of, and obstacles to, territorial integration in Europe [19].

Borderlands are places where diverse ideas and values come into contact, which sometimes concur and other times conflict. This situation stimulates the communities inhabiting them to connect (hybridize) and search for new ways of thinking and acting [18]. In pursuit of this, the process of forming Euroregions on the borders within the current EU was initiated in the 1950s. Euroregions are defined as a type of formalized and institutionalized cooperation between subnational authorities and (potentially) private actors who are located close to a border in two or more countries [20]. They are appointed to support the processes of creating a community based on ties that meet the needs of their participants and border residents [21]. As a result of their activities, social capital is created to increase the efficiency of regional and local communities [22].

One of the main ways to implement the functions of Euroregions is to create suitable conditions for the emergence, dissemination, and consolidation of cross-border partnerships that implement joint activities in various areas of the life of local borderland communities. In this way, on the one hand, the actions of EU regional policy concerning the border are realized through the activities of partnerships; on the other hand, this activity takes place while maintaining the freedom of the partners in formulating their goals and the ways of achieving them.

We assume that the effectiveness of Euroregions as social networks can be measured by their activity in initiating partner projects, and the frequency and nature of cross-border contacts between residents of the border regions [23, 24]. Conversely, shaping relationships between the partners of joint cross-border projects is an essential element in the development of networks [25, 26]. Relations between network actors have an impact on the thematic scope, durability, and effectiveness of cooperation, as well as its nature and strength. The use of the network paradigm in the analysis of Euroregions allows us to pay attention to the conditions and consequences of their construction (distribution of positions, density, diversity) and the goods supplied to each other (nature and direction of flows) [27].

The Euroregion features mentioned above, lead us to perceive the regions not only as social networks, but also as governance networks. The activities of Euroregions as governance networks are in response to the particular living conditions created by the border. They serve to meet local needs, and overcome special restrictions and differences through adaptation and negotiation, accompanied by daily contact and the exchange of tangible and intangible goods. According to Marcus Perkmann [28], the participation of civic sector entities in the implementation of EU programs is a key condition for sharing responsibility for the functioning of the public sphere in Euroregions. This enables trusted actors with high competences and social capital to be found and engaged, which is necessary to achieve the goals of strategic programs. Besides, the sustainability of cooperation is supported by the creation of informal networks [23, 24]. Personal relationships, on the other hand, promote effective collaboration between organizations and the reproduction of role models [29].

Other authors have addressed the problem of the Euroregions' participation in political decision-making processes. Attention has focused mainly on particular aspects of the functioning of the public sphere (e.g. transportation) [19, 30]. Sara Svenson [31] presented the determinants of local government contact networks within cross-border Euroregions, while also maintaining the theory that although Euroregions lack executive power, their participation in all stages of the decision-making process is possible (agenda setting, policy design and formulation, adoption, implementation, evaluation, and reform) [32].

In our opinion, Euroregions meet the conditions for governance networks. At this point, we formulate a thesis that *the SNBE is a stable set of interdependent public, private, and social entities, operating on the basis of stable relationship patterns, centered around needs and social problems within the implementation of public policies, using both the resources of individual network participants and those created as a result of its operation, thus implementing the principles of a governance network.* The next part of the article contains a description of the research procedures we used to collect data, on the basis of which it is possible to check the validity of the above thesis.

3 Sourcing Foundation of the Analysis

The presented analysis is based on the results of research conducted in the SNBE during 2019–2020 by applying two methods: social network analysis and unstructured interviews. The goal of the SNA was to present the structure of the Euroregion as it was in 2007–2013. To reconstruct it, we used data on 57 projects that were implemented under Interreg IVA, and published on the European Funds Portal. The data contained information on the composition of partnerships, lead beneficiaries, the thematic scope of projects, among other things. The rationale for choosing such a data source was, (1) the availability of information on projects implemented by the Euroregional network under local public policies, and (2) the ability to access data on completed projects and therefore assess their implementation, relevance to local communities, and the degree of satisfaction felt by residents. During our research, newer projects were still being implemented as part of the Interreg VA, in force during 2014–2020, which prevented us from accessing information that was of interest to us. Eighty organizations from the SNBE participated in the projects. Examples of these organizations are mentioned in the presented text, and in Fig. 1, they are identified by numbers. Each organization's associated number can be found in the Table 1. The list of the participants in partnerships determined the direction taken in seeking participants for the unstructured interviews.

In total, 11 interviews were carried out with persons representing the beneficiaries of leading projects. In this way, we managed to information on the course of 34 projects. Eighteen leaders were entities from Poland, and 16 from Germany. The interviewees were people selected by the surveyed organizations as having the most knowledge about the course of the projects. The collected information primarily concerned the principles of partner cooperation, the methods of communicating and making decisions, as well as the methods used to inform the local community about projects and get feedback on the residents' satisfaction with the results. Based on the interviews, we conducted qualitative analyses focused on the study of networking patterns and principles [33]. The audio recordings and the interview transcripts are stored at the University of Zielona Góra. In this text, the interviews, which include the premises for the conclusions being drawn, have been tagged with the letter 'I' and a number (e.g., I1).

4 Circumstances of Cross-Border Cooperation in the Polish-German Borderland

A border region's features create the context for the Euroregion's functioning. The Polish-German border region represents a new borderland type [34]. Apart from the relatively short period in which Polish-German relations were created, it demonstrates the contact nature of the border resulting from cultural, religious, and linguistic diversity, as well as the coexistence of communities distant from each other due to their level and quality of life. Four stages can be distinguished in cross-border contacts between Poland and Germany: (1) the opening of the border in 1972, beginning a period of intensive contact (lifting the visa requirement resulted in several dozen million people crossing the border on average) and the borderland's process of transition from administrative to social [35]; (2) Poland's unilateral closure of the border in 1980, which limited links between organizations and institutions, ultimately stopping them in the mid-1980s; (3) reopening of the border in 1989, thus initiating intensive processes that shaped Polish-German relations [36]; and (4) completing the process of border opening through Poland's accession to the Schengen Agreement in 2007.

In 1990 the Border Treaty[2] was established and in 1991 the Treaty on Good Neighborliness and Friendly Cooperation[3]. These agreements were the basis for the development of cross-border cooperation and Euroregions along the Polish-German border. However, the social conditions of cross-border cooperation have changed rapidly, and its leaders have faced the specificity of recognizing the nature of the state border and its reception by the residents, as well as the low level of confidence in social relations within the Euroregion [37] as a result of a common war and post-war past. The effect of closing the border was to consolidate stereotypes and the dislike and distrust of neighbors from across the Oder. At the beginning of the 1990s, the foregoing barriers to cross-border cooperation also include the polarization of the social structure of borderland residents, indicated by a significant disproportion in income and standard of living. In particular, the perception of the asymmetry in standards of living along the border was one of the reasons Poles reached for innovative methods of achieving economic resourcefulness [38], including intense illegal phenomena (theft, smuggling, prostitution).

[2] Treaty between the Federal Republic of Germany and the Republic of Poland, 1990.

[3] The Polish-German Treaty of Good Neighborliness and Friendly Cooperation, 1991.

The response to these challenges was to establish the SNBE. The Euroregion was founded in 1993 as a result of an agreement between the Polish Association of Municipalities, *Euroregion Sprewa-Nysa-Bóbr* and the German association *Euroregion Spree-Neisse-Bober*. The procedure for its creation had the features of a local government model. Its primary goal has been to undertake comprehensive actions along the Polish-German border, economic and cultural development, and the continuous improvement of living conditions of the inhabitants, with particular emphasis on attempting to eliminate disproportions in their economic situation. For this purpose, the Euroregion began to raise funds to implement EU projects during the pre-accession phase. This was the first stage of the Euroregion's operation, and lasted until 1995. Access to EU funds began in 1995 under the PHARE CBC program, which continued until 2003. This was the second phase in the development of the Euroregion's operation, in which we observed substantial growth and the development of a large number of partnerships. From 2005 (the last stage: the cooperation's consolidation), the use of funds from structural programs began. These were Interreg IIIA 2005–2006, Interreg IVA 2007–2013, and Interreg VA 2014–2020. This was a particularly crucial stage for initiating partnerships arising from cross-border projects and was determined by two main factors: Poland's accession to the EU, and the possibility of financing projects under the Interreg program[4]. As a result of implementing new possibilities for using funds for infrastructure projects, the accession to the Interreg program has, additionally, changed the way tasks and objectives set for the Euroregion are implemented. In addition to statutory activities, large-scale participation in the management of EU funds began with issuing opinions on the financing of 'soft' projects, focusing on levelling disproportions, but also shaping social ties on both sides of the border, and building network projects, allowing for the creation of new infrastructure solutions.

5 The Spree-Neiße-Bober Euroregion as a Governance Network

The purpose of the Euroregion is to implement public tasks. In network terms, an essential aspect of this activity is its efficiency in performing the functions assigned to it, the adequate allocation of resources, and directing the participants' activities in a way that guarantees their proper implementation. With such tasks formulated, it is vital to implement an appropriate management process. This is about creating opportunities for an intensive exchange of opinions and views, and for actively dealing with everyday problems in a given community. In diagnosing this process in the SNBE, we found it important to analyze the goals and tasks of the partnerships, the principles of cooperation they adopted, the principles of communication used, and the methods for making decisions. It was also important to analyze the diversity of the network participants and the challenges of collaboration. This chapter consists of three parts relating to the above-mentioned features of the network under examination. These are: network structure, cooperation principles, and decision-making processes.

[4] The development of the cross-border cooperation network in the SNBE was described in detail in previous publications of the authors of this article [23, 24].

5.1 Network Structure

During 2007–2013, the cooperation network in the SNBE consisted of entities of various statuses, goals, and functions, formally autonomous but operationally interdependent on each other [7]. An analysis of documents showed that it was created by representatives of local government and the public institutions that were subordinate to them, social organizations, and private enterprises. Each partnership had at least two entities, which were diversified in terms of origin (Poland, Germany), and in some cases, in terms of the sectors they represented (public, civil, private). *Disconnected partnerships* were found in these types of networks. In general, there were pairs of entities, trios, and larger systems. Most often, these were event partnerships implementing one joint project. The composition of the network is presented in Fig. 1.

Local government entities (31 entities) and non-governmental organizations (30 entities) were most frequently represented in the SNBE cooperation network. The network also included 16 public institutions subordinated to local governments, and three private enterprises. Private enterprises were found only in the German part of the network. This is the result of differences in the countries' legal systems, and difficulties in implementing public-private partnerships in Poland. Insofar as local government institutions assumed the role of cooperation leaders (often alternately), non-governmental organizations and

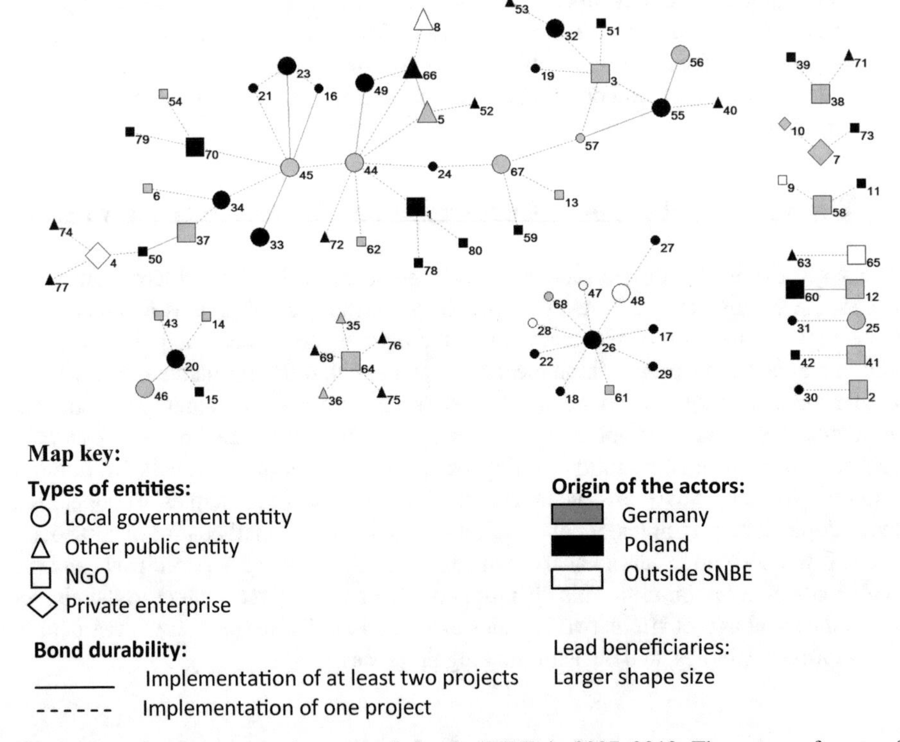

Fig. 1. Cross-border cooperation network for the ESNB in 2007–2013. The names of some of the entities that are indicated by numbers can be found in Table 1.

private enterprises were assigned expert tasks related to the care of the substantive content of tasks to be implemented, and their courses. The possibility of utilizing the knowledge and experience of their participants in the event of sudden problems in the implementation of project tasks was also considered particularly valuable (I1, I4, I11).

The leading beneficiaries of the 39 projects were local government entities, 12 non-governmental organizations, three other public institutions, and three enterprises. Adoption of the leadership role by local governments resulted from the organizational and financial resources they managed, compared to the resources of other types of entities, their size and scale allowed them to maintain the organizational solutions necessary to support projects for two to three years (relevant departments and jobs). Partnerships also operated within the network, of which social organizations were the leading beneficiaries.

As part of our analysis, we also found three project partnership cases led by private companies. An example of this was the cooperation between the NGO Western Chamber of Commerce and Industry, and two private enterprises: EGC Development Agency Cottbus GmbH and Centrum für Innovation und Technologie GmbH, which was the project leader. This collaboration's purpose was to support the development of networks of small and medium enterprises, and research units.

The numbers used in Fig. 1 and indicated in this paper are listed below along with the names of the entities and actors they represent.

Most often (35 times) the lead beneficiary's obligations were implemented by entities from Germany. Twenty-two leading beneficiaries came from Poland. This result may undermine the conclusion in the literature that German partners demonstrate lower interest in cross-border cooperation. At the same time, German entities have a more exceptional ability to cover their contribution and a greater orientation towards the implementation of infrastructure projects that are measurably beneficial for their local communities. These entities from Germany are not, however, strongly interested in the implementation of soft projects. This is evidenced by our observation of a much lower proportion of leading beneficiaries in the implementation of 'soft projects' from the Small Projects Fund (SPF)[5].

Organizations on the Polish side of the Euroregion were definitely more active in searching for partners and preparing project applications than German organizations (in 2007–2013 under the SPF, the Polish part of the SNBE carried out 1,755 projects, for a total amount of € 11,765,614.35, while the German part had 1,010 projects for the amount of € 4,047,486 [23, 24]). Our collected data indicate that the part of the EU structural funds dedicated to border regions is an important source of funding for activities aimed at achieving the objectives of various local government public policies, and thus stimulates efforts to obtain them.

In the analyses presented here, we found associations between entities became permanent when they jointly implemented more than one project. Bonds of this kind were characterized by partner homophily. Stable relationships were maintained, primarily, between local government entities. The exception was the partnership between the *Spree-Neisse-Bober* Euroregion e.V. NGO (12), and the Association of Municipalities of the Republic of Poland Euroregion *Sprewa-Nysa-Bóbr* (60) – the German-Polish organization that created the SNBE. Two features primarily favor the maintenance of relations

[5] European Regional Development Fund as part of the Cross-border Cooperation Operational Programme – Poland (Lubuskie) – Brandenburg 2007–2013.

Table 1. Names of participants of the network

Name	No.	Name	No.
Association of Municipalities of the Republic of Poland Euroregion Sprewa-Nysa-Bóbr	60	Lubsko	23
Brody	16	Marina Winterhafen Sportboot Frankfurt (Oder) e.V. association	61
Bytom Odrzański	17	Marketing und Tourismus Guben Association e.V	43
Centrum für Innovation und Technologie GmbH	7	Nowa Sól	26
Chamber of Commerce and Industry of Cottbus	37	Organization of Employers of the Lubuskie Region	50
Cottbus	44	Polizeisportverein 1893 Forst (L.) e.V.	54
EGC Development Agency Cottbus GmbH	10	Provincial Sports and Recreation Center in Zielona Góra	70
Eisenhüttenstadt	68	Sports Club of Zielona Góra	79
Forst	45	Spree-Neisse-Bober Euroregion e.V. NGO	12
Frankfurt [Oder]	48	Sulechów	29
Górzyca	18	Technical University of Cottbus	5
Guben	46	Trzebiechów	32
Gubin	19	University of Zielona Góra	66
IHK (Industrie- und Handelskammer Cottbus)	37	Western Chamber of Commerce and Industry	73
Jasień	21	Zielona Góra	49
Krosno Odrzańskie	22		

between territorial self-governments and the above-mentioned social organizations. The first of these were organizational and financial resources. Actors that maintained lasting relationships with partners have, within their structures, units or positions responsible for continuing cooperation with foreigners. A similar explanation was related to the dominance of local government entities among the beneficiaries of leading projects. The second feature was in taking action based on long-term strategic documents, in which cross-border cooperation was one of the tools for achieving the goals. Besides, the sustainability of relations between both parts of the Euroregion is inscribed in the essence of its functioning. It is a necessity arising from the obligations contained in its statute, and structural funds support this sustainability. Part of these funds are transferred to Euroregions as SPFs. The event nature of the remaining partnerships resulted, primarily, from deficits in organizational and financial resources. In the case of NGOs, an additional factor was the spontaneity of their activities and the dominance of the project financing method.

5.2 Cooperation Principles

An analysis of the goals of the partnerships indicates their focus on providing public services and solving social problems. The main purpose of partnership cooperation was to improve the quality of life in the Euroregion. The analyzed projects concerned

1. Communication projects: the construction of bridges, roads, and bicycle paths (e.g., cooperation between the cities of Brody, Forst, Lubsko, and Jasień, and between Gubin and Guben);
2. Projects related to the development of tourism, such as renovating tourist information centers, revitalizing the downtown rivers and quays' environment (cooperating cities: Zielona Góra and Cottbus, and Forst and Brody), including water tourism (cooperating communes: Nowa Sól, Bytom Odrzański, Sulechów, Krosno Odrzańskie, Górzyca, cities: Eisenhüttenstadt, Frankfurt [Oder] and the Marina Winterhafen Sportboot Frankfurt [Oder] e.V. association);
3. The revitalization of leisure time places, for example, the revitalization of parks and monuments, and the planetarium building (cooperating cities: Brody and Forst, Cottbus and Zielona Góra, and the Technical University of Cottbus and the University of Zielona Góra);
4. Development of economic cooperation, seen as raising workforce competences, organizing marketing centers, and promoting craft (Chamber of Commerce and Industry of Cottbus, Organization of Employers of the Lubuskie Region, Gubin and the Marketing und Tourismus Guben Association e.V., and Gubin and Guben),
5. Creating cross-border neighborly relations: construction and renovation of Polish-German meeting centers (implementation: commune of Trzebiechów, Zielona Góra, and Cottbus); and
6. Sport and recreation promotion: expansion of sports facilities (cooperating cities: Gubin and Guben, as well as the Provincial Center of Sport and Recreation and the town of Forst).

The tasks implemented in the partnerships concerned infrastructure changes. Soft projects financed from the SPF were also carried out on these occasions. These were used to recruit participants for meetings organized as part of structural projects, such as training, international sports competitions, and classes at meeting centers, and were not only aimed at implementing the indicators written in projects, but their task was also to shape new patterns of behavior, establish relationships between borderland residents, and tame otherness.

The participants of the study were dominated by a belief in the high managerial and executive potential of the partners. All interviewees talked about the transparency and openness of the decision-making process, having competent project staff, having a high capacity to mobilize local resources for project implementation, and adapting project tasks to the needs of residents (I1–11). The experience gained from long cooperation was indicated here, providing the possibility for new projects to be quickly developed, and for the effective distribution of roles and tasks (I2–4, I5–8). The adoption of such a standard in partnerships could indicate an attitude of conservatism. This served to consolidate and replicate good collaboration patterns. The frequency of contacts in infrastructure projects depended on the location of the investment. Due to these locations, we distinguished two types of projects. In the first, the investment was implemented on each partner's territory, for example, on the border or simultaneously on both sides (e.g., the construction of a bridge or marina). Such projects required more frequent partner meetings, usually once a week (e.g., I4, I5). This was mainly due to differences in construction laws between Poland and Germany. The effect of this was to necessitate ongoing arrangements by the partners as to the next stages of the investment. The second type of project focused on activities on one side of the border, usually at the headquarters of the lead beneficiary. In these cases, meetings were arranged in a monthly (I2, I3) or quarterly (I3, I10, I11) regime. In all conversations, however, it was emphasized that meetings were always held as needed, regardless of the previously agreed schedule. In urgent or difficult situations, cooperation was strengthened. The meeting places were the partner headquarters or places where the task was being implemented, for example, a bridge construction site, the site of the reconstruction of an information center, or the SNBE headquarters. They always shared information about the progress of works, difficulties, and ways of solving them, even when the implemented infrastructure solutions concerned only one side of the border (I1–11). For example, informant 11 claimed: 'The cooperation consisted of the fact that there were these meetings, there were contacts, there were emails with the German side, i.e., mainly with the person responsible in the Forst City Hall as well as in this PSV Club.' While informant 1 said:

> Our main task was carried out on the Polish side. However, together the history of the place was sought and an attempt was made to search several archives. Partners from Germany were mainly involved in this – they were only available on the German side. The Polish burned it down during the war. There was also no cooperation in renovating the building. Everything was done following Polish law and by Polish companies. But we informed each other about everything. [...] And when we found the wall paintings and were looking for additional funds for conservation works.

Meetings were mostly spoken in German. Partners from Poland demonstrated their communicative level, but professional translators were used for some of the specialized translations. The meetings were attended by a permanent core group (leaders from both sides) and additional people were invited as needed. These constituted the remaining parts of the crews of the cooperating organizations as well as other experts, as in the case of the construction of a bridge in the Oder-Spree and Krosno (I5) counties, or the construction of a meeting center (I6). This indicates the great ability of the network to integrate the resources of its participants, serving both public interest and the particular purposes of its participants. An example of such a situation was the cooperation between the Provincial Sports and Recreation Center in Zielona Góra and the city of Forst, together with riding associations from Poland and Germany (Sports Club of Zielona Góra and Polizeisportverein 1893 Forst (L.) e.V.) In one of these projects, when it turned out that the amount allocated for a fence was insufficient (a fence for competition parkour in equestrian tournaments), consultations with association representatives allowed a cheaper model of fence to be developed, but that still met the requirements of the games. This is what the informant 1 told about this situation:

> Yes, that's exactly when the people from the association were at the meetings. And it was important. These were typically construction meetings, including [those] associated with this fence to solve the problem, but these were Polish-Polish meetings. Nevertheless, they were all the time. [...] In this type of talk about problems in implementation, the core [group] of the activity was the main one, and the remaining people were invited as needed.

There were also difficulties in cooperation within the network we examined. The three most important were imbalances in levels of interest, procedure, and organization of work. However, these did not constitute complete obstacles to the implementation of tasks; instead, they were challenges that the partners were aware of and able to deal with. In their opinion, however, these difficulties limited the dynamics of cooperation and led to public needs going unmet, and unused resources on the network.

The imbalance in levels of interest concerns one party's willingness to implement projects being less than the other, which is characteristic of the German side. This is the result of differences in the socio-economic development of the Polish-German borderland and less need for infrastructure changes on the German side. In the qualitative research, information about the imbalance in levels of interest in working in partnerships comes from Polish network participants. They spoke of difficulties in finding partners from Germany who had an 'interest' in project activities and were therefore less willing to cooperate. There are partnerships in the network under analysis in which Polish representatives testified that their partners never refused to sign a partnership agreement with them, and also those who reported a lack of interest in their offer of cooperation among actors from Germany. An example of this is the lack of expansion of water infrastructure in the 'Odra for tourists' project, or the termination of cooperation on a sports association. The research participant 4 confirms these observations by pointing to the difficulties in involving Germans in projects which, in their opinion, will not benefit them:

> Interreg projects give the chance for this soft sphere to function and to cooperate. But I will also say so critically. Germany cooperates as long as they have an interest and see it as an interest. Because it is the interest of two parties, after all. If we are to talk about it this way, then this business should always be for the benefit of two parties. And it is challenging today to convince the German side to carry out further tasks for our Odra river actions, because their side does not see these benefits.

Procedural imbalance concerns different operating principles for functioning organizations in partnerships. These differences are particularly evident when comparing the organizations' operating policies on both sides of the border. Participants from Poland point to the existence of many documents regulating proceedings in administrative matters in Germany that do not exist in Poland. Despite the high flexibility of Polish network participants and their quick response to tasks, this extends the number of procedures used, and slows down the execution of tasks; for example, due to this issue, it took ten years to prepare a bridge construction project (Krosno County and Oder-Spree County). The informant 4 recalled differences in planning and strategic programming:

> The German side looks like it deals with large projects differently. They often have specific concepts, specific programming, and strategic documents. We were very lame in this stage in those years. We had ready ideas. Ready ideas and ready project documentation. However, some programming, some strategies, some attachments to concepts were new to us.

While informant 3 said about the differences in setting directions for social and economic development:

> One of the discrepancies concerned different development goals of Lubuskie Province and Brandenburg. The Polish side wanted to support tourism. On the German side, tourism was not so important. On the German side, the metal industry (metalworking) was more important. For this reason, a compromise had to be found.

And informant 4 also reflected the prolongation of decision-making periods due to the large number of regulations and documents:

> For Germans, the period for making decisions was so large that we took a team and drove to them and settled matters. We took Mayor Tyszkiewicz as a leader, we drove, and we dealt with the matter at the base, which in Germany, among our partners, aroused sympathy. Maybe not necessarily adapting to our principles, but liking that we approach the topic so ambitiously.

The third imbalance, related to the organization of work, refers to a different way of working and a way of seeing project situations. It also implies adopting different performance strategies for approved tasks. Polish network participants emphasized the excellent organizational ability of their German partners, their timeliness, and their compliance with rules once they had been established. The German participants in the partnerships pointed to the innovation and spontaneity of Poles as well as their great determination to act. As informant 4 recounted: 'But I would say we [Poles] were more creative and they [Germans] were more literal in understanding some things. And that had to be reconciled.' While informant 3 claimed: 'In Germany, preparations for ventures start 3–4 months earlier. This is the normal rule. The Polish side usually starts shortly before the deadline, spontaneously, it gives a sense of tension, but after 15 years, IHK [Chamber of Commerce and Industry] got used to it.'

Providing instruments to deal with such situations is one of the goals of the Euroregion's operation.

5.3 Decision-Making Processes

In all partnerships, decision-making was carried out according to the scheme/rules that resulted from the conditions of applying for and using funds from Interreg IVA (I1, I9). One of the necessary conditions for receiving funds was the implementation of three out of four partnership criteria: joint project preparation, joint financing, joint staff, and joint implementation. In the partner projects of the Cottbus and Zielona Góra municipalities, these criteria were implemented as follows. Representatives of both sides worked on the preparation of the project. Appropriate units were established in Zielona Góra (Department of European Funds) and Cottbus (Mayor's Office) to prepare and submit applications, develop annexes, and settle the project. The city hall in Zielona Góra was represented by the head of the above department, and the city hall in Cottbus by the mayor and the head of the mayor's office. The relevant departments in these offices were responsible for substantive issues. The lead beneficiary had power of attorney only to prepare the application document; other decisions were made jointly. During the project, each party made decisions about the project's implementation independently. But in the event of changes in the content of the submitted application, the relevant partner (e.g., Zielona Góra) was obliged to apply to the lead beneficiary (e.g., Cottbus) for appropriate corrections, who then considered the case and potentially approved the changes. The application for change was examined in partnership. It was possible to make two changes to the project in this way during its duration. Receipt of funds followed the development of a report on the activities carried out by the lead beneficiary, with confirmation by the partner(s). On this basis, certification and the payment of funds took place.

Negotiating decisions and seeking consensus in managing cross-border projects was another factor in the shaping of practical cooperation rules [11]. This is evidenced by the fact that, although the formal, legal, and procedural problems caused the partners difficulties in implementing the projects, they did not weigh on their implementation or the satisfaction of the beneficiaries. Importantly, new projects implemented in the same partnerships have been and are being submitted on the basis of the first ones. This means that the analyzed network has a relatively high capacity for self-regulation and learning. This allows it to continue working while, at the same time, avoiding mistakes. Additional decision-making methods were also used in this network. These included giving opinions on solutions (I5), and establishing agreements and monitoring committees (I4). As informant 4 reflected:

> The project was far reaching in the formula of democratic principles. We even had a body like the monitoring committee of the project. However, reconciling the interests of nine communes [...] believe me, as a leader I had a communication problem, because it was to embrace these nine communes and democratically make every decision; it was not easy. This body was created. And sometimes we forced some things very hard on this committee in voting.

Most often, however, participants talked about the unanimity resulting from many years of cooperation in partnerships. Therefore, we can say that the examined network has principles conducive to shaping dialogue and conciliatory attitudes, while reducing those that are confrontational, indicating a particular situation in the implementation of public tasks in the Euroregion.

The social sense of creating public policies lies in the fact that this process itself should be a necessary part of the mobilization of the community for successful strategic goals to be achieved. To make this possible, it is crucial to participate in the debate on the purposes and principles of the operation of broad social groups and individual citizens involved in the process. In the SNBE collaboration network, project ideas were usually not the result of consultations with residents. Instead, they were the result of analyzing the municipalities' strategic documents and understanding the needs of local governments. Project leader 4 said: 'Tasks are written in strategies. We do not ask residents in person, but there are diagnoses in which residents told us what they would like.'

The representatives of offices contacted residents at the design and implementation stage, however, these were not regular meetings but rather occasional contacts (I4). It was also pointed out that representatives sometimes had a private circle of people whose opinions were considered reliable and sufficient (I6). The interviewees also referred to social analyses conducted for the development of public strategic documents in which residents informed the authorities about problems and expectations related to their immediate environment (I1, I4). Promotional activities included meetings with the media (I1–11), interviews and press conferences (I1, I4, I5, I9–11), meetings with residents (I5), the use of information boards (I1–11), and placing information on the offices' websites and those organizations cooperating with them (I1–11). Thus, we can say that the dominant form of contact between project partners and residents was informal, unsystematic, and unorganized. However, this did not affect the overall reception of the projects. During the interviews, press mentions of projects were shown, photos of project results, as well as information on the ways resident used them. What was also pointed out was the specific, positive changes within the Euroregion regarding (1) the establishment of new enterprises, the development of agritourism, the economic development of the region in relation to the construction of roads and bridges (I1, I2, I5); (2) increasing the number of tourists by renovating monuments, waterfronts, parks, and entire tourist routes (I4, I7, I8, I11); (3) intensifying cooperation in the field of sport (I6, I11); and (4) establishing cross-border relations between the inhabitants of each region concerning the participants of training and classes in meeting centers (I3, I7, I11).

6 Conclusion

Euroregions as social networks are created to solve a common problem (or problems) that cannot be resolved by mobilizing the members of the network's own resources due to the residents of local communities living along the border belonging to at least two political systems, and the impact this has on their lives. From this follows the interdependence of the entities operating on the border. This interdependence entails the need to search for new ways of achieving goals, for example, the development of institutional procedures, or creating relationships. The network co-management model highlights the central importance of interactions between actors of different statuses in the process of consulting, agreeing, and deciding on common issues. The shape and consequences of these interactions are determined by the ability of the partners to control network relationships. The approach to network coordination specific to public co-management is characterized by the diversity of the cooperation partners, the use of negotiating decision-making methods, and the decentralization of cooperation.

In this article, we have analyzed the functioning of the cross-border cooperation network of the SNBE as a governance network. In conclusion, we tend to accept the thesis that the cooperation network in this Euroregion has met the conditions for a governance network. The results of the presented research indicate that the shape of the SNBE network and the rules it functions by are the result of both political and bottom-up impacts.

The network consists of many entities with various statuses, goals, and functions, formally autonomous but operationally interdependent. An important factor in creating partnerships was the homophily of the actors, in this case, the type of organization. Moreover, contrary to what might have been expected, this homophily was conducive to the durability of ties and the creation of activities, which was manifested in undertaking subsequent joint projects. Unlike other types of network actors, local governments had adequate financial resources and staff to ensure the continuity of international cooperation. The most advanced level of the institutionalization of local government activity also seems to be of key importance, as it includes the rules for establishing cooperation with other units, in which international cooperation has a special place.

The analyzed network uses a variety of decision-making mechanisms to reach a consensus. The network also demonstrates its ability to integrate the resources of its participants, serving both the attainment of public goals and the specific goals of the network participants. However, difficulties in cooperation do appear within it, taking the form of imbalances regarding interest in cooperation, applied procedures, and work culture. These are the result of the participation of various entities in the implementation of joint action programs. This is a key condition for taking joint responsibility for the functioning of the Euroregions' public spheres; providing instruments to deal with such situations is one of the goals of the Euroregion's functioning – sometimes, however, it is a challenge to achieve.

The SNBE is the result of matching functions, tasks, and responsibilities to the resources of the network and the possibilities of the partners' activities. As a result, it developed a new dynamic action, based on the participation of many different entities in solving public affairs and involvement in causative activities. The SNBE cooperation network is characterized by an ability to self-regulate and learn, which is associated with a replica of good collaboration patterns, and to develop the ability to respond to imbalances related to interest in cooperation, applied procedures, and work organization. This will determine the implementation of public services and local policymaking in the future.

References

1. Knoke, D.: Changing Organizations: Business Networks in the New Political Economy. Boulder, CO, Westview (2001)
2. Granados, F.J., Knoke, D.: Organized interest groups and policy networks. In: Janoski, T., Alford, R.R., Hicks, A.M., Schwartz, M.A. (eds.) The Handbook of Political Sociology. States, Civil Societies, and Globalization, pp. 287–309. Cambridge University Press, Cambridge (2005)
3. McPherson, M., Smith-Lovin, L., Cook, J.M.: Birds of a feather: homophily in social networks. Annu. Rev. Soc. **27**, 415–444 (2001). https://doi.org/10.1146/annurev.soc.27.1.415
4. Folke, C., Colding, J., Berkes, F.: Synthesis: building resilience and adaptive capacity in social–ecological systems. In: Berkes, F., Colding, J., Folke, C. (eds.) Navigating Social–Ecological Systems: Building Resilience for Complexity and Change, pp. 352–387. Cambridge University Press, Cambridge (2003)
5. Bodin, Ö., Crona, B., Ernstson, H.: Social Networks in Natural Resource Management: What Is There to Learn from a Structural Perspective? Ecology and Society (2006). http://www.ecologyandsociety.org/vol11/iss2/resp2/ Accessed 05 May 2020

6. Pappi, F.U., Henning, C.C.A.: Policy networks: More than a metaphor? J. Theor. Politics **10**, 553–575 (1998)
7. Kenis, P, Schneider, V.: Policy networks and policy analysis: Scrutinizing a new analytical tool box. In: Marinand, B., Mayntz, R. (eds.) Policy Networks: Empirical Evidence and Theoretical Considerations pp. 25–62. Westview Press, Boulder, Co. (1991)
8. Knoke, D.: Political Networks: The Structural Perspective. Cambridge University Press, New York (1990)
9. Klijn, E.H., Koppenjan, J.: Governance Networks in the Public Sector. Routledge, New York (2016)
10. Klijn, E.H., Steijn, B., Edelenbos, J.: The impact of network management strategies on the outcomes in governance networks. Public Admin. **88** (4), 1063–1082 (2010) https://doi.org/10.1111/j.1467-9299.2010.01826.x
11. Mandell, M., Keast, R.: A new look at leadership in collaborative networks: process catalysts. In: Raffel, J., Leisink, P., Middlebrooks, A. (eds.) Public Sector Leadership: International Challenges and Perspectives, pp. 163–178. Edward Elgar Publishing, United Kingdom (2009)
12. Piattoni, S.: The Theory of Multi-level Governance: Conceptual, Empirical, and Normative Challenges. Oxford University Press, New York (2010)
13. Bieniada, R.: Regionalizm i regionalizacja w definicji. Wybrane problemy teoretyczne [Regionalism in Definition. Selected Theoretical Problems]. Kwartalnik Naukowy Ośrodka Analiz Politycznych Uniwersytetu Warszawskiego 'e-Politikon' [e-Politikon. Political Science Quarterly of the Centre for Political Analysis of the University of Warsaw] (6), 281–299 (2013)
14. Klatt, M.: Mobile regions: competitive regional cocepts (not only) in the Danish-German border region. In: Andersen, D.J., Klatt, M., Sandberg, M. (eds.) The Border Multiple: The Practicing of Borders between Public Policy and Everyday Life in a Re-scaling Europe, pp. 55–73. Aldershot and Burlington Publisher, Ashgate (2012)
15. Paasi, A.: The resurgence of the 'Region' and 'Regional Identity': theoretical perspectives and empirical observations on regional dynamics in Europe. Rev. Int. Stud. **35**, 121–146 (2009). https://doi.org/10.1017/S0260210509008456
16. OECD/SWAC, Cross-border Co-operation and Policy Networks in West Africa, West African Studies, Paris: OECD Publishing (2017) http://dx.doi.org/10.1787/9789264265875-en
17. Kaczmarek, T.: Struktury terytorialno-administracyjne i ich reformy w krajach europejskich [Territorial and Administrative Structures and their Reforms in European Countries]. Wydawnictwo Naukowe UAM, Poznań (2005)
18. Sohn, C.: Cross-border regions. In: Paasi, A., Harrison, J., Jones, M. (eds.) Handbook on the Geographies of Regions and Territories, pp. 298–310. Edward Elgar, Cheltenham (2018)
19. Durand, F., Nelles, J.: Cross-border governance within the Eurometropolis Lille-Kortrijk-Tournai (ELKT) through the example of cross-border public transportation. CEPS/INSTEAD, Working Papers 16, (2012), https://www.liser.lu/?type=module&id=104&tmp=2652 Accessed 07 May 2019
20. Perkmann, M.: Institutional entrepreneurship in the European Union. In: Perkman, M., Sum, N.-L. (eds.) Globalization, Regionalization and Cross-Border Regions, pp. 103–124. Palgrave Macmillan, New York (2002)
21. Allmendinger, P., Haughton, G., Knieling, J., Othengrafen, F. (eds.): Soft Spaces in Europe: Re-negotiating Governance, Boundaries and Borders. Routledge, London (2015)
22. Andersen, D.J., Klatt, M., Sandberg, M. (eds.): The Border Multiple: The Practicing of Borders between Public Policy and Everyday Life in a Re-scaling Europe. Aldershot and Burlington Publisher, Ashgate (2012)
23. Frątczak-Müller, J., Mielczarek-Żejmo, A.: Cross-border partnerships – the impact of institutions on creating the borderland communities (the case of Spree-Neiße-Bober Euroregion). Innov. Euro. J. Soc. Sci. Res. **29**(1), 77–97 (2016)

24. Frątczak-Müller, J., Mielczarek-Żejmo, A.: Networks of cross-border cooperation in Europe – the interests and values. The case of Spree–Neisse–Bober Euroregion. Euro. Plan. Stud. **28**(1), 8–34 (2019). https://doi.org/10.1080/09654313.2019.1623972
25. Strihan, A.: A network-based approach to regional borders: the case of Belgium. Reg. Stud. **42**(4), 539–554 (2008)
26. Perkmann, M.: Construction of new territorial scales: a framework and case study of the EUREGIO cross-border region. Reg. Study **41**(2), 253–266 (2007). https://doi.org/10.1080/00343400600990517
27. Borgatti, S.P., Halgin, D.S.: On network theory. J. Organ. Sci. **22**, 1157–1167 (2011)
28. Perkmann, M.: Policy entrepreneurs, multi-level governance and policy networks in the European policy: The case of the EUREGIO. Lancaster University, Lancaster (2002). http://www.comp.lancs.ac.uk/sociology/papers/Perkmann-Policy-Entrepreneurs.pdf Accessed 01 May 2019
29. DeMarco, T., Lister, T.: Czynnik ludzki. Skuteczne przedsięwzięcia i wydajne zespoły [Human factor. Successful ventures and efficient teams]. Wydawnictwo Naukowo-Techniczne, Warszawa (2002)
30. Dörry, S., Decoville, A.: Transportation policy networks in cross-border regions. First results from a social network analysis in Luxembourg and the greater region. Euro. Urban Reg. Stud. **23**(1), 69–85 (2013)
31. Svenson, S.: The bordered world of cross-border cooperation: the determinants of local government contact networks within Euroregions. Reg. Federal Stud. **25**(3), 277–295 (2015)
32. Babu, S.C.: Policy processes and food price crises: a framework for analysis and lessons from country studies. In: Pinstrup-Andersen, P. (ed.) Food Price Policy in an Era of Market Instability: A Political Economy Analysis. Oxford University Press, Oxford (2014)
33. Fowler, J.H., Christakis, N.: Cooperative behavior cascades in human social networks. Proc. National Acad. Sci. **107**(12), 5334–5338 (2010)
34. Babiński, G. Pogranicza stare i nowe. Ciągłość i zmiana procesów społecznych [New and Old Borderlands. Continuity and change of social processes]. In: Krzysztofek, K., Sadowski, A. (eds.) Pogranicza etniczne w Europie. Harmonia i konflikty [Ethnic Borderlands in Europe. Harmony and Conflicts], pp. 69–92. Wydawnictwo Uniwersytetu w Białymstoku, Białystok (2001)
35. Kurcz, Z.: Powrót pograniczy. Entuzjazm, normalność, problemy [The Return of the Borderlands. Enthusiasm, Normality, Problems]. In: Frysztacki, K., Sztompka, P. (eds.) Polska początku XXI wieku: przemiany kulturowe i cywilizacyjne [Poland at the beginning of the 21st century: cultural and civilization chan ges], pp. 461–482. Polska Akademia Nauk, Warszawa (2012)
36. Council of Europe. European outline convention on transfrontier cooperation between territorial communities or authorities. European Treaty Series No. 106. Council of Europe, Strasbourg (1982)
37. Greta, M., Stanisławski, R.: The significance of Euroregions in the process of achieving social and economic cohesion in the European Union. In: Stanisławski, R., Greta, M., Maciaszczyk, A. (eds.) Structural funds for the socio-economic development of the Lodz Region in the perspective of experiences of Polish regions, pp. 213–225. Politechnika Łódzka Press, Łódź (2008)
38. Brym, M.: The integration of European Union borderlands: a case study of Polish opinions on cross-border cooperation along the Polish-German border. Univ. Fraser Valley Res. Rev. **3**(3), 15–33 (2011), http://journals.ufv.ca/rr/RR33/article-PDFs/2-brym.pdf Accessed 26 May 2019

Planning of Needs and Collective Actions. The Functioning of 'Mixed Networks' in Urban and Social Integration Process

Maria Camilla Fraudatario(✉)

University of Naples Federico II, Naples, Italy
mariacamilla.fraudatario@unina.it

Abstract. In such a singular context as the city of Naples in which the inclusion process of many refugees and asylum seekers is often weak and barely supported by the local institutions (for example in housing and educational policy), participatory planning among social groups might be the most efficient horizontal practice for answering to this issue. The starting point of this study is to consider the places of the city like devices in which relationships can be produced and transferred among citizens. Hence this contribution presents the main results achieved under a research-action project, This Must Be the Place (TMBP), that has aimed to work alongside with beneficiaries on creating social inclusion and to extend outcomes to many other actors through a rising effect. Social Network Analysis approach has been used to examine how individuals with different backgrounds play a relevant role within a network to facilitate inclusion in the urban context. Within the scope of the TMBP project, the following dimensions were examined: (a) reconstruction of the network of project participants and its evolution at different stages of the project; (b) understanding whether the relationships born within this facilitating context have been transferred outside the project (e.g.: in the places of the city and the daily life of participants); (c) understanding how the relationships have become essential resources to support the process of social and urban integration. This work shows that considering the interrelationship between urban space and horizontal network structures provides a strong basis for understanding and facilitating immigrant's inclusion.

Keywords: Integration process · Urban territory · Social networks analysis

1 Social Capital as Limit and Opportunity for Social Inclusion

Immigration policies and integration strategies are often linked to poorly coordinated practices among them, which produce uncertain outcomes, sometimes far from the real needs of the specific target group. This is partly due to governance models that are deficient in defining their missions. Furthermore, they employ very weak instruments to produce satisfactory results in terms of inclusion (first), and integration (second) in host contexts [1]. The weakness of these integration models (e.g. assimilation and

A. Antonyuk and N. Basov (Eds.): NetGloW 2020, LNNS 181, pp. 266–282, 2021.
https://doi.org/10.1007/978-3-030-64877-0_18

multiculturalism approach) lies in the fact that all measures would aim to a process of naturalisation into a 'normal' way of life and identity close to the majority [2], shifting the burden of adjustment onto the individual migrant.

Based upon these limits, it is necessary to reconsider integration as an interactive process involving the whole parts of the social system [3]. The integration – probably one of the most difficult challenges for host societies – can only be achieved, over time, by way of interaction between actors both individual and collective. Therefore, it is possible to assert that inclusion pathways and *integration trajectories* take shape in urban territories and they are the result of a long and well-structured process involving both immigrants and the entire city environment made up of citizens, institutions, cultures, and relationships. Following this approach, one can deduce some considerations. The first one is that context dynamics directly affect the processes of inclusion, but vice versa too, i.e. immigration phenomena show a transformative power acting on many areas: institutional-political, economic, socio-cultural, and identity [4]. Secondly, no intervention can be designed without knowing the real needs of the most disadvantaged populations. Hence it is a priority to involve new actors as well as protagonists for whom the centrality of social need is highlighted [5].

Lastly, interaction in all these processes refers to the relational element. As many studies have highlighted, the achievement of immigrants' integration in reception contexts is supported by the density and quality of personal social networks [6]. Every migratory event occurs in a relational system which connects several territories; in each of these places, the migrants constantly 'feed' on old ties and build new ones, which become essential resources of support [7–9]. Therefore, social capital produced in a host place determines advantageous effects in terms of assimilation and social mobility [10, 11].

However, as Putnam points out, the relational resources can produce positive, sometimes perverse, and murky effects on immigrants' inclusion. This distinction is identified by the author in the concepts of *bonding social capital* and *bridging social capital*. In the first case, it is an exclusive form of social capital that has its source within a small group without showing openness to outward connections, and conversely lacking in flexibility to include other groups. The bonding social capital could be the result of a forced choice of immigrants both due to the lack of acceptance in the urban context (from the bottom), as the new arrivals mark moments of crisis between old and new inhabitants [12], and institutional carelessness (from the top), perpetuating their irregularity and clandestine state. Because of a deficit of quality social capital, a series of 'place effects' [13] display in cascade. It happens that a diminished sense of group identity and absence of identification with the country where one lives generates some traits of digression, such as marginality and social exclusion, making the immigrants more exposed to the risk of forming ethnic enclaves [14] or even ghettos. Sometimes the non-integration is linked to the temporary nature of migratory projects. Just look at *middleman minorities* [15], which create an enclave economy also supported by *chains of employees* (people recruited into the same community) [16]. This not only supports closed group identity but also responds to migrants' desire to return to their homeland once they have earned adequate financial resources. Benefits from clustering might facilitate mutual-aid resources among the members, increasing traits of their own culture. This solidarity social capital as well defined by Pizzorno [17] works into a limited ethnic network and never meets the social capital of the local community. For utility, the social capital of the first case is defined as 'micro' and the second one as 'macro'. The gap generated by the strong polarisation of the two social capitals could be an index of the lack of immigrant integration in the urban context. The effects generated by micro-social capital are certainly positive within the group, but they cannot be expressed with the same positive effect in the external context. For this reason, the social capital has a situational character; indeed, whether an action is an opportunity on the one side, it could be a limit on the other [18].

An alternative proposal would be to encourage the creation and diffusion of more open and inclusive social capital (bridging). This formula would benefit both the immigrants and the locals, increasing integration levels in the host societies, and helping to uproot any form of fear or intolerance. As underlined by many studies, being part of a large social network facilitates the success of immigrants in economic and social integration [19–21]. Besides, it follows that high levels of bridging social capital would contribute to improving the quality of life in the places in which it is practised.

In accordance with this premise, this contribution aims to prove the importance of the mixed network, meant as interethnic relationships, in the integration process of immigrants, starting from the results achieved in an experimental project in Naples, This Must Be the Place (TMBP). The promotion of more active citizenship practices at the local level could provide a key resource, contrary to what is claimed by several studies on the exclusive role of ethnic clusters in the various stages of immigration [22]. In this sense, the role of urban contexts (micro-level) in addressing the challenge of social inclusion should be empowered, since cities are the shared places where intercultural relations can be built and nurtured [23].

2 Urban *Containers* of Social-Innovation Practices

Social inclusion processes cannot be analysed by separating them from the urban contexts in which they take shape and from which social, economic, and cultural actions are influenced. In this case study, the context is the city of Naples which is a peculiar urban reality model. Naples has always been a place of contradiction: on the one hand urban regeneration and social inclusion, on other hand resistance of the ancient elements.

Spontaneous social, cultural, and economic diversity has aroused interest in testing new shapes of social-urban innovations. Moreover, it has allowed the city to be considered as a *natural* research laboratory. Our interest area corresponds to the *ancient core* of Naples also known as Decumani (Fig. 1), where physical places could become *devices* of inclusion or exclusion. Characteristic of the area is the partitioning street of Via Duomo which, dividing the Decumani perpendicularly, delimits an 'upper' and regenerated area, and a 'lower' and degraded area. This peculiarity of the Decumani suggests a segregated interpretation of urban space. The first form of segregation is horizontal and recalls the dual nature of the historical centre.

In the *upper* area (left part of the map on Fig. 1), the presence of universities, aggregation places, important monuments, and archaeological sites attract students, workers, young Neapolitans, and tourists. Moreover, the commercial vividness has contributed to giving specificity to some streets in the historical centre and has also fed local economy, and a spontaneous urban and social regeneration.

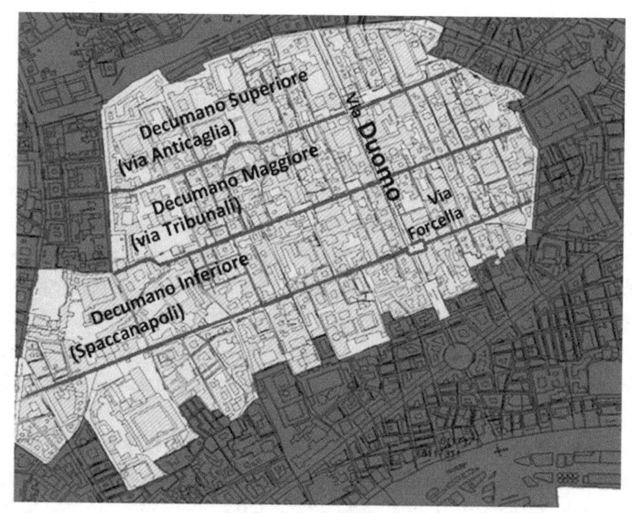

Mappa, Ns elaborazione

Fig. 1. The area of the ancient nucleus of Naples, also known as the Decumani. The Decumani are the main three streets dating back to VII B.C.

A completely opposite scenario arises in the *lower* area (right part of the map), where greater structural critical matters and social weaknesses emerge. Walking along the streets, like Via Forcella or the end of Via Tribunali, anyone can see the precariousness of the buildings. Some of them are historic and still ruined by the earthquake of the 1980 [24].

In this architectural setting, the Decumani reinvent themselves thanks to the chaotic life of its inhabitants. The streets are livened up by daily practices. Coffee bars become a meeting point for the men or the young people; some businesses gravitate around the black market[1] such as smuggling of the cigarettes.

In these places where the micro-criminal phenomenon is recurring, urban safety is maintained by the spontaneous social control of the shopkeepers and residents, in a form of *eyes on the street* [25].

Another shape of segregation develops vertically and can be observed in many buildings in the historic centre. Families from different social backgrounds live in whole buildings, but usually, Neapolitan families belonging to lower classes, or immigrants who have recently arrived in the city, are living on the ground floors (the so-called *bassi*). Throughout the 1970s, while in many other Italian cities there was a progressive spill out from the historical centres by the wealthier families, in Naples this movement never took on mass characteristics. In addition to that, the lack of a total functionalisation of the housing stock, typical in gentrification phenomena [26], has created a very composite social stratification. In fact, in many streets of the centre, alongside buildings renovated and inhabited by professionals or used as staying places for tourists, old buildings welcome Neapolitan families who have lived there for generations, foreign families

[1] The houses on the ground floor are functionalized as shops during morning.

with a migration history consolidated over time (Peruvian, Ceylonese, Ukrainian, and Bulgarian), and off-site students.

The brief illustration of the urban context suggesting to the reader an immediate idea of the socio-structural composition also allows us to intuit how this is an essential prerogative for social movements arising from the bottom. In a territory like Naples, where proximity between social classes is the characterising element of daily interactions, several initiatives have been carried out with the common good as their sole purpose. It is worth recalling some initiatives that have recently been able to start a real process of urban and social regeneration. Many groups of citizens from different social backgrounds, together with local associations, have led the recovery and re-use of some degraded sites[2] [27]. In accordance with Collins' theory, therefore, one could say that spatial proximity is a necessary condition for interactions that generates common emotions [28] and lead to innovative social changes.

The central role of urban agglomerations as incubators of sociality also refers to the broad concept of the *right to the city* [29], and to the appropriation of public space by those who had been deprived of it. Cities as devices of exclusion would reproduce the dynamics of inequality and injustice to the detriment of the most disadvantaged social classes [30]. Suffice it to say that denied accessibility to some basic essential services such as healthcare, education or a worthy and suitable housing becomes a more pressing need especially for immigrants who have not yet regularised their presence in the host context.

These sorts of social needs were intercepted by the action-research project This Must Be the Place, for which the objective was to create an alternative model of welcoming and participatory integration. The project involved a group of immigrants leaving a reception centre and a group of university peers to intervene with project hypotheses on three specific spheres of need: educational rights, housing rights, and social relations. All these aspects are interconnected and affect the outcomes of integration in the host context.

At the initial stage of the project (May–July 2018), 15 meetings were held to encourage acquaintance/friendship among all participants. Some days were dedicated to focus groups, where the participants had the opportunity to reflect on the needs they shared. On other days, the participants took part in some urban exploration events to stimulate the knowledge of the territory with living places considered significant for emotional reasons or because they are often frequented.

Creating new spaces of sociality that can foster more than a simple opportunity for migrants and locals to meet on the one hand increase levels of social cohesion within the urban contexts, and, on the other hand, allows to establish and support open and inclusive relationships. Social relations represented an objective of the project because they are also important to achieve full socio-cultural, linguistic, housing, and work integration in the host context [31, 32].

[2] These actions include the recovery of the former Conservatory of Santa Maria della Fede, later renamed Santa Fede 'Liberata' which now hosts several activities such as film forums, folk festivals, and neighbourhood meetings. This illustrates how a space perceived as a simple urban void can become a harbinger of social change.

As regards the other two needs, education and access to housing, complex areas of intervention have been outlined.

The problems concerning education are linked to the recognition of foreign education level and, consequently, to the transfer of all skills acquired in a 'quality' labour market. Moreover, despite the propensity of refugees and/or asylum seekers to continue their studies, access to the education system poses countless barriers. In the most fortunate cases, one proceeds with integration into a lower class (normally corresponding to a lower class in a middle or high school), which in any case generates discrimination for all immigrants with high qualifications. The objective, therefore, has been to create a 'school network' to improve the current state of many students and all those who want to follow training courses.

For the second social need, a study on housing problems has been launched, also considering transformative phenomena that are affecting the Neapolitan city. Recently, the tourist boom has led many owners to rent rooms or flats for short or long stays in the most central areas. This has weakened real estate demand, which until recent years was mainly aimed at immigrants and off-site students and has also triggered an increase in rents on the few properties available for this target group.

The housing emergency is heightened for young people leaving the reception centre who, still in job insecurity condition and poorly supported by territorial networks, do not have access to the local property market. Hence, it has proceeded in two ways: a) strengthening existing services (e.g. 'housing help desk'), b) identifying potential housing spaces through the mapping of the territory, in which the self-recovery strategy could develop a fair and sustainable housing model.

Ultimately, this project, which presents itself as a first Italian experiment, reflects how collaboration between peer groups is an essential resource for an appropriate response to pressing social needs; and how this participatory integration model stands for a first step towards the construction of a shared and plural society.

3 Objectives and Methods of the Research

Fundamentals and methods of Social Network Analysis have been widely used in migration studies to understand the dynamics of the ethnic network in supporting the different stages of immigration (from arrival to first inclusion in the host context). While the literature has carried out an in-depth analysis of the benefits of family and ethnic linkages, few studies have analysed interethnic relationships in the inclusion process [22]. Starting from this lack in the literature, this contribution mainly uses the SNA approach to examine the leading role of an inclusive network of individuals with different backgrounds to facilitate inclusion in the urban context. Moreover, a mixed-use of different analysis techniques has also provided richer results for supporting the sociological interpretation [23].

Within the TMBP project, following dimensions were examined: (a) reconstruction of the network of project participants and of the evolution of relations at the project's different stages; (b) understanding whether the relationships born within this facilitating context have been transferred even outside the project (e.g.: in the places of the city and the daily life of participants); (c) understanding how the relationships have become essential resources to support the process of social and urban integration.

It should be pointed out that only the sphere of need related to social relations is taken into account here since the other two social needs (housing and education rights) require a long time of achievement before significant results in terms of the impact of the project can be gathered.

The objective of the research has been pursued by involving human resources in a dual role both as participants in the project and as undeclared researchers in order not to threaten the creation of natural relationships between the groups.

Concerning the objective (a), a survey was used to collect directed relational data at three different times: time 0, corresponding to May 2018 (project starting stage); Time 1, November 2018, a period in which many activities were carried out (training meetings, focus groups, and urban explorations); Time 3, July 2019 (end-stage). The survey was carried out on all participants (33). Starting from the second stage, the technical interview was used to understand the reasons why some participants left the project and to better support the explanation of the networks.

Regarding the objectives (b) and (c) introducing the second part of the analysis, both surveys and interviews with some participants were used to get results in terms of the social inclusion process. At all stages of the research, the participant observation technique supported comprehension of some relational dynamics.

The following chapters will show the network structures at the three project stages (see Sect. 4.1), and the results of both the survey and the interviews in relation to the generic connection between mixed relationships and social and urban inclusion (see Sect. 5).

4 Analysis

4.1 The Network of Project Participants

The first network was calculated with the information collected at Time 0, the initial stage of the project. The network structure (Fig. 2) clearly shows this first stage, in which the participants join the social cause of the TMBP. At the beginning, two groups (university students and beneficiaries of the reception centre) are divided into separate networks. The only ties to connect them were the Marta–Adam ties, since Marta knew Adam previously. There are 33 nodes in the network and 181 ties, of which 25.96% are classified by participants as friendly relations and 74% as acquaintance relations. The density of the network is low at 0.216.

An area of major connectivity coincides with the group of the students, in which the prevalence of acquaintance relations is noted, unlike the group of the beneficiaries which, despite a few acquaintance connections, already has stronger relationships within it (orange ties). This is because the groups move in two extremely different environments. While the university context offers the possibility of meeting countless people, it is also true that most relationships are limited to education. Therefore, such ties are weak, in Granovetter's pure meaning [33]. The reception centre is instead the first integration context for immigrants who often arrive alone after long and difficult migration routes. Sharing similar experiences, as well as the adjustment to the host society, affects their need to quickly recreate meaningful relationships.

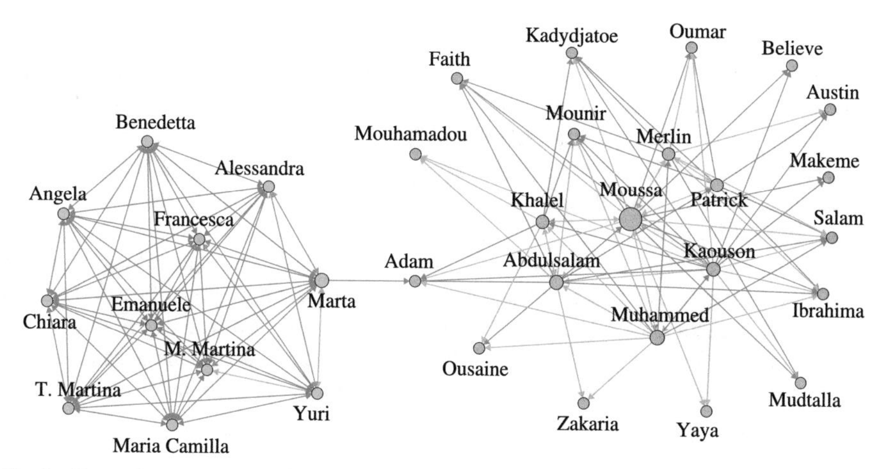

Fig. 2. General network at the first project stage. Henceforth, the beneficiaries of the reception centre are represented with orange nodes and university students with green nodes. Size of a node according to Betweenness centrality. Orange ties represent friendly relations and grey ties represent acquaintance relations.

Some measures of centrality were calculated to better characterise the nodes of this first network. The first refers to the nodes' In-degree, deriving from the simplest measure of the centrality of a network, the Degree centrality. The aim was to immediately identify which and how many actors were considered important by all the alters, within their relational system. In the group of students, these values were rather homogeneous (variation between 11 and 9), and not indicating any prominence role. The same value recorded (11)[3] is mainly due to the attendance of the same university course (Sociology, Federico II), with exception of Martina T., Angela, Benedetta, Chiara (9) who came from different courses of the Orientale University, and not all of them knew each other. In this case, some meetings organised within the project only with the students brought them in touch with each other.

In the group of beneficiaries, the following individuals distinguished themselves for high values of In-degree: Moussa (value 11) and Ibrahima, Mounir, and Salam (value 9). It is important to observe how these last three individuals were defined as 'popular' as they are also intercultural mediators of the reception centre. This role makes them well-known because they work in linguistic and cultural assistance with newcomers. Although Moussa is a beneficiary of the centre, he records the highest value because, as told by the other participants, he was the "*first-person met in the centre*". Therefore, if the popularity in the first three cases is due to the institutional role played, in the second case it is due to a greater personal capacity for socialisation.

Betweenness was the second measure of centrality calculated on the whole network and made it possible to identify the subjects who, occupying a strategic position within the network, play an intermediary role among all the others. Moussa has the highest Betweenness centrality (equal to 38.8 points). The difference recorded between this

[3] This value refers to Emanuele, Marta, Francesca, Maria Camilla, Alessandra, M. Martina, Yuri.

value and all subsequent nodes is very significant: Muhammed (10.6), Marta (9.12) Abed (9.0). About 40% of the individuals in the network pass through Moussa to reach all the others. As it has been observed during the first project-meetings, attendance of many beneficiaries was initially discontinuous, but thanks to Moussa who was very included in several micro-groups of the reception centre, it was possible not only to receive information about them but also to pull their participation at least until the second stage of the project.

The interesting aspect that emerges so far is that these synthetic values somehow obscure the presence of the intercultural mediators into the network. Although the latter have proved popular on In-degree values, they do not play the role of brokers within the network. This reflects in any case a substantial weakness in the role of intercultural mediators, due to the lousy attention to these figures by those who institutionally work on foreigners' integration issues.

4.2 The Inclusive Network

In the second survey (Time 1), the network structure is much more complex than the previous one (Fig. 3). Most nodes are connected to each other. It is therefore a very dense network (0.548).

There are always 33 nodes in the network, but its density increases (439 links vs. 181); above all, friendship ties increase from 25.96% in the first survey to 31.89% in the second. This shows that in recent months several friendships have been established.

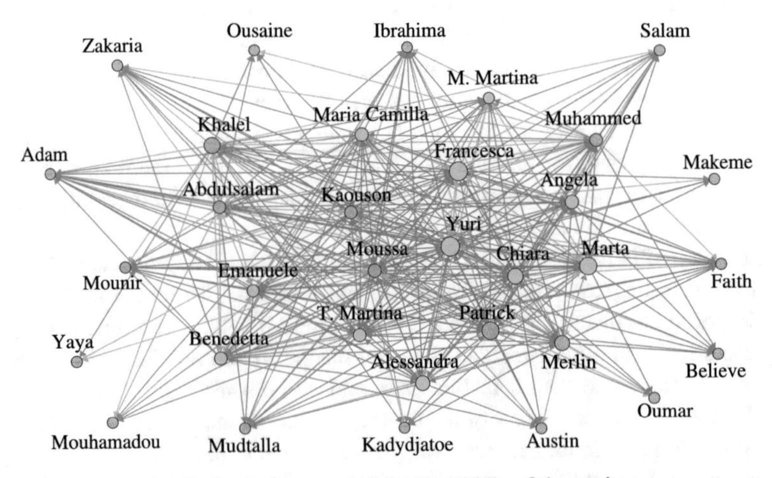

Fig. 3. Inclusive network in the middle of the project

They replace the intercultural mediators who took part in the project to facilitate social interaction between participants with a migratory background. Already starting from this second stage, a greater 'mixture' can be noticed, because of a certain interaction between the groups. Indeed, horizontal network expressed by centrality measures confirms an equal role played by each of them. Up to now, the project has been able to create more strong and transversal relationships between the groups.

4.3 The Mixed Network

In the last stage coinciding with the period of June–July 2019, the research has allowed both to collect the data to understand how the relationships have evolved, and to understand through interviews the causes that would have generated a discontinuous attendance and, in a few cases, the project's failure by some beneficiaries. This was an expected outcome and partially anticipated by the network analysis at the second stage.

As told by six beneficiaries, the discontinuity of attendance in the project activities would have been linked to the working needs, as well as to the increasingly urgent need to naturalise one's legal status in Italy. Towards the end of the reception period at the centre where they resided, the six beneficiaries embarked on job careers, sometimes precarious, and often moving within the regional context of Campania. The difficulty to reconcile working time with other activities emerges in all the interviews. Therefore, the time dimension is experienced differently by asylum seekers. In almost all the interviews, perception of living in a time of continuous waiting and uncertainty [19], in a form of eternal temporary [34], becomes pressing. Suffice it to say that from the moment the personal status of an asylum seeker is formalised until the end of the procedure, it takes years to get an answer as to whether or not it is possible to start on a settlement visa in Italy.

In five other cases, relocation to other cities in northern Italy or even to other countries has led to a concrete exit from the project. These are beneficiaries who, once they had obtained a settlement visa, moved elsewhere to look for better job opportunities in the most dynamic economic contexts, or to reunite with their families.

The structure of this third network confirms the personal dynamics of the beneficiaries described above (Fig. 4).

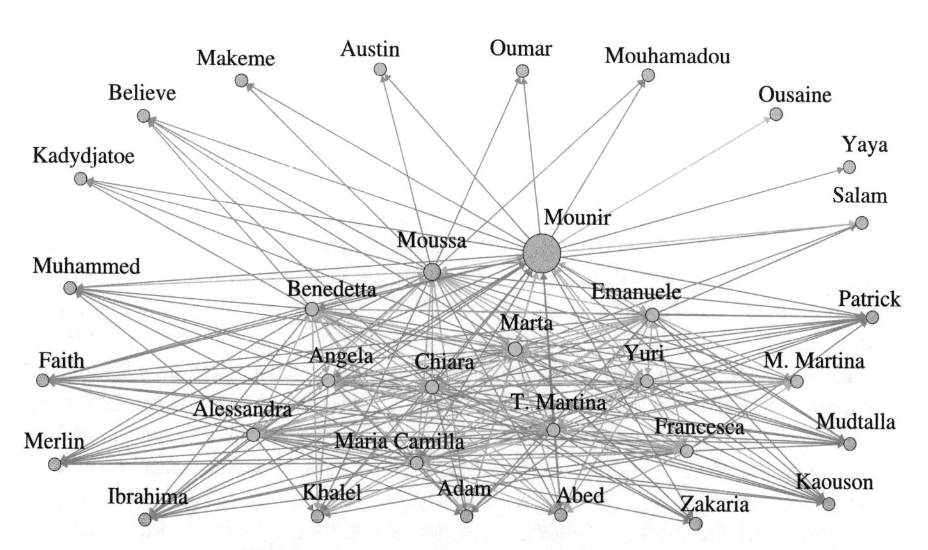

Fig. 4. Mixed network at the end of the project

Looking again at the network, one can also see how the content of ties changes as compared to the two main areas both in the periphery (lower density) and in the core (higher density). In the core, there are the relations defined by the interviewees as friendly (56.8% orange ties) that extend even outside this context (see next chapter), while the periphery mostly consists of acquaintance relationships (43.2%) created within this common context but never increased in significant mutual relations.

Even for this network, the same measures of centrality used previously have been calculated, and for convenience summarised for each stage (Table 1).

Table 1. Betweenness centrality in comparison

T1		T2		T3	
Moussa	38.8	Yuri	27.4	Mounir	41.9
Muhammed	10.6	Francesca	25.9	Moussa	30.9
Marta	9.12	Marta	21.3	Marta	19.9
Abed	9.0	Patrick	21.1	Benedetta	7.4
Khalel	6.5	Khalel	17.6	Emanuele	4.2
Patrick	5.0	Chiara	16.4	Martina T.	3.1

As can be seen starting from the most significant values of Betweenness centrality shown in the table, at Time 3 the intermediation function is anchored to three nodes of the network (Mounir, Moussa, and Marta).

Starting from the third stage, one can see an analogy with the Betweenness centrality values of the initial stage, as opposed to the second stage in which many students and beneficiaries have recorded similar values, reflecting in this way an equal centrality within the network. This variation in growth can certainly be explained by the fact that there have been several cases of progressive distancing from project activities and spill out. After the exit of a significant part of foreign participants, the relevant scores were centralised in the only beneficiaries remained, Mounir (41.9) and Moussa (30.9). Following them are Marta (19.9) and others with less relevant scores (Benedetta 7.4, Emanuele 4.2, and Martina T. 3.1). This dynamic is clearly visible also in the graphic rendering of the network. Above all, the first three subjects play a bridge role between the internal area of the greatest connection and the external area where few nodes with a visibly reduced number of ties are located. They have been able to maintain strong relationships that would otherwise have been broken more than others. In the Mounir and Moussa cases, the common experience in the reception centre meant that friendships were maintained between the beneficiaries. Instead, as has been observed, Marta's role is linked to her ability in involving persons in leisure activities or seeking touch with them regardless of the project context. From these last results, some fundamental considerations arise. Above all, the project has worked as an incentive to create an opportunity for meeting in which the sharing of experiences, needs, and feelings has certainly increased relationships of trust. In any case, the importance of the 'contact space' is also true, since many relationships need specific places to support themselves at the beginning. It could be argued that physical proximity is also a necessary condition for social proximity.

However, the third network overall shows a strong relational potential that could be a crucial resource in terms of social integration. As a result of this case study, it is important to consider the quality of ties but also the time variable to observe the changes in the network structure. Interethnic relations and time dimensions are necessary to support the social inclusion process. Starting from the first survey in which the network is represented by separate groups (Fig. 2), the project as a facilitating context helped develop inclusive relationships (Fig. 3) until stabilising them in the last network. Despite many dynamics that occurred in the personal lives of participants, the strong relationships were still alive and preserved in clusters.

The last section of the survey focuses on the type of impact that social relations have generated on the inclusion process, in terms of the daily life of the participants and the places in the city as containers of all social practices.

5 Friendship, Free Time, and City

Friendly relations born in this facilitating context have found new expression in other spaces of the city and especially in the daily life of the participants. About 60% of the students and beneficiaries said they were satisfied and surprised to have expanded their relational sphere to include people with backgrounds other than their own. From the project space to intervene in common social needs, the participants began to spend more time together, meeting beyond the project's time frame and carrying out plenty of leisure activities.

As regards the activities carried out together (Table 2), the answers obtained from the survey were gathered into seven categories (University activities, Cinema, shows, and cultural events, Lunch break, Night out, Happy hour/Aperitif, Dinner at a friend's house, Day trip).

Table 2. The main leisure activities carried out together by participants

Leisure activities	Frequency	Relative frequency (%)
University activities	7	14%
Cinema and cultural events	6	13%
Lunch break	9	19%
Night out	7	15%
Happy hour/aperitif	7	15%
Dinner at a friends' house	5	11%
Day trip (seaside, picnic)	6	13%
Total	**47**	**100%**

The activities most frequently carried out refer to lunch breaks together (19%), and study and training activities (14%). The other top categories are activities strictly related to leisure time and that gather the interests of a more heterogeneous group: aperitifs and evening outs occur in 15% of cases.

Other activities are linked to trip days, such as spending a day at the beach or in the park, and go to the cinema, shows and cultural events (13%). Dinners at a friend's house represent 11% of the frequencies.

From these first results, it is possible to draw some synthetic considerations. Organising for study or lunch breaks has become a real daily practice. As has been observed in the last year, while in the first stage these activities represented an opportunity for students from Neapolitan universities to meet, in a more mature stage they have become increasingly inclusive. Many beneficiaries decided to meet students in university libraries to improve Italian language learning thanks to their support.

For many, the location of the Universities in the *historical heart* of Naples allows them to spend more time in these areas also in the afternoon and evening hours. This makes the historical centre a place of aggregation often practised also because of its strategic position as it is easily reachable both on foot and by public transport.

The close link between local space and social relations is still prominent. The need for space for individuals is fundamental not only for the construction and expression of their socio-cultural identity but also because it is precisely in this space that systematic contact can be established with others [35]. Social relations and the norms of reciprocity through the individuals are nourished *in-and-between* places of daily life.

It was interesting to measure the importance of friendly relations to understand how they support social and urban inclusion. Interviewees were asked to assign a value from 1 (not at all important) to 5 (very important) on the measuring scale, to eight items as shown in Fig. 5. From the bar chart, one can see that significant proportions of participants believe that the friendships established within the project were important to *receive emotional support* (45%), *to meet new friends* (38%), and *get to know the city better* (37%). This dynamic was not unilateral as also the Neapolitan participants found a wider knowledge of the places, and common practices lived by migrants in the Neapolitan territory than in the past. The idea of a network that works as a duplicator of relationships, i.e. a network able to include new relationships based on trust fundamentals (*bridging social capital*), clearly emerges.

The importance of these relationships is also linked to the *exchange of university information* (33%) and the and *access to several services* (7%). Therefore, the growth of friendship and trust relations has been a flywheel for greater integration in the urban context. An exemplary case like this refers to a Syrian asylum seeker who managed to access the information needed to continue his university studies in Italy by overcoming bureaucratic and legal barriers, also thanks to the support of the network as well as project-managers.

Only 13.3% of the interviewees said that these relationships were slightly important *for finding a job or a home*. However, this is a result to be contextualised with the statistical units for which the presence of locals or commuters pushes on rather different needs from those of the beneficiaries. The interviews carried out show an opposite and positive scenario for the integration outcomes. For example, in the case of a beneficiary of African origin, the friendships established within the project were essential to deal with the subsequent steps leading to the exit from the reception centre. Moussa managed

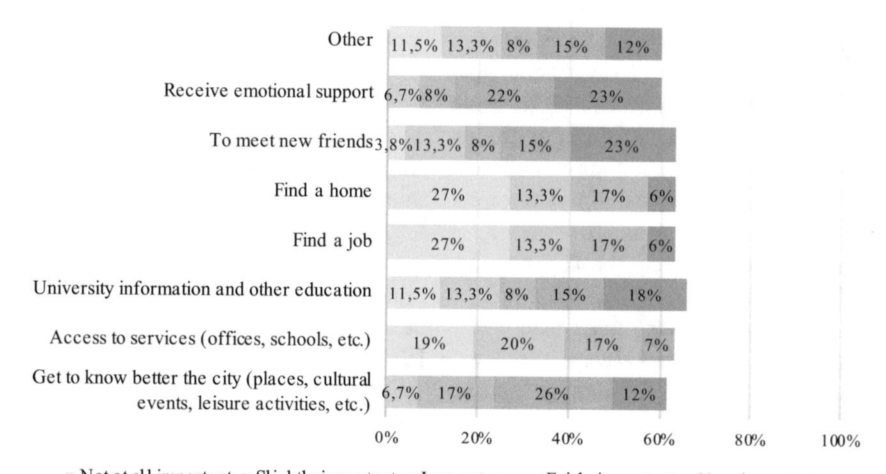

Fig. 5. The importance of social capital on eight integration items

to find a job first and then a house where he lives with other university students, with the support received from some students:

> They have done so much for me. They knew I needed a home, but without a job, I could not do anything. So, they helped me write a curriculum vitae and look for a job. Finally, we found Barrett's [coffee bar]. When I have been hired, I called Martina and Benedetta again and I said that I could have afforded a house and we searched together [...] now I live in Via Forcella with two students and a Spanish guy in Erasmus in Naples. (Moussa, 15th July 2019)

Ultimately, Moussa's job and housing integration supported by the project network leads to consider the social capital from mixed relationships as a 'key resource' in integration processes. As widely stated in the relevant literature [33], people who move in relational systems other than their own offer the possibility to access new information to reinvest in successful professional and social mobility.

6 Conclusion

The TMBP project as the first Italian experiment aims to reflect how collaboration between peer groups is an essential resource for efficient response to pressing social needs. Moreover, rethinking a new model of inclusive and participatory integration could help to overcome an obsolete institutional way of proceeding, based on less consistent top-down approaches, and consider the characteristics of social groups and urban contexts.

This contribution triggers some synthetic considerations on two specific dimensions. The first one recalls the methodological approach adopted, the second the actual results achieved within the project.

As regards the methodology used to analyse the importance of social relations in integration processes (one of the spheres of needs), the SNA approach proved to be particularly productive. The networks not only showed their evolution compared to the three moments predetermined by the research, but they also enabled foreseeing some dynamics taking shape within the groups. Therefore, the SNA lent itself as a useful approach both for the monitoring and for the impact study of the project.

The social capital resulting from interethnic relations has shown to have a rather dynamic and processual nature in creating and supporting itself over time, and a situational nature that takes different forms, each of them productive for a specific purpose [18].

Concerning the results of the project, an undoubted strength that has characterised all the stages of the work transversally was linked to the constant leadership of the participants, whose role has always been active. This was made possible by a positive spirit from the very beginning, suitably reinforced by horizontal relations not only between students and beneficiaries but also between the project-managers. Further evidence of the achievement of one of the project objectives refers to the creation of strong friendships. The growth of friendly and trusting relations has been an important resource to undertake a gradual path of integration in the Neapolitan context. This has opened a new opportunity not foreseen at the planning stage, which instead initially focused on other objectives. Housing and Education rights are still ongoing, also because of the structural weakness that characterises the territorial network. Although institutional and third sector actors were included for these latter needs, they did not prove to be sufficiently ready for the prompt implementation of the more onerous interventions (e.g. housing problem).

However, a very important result was the determination of some participants to continue the path undertaken beyond the official closure of the TMBP. The group of participants showed great resilience and ability to self-organise in facing new challenges. With the spontaneous aggregation of other people and associations in the area, it was decided to carry out some workshop activities aimed at involving citizens in a valuable intercultural exchange. For this purpose, space has been identified in the historical centre of Naples, upon a municipal concession, and several activities have been launched: language tandems in Arabic and French with the involvement of foreigners, a talking library (reading workshop on the intercultural themes), and a dance laboratory. In addition to that, monthly events were organised such as docufilm, musical evenings, and urban explorations, which made it possible to attract many people and significantly expand the network.

This experience shows how both the city and the social groups active in the urban context play an essential role in achieving innovative social change, and how they also represent a first step towards the building of a common and plural society.

References

1. Villa, M., Emmi, V., Corradi, E.: Migranti: la sfida dell'integrazione [Migrants and integration challenge], CESVI-ISPI, Ledizioni, Milan (2018)
2. Carrera, S., Wiesbrock, A.: Civic integration of third-country nationals: nationalism versus Europeanisation in the common EU immigration policy. Centre for European Policy Studies, Brussels (2009)

3. Berry, J.W.: Integration and multiculturalism: ways towards social solidarity. In: Papers on Social Representations, vol. 20, pp. 2.1–2.21 (2011)
4. Zanfrini, L.: Sociologia delle migrazioni [Sociology of immigration], Laterza, Bari (2007)
5. Crescenzi, M.: Social Innovation e social Business. Graphofeel, Rome (2012)
6. Kindler, M., Ratcheva, V., Piechowska, M.: Social networks, social capital, and migrant integration at local level. IRIS Working Paper Series, no. 6, University of Birmingham (2015)
7. Tilly, C.: Trust networks in transnational migration. In: Sociological Forum, vol. 22, no 1, Wiley and Springer, (2007)
8. Lubbers, M., Molina, J.L., Lerner, J., Brandes, U., Avila, J., McCarty, C.: Longitudinal analysis of personal networks. The case of Argentinean migrants in Spain. Soc. Netw. **32**(1), 91–104 (2010)
9. Bilecen, B., Sienkiewicz, J.J.: Informal social protection networks of migrants: typical patterns in different transnational social spaces. Popul. Space Place **21**(3), 227–243 (2015)
10. Putnam, R.: Bowling Alone. The Collapse and Revival of American Community. Simone & Schuster, New York (2000)
11. Coleman, J.: Social capital in the creation of human capital. Am. J. Sociol. **94**, S95–S120 (1988). https://doi.org/10.1086/228943
12. Hirschman, O.A.: Exit Voice and Loyalty: Responses to Decline in Firms, Organizations, and States. Harvard University Press, Cambridge (1970)
13. Paba, G.: Il territorio delle Piagge come risorsa fisica e sociale della città di Firenze [Piagge territory as physical and social resources of Florence city]. In: Marcetti, C., Solimano, N. (eds.) Immigrazione, convivenza urbana, conflitti locali [Immigration, urban coexistence, local conflicts], Angelo Pontecorboli, Florence (2001)
14. Douglas, S.M.: Social class and ethnic segregation: a reconsideration of methods and conclusions. Am. Sociol. Rev. **46**(5), 641–650 (1981)
15. Bonacich, E.: A theory of middleman minorities. Am. Sociol. Rev. **38**(5), 583–594 (1973)
16. Werbner, P.: Metaphors of spatiality and networks in the PluralCity: a critique of the ethnic enclave economy debate. Sociology **35**(3), 671–693 (2001)
17. Pizzorno, A., Trigilia, C.: Il capitale sociale. Istruzioni per l'uso [Social capital. Instructions for use], il Mulino, Bologna, (2001)
18. Piselli, F.: Capitale sociale: un concetto situazionale e dinamico [Social capital: situational and dynamic concept]. In: Bagnasco, A., Piselli, F., Pizzorno, A., Trigilia, C. (eds.) Il capitale sociale. Istruzioni per l'uso [Social Capital. Instructions for use], il Mulino, Bologna (2001)
19. Rainer, H., Siedler, T.: The role of social networks in determining migration and labour market outcomes. Econ. Transit. **17**(4), 739–767 (2009)
20. Carrington, W.J., Detragiache, E., Vishwanath, T.: Migration with endogenous moving costs. Am. Econ. Rev. **86**(4), 909–930 (1996)
21. Massey, D.S., Garcia-Espana, F.: The social process of international migration. Science **237**, 733–738 (1987)
22. Morosanu, L.: Mixed Migrant Ties Social Networks and Social Capital in Migration Research. Technical Report, CARIM Analytic and Synthetic Notes, Migration Summer School - Best Participant Essays Series (2010)
23. Ryan, L., D'Angelo, A.: Changing times: migrants' social network analysis and the challenges of longitudinal research. Soc. Netw. **53**, 148–158 (2018)
24. Zaccaria, A.M.: Dentro il cratere. Il terremoto del 1980 nella memoria dei sindaci [Inside the crater. The 1980 earthquake in the memory of the mayors]. In: L'Italia e le sue regioni [Italy and its regions], Treccani, Rome (2015)
25. Jacobs, J.: The Death and Life of Great American Cities. Random House, New York (1961)
26. Schmoll, C.: Aux marges de la forteress Europe. L'immigration en Italie [At the fringe of the Europe fortress. Immigration to Italy]. In: Liauzu, C. (ed.) Tensions mediterranées [Mediterranean tensions], L'Harmattan, Paris (2003)

27. Zaccaria, A.M., Delle Cave, L.: Reti di volontariato nel centro antico di Napoli [Charity network in the ancien centre of Naples]. In: Olivieri, U. (ed.), Lavoro, Volontariato, Dono [Work, Voluntary, Gift], Milella, Lecce (2017)
28. Collins, R.: Interaction Ritual Chains. Princeton University Press, Princeton (2004)
29. Lefebvre, H.: Spatial Politics, Everyday Life and the Right to the City. Routledge & CRC Press, London (2012)
30. Nicholls, W.J., Uitermark, J.: Cities and Social Movements: Immigrant Rights Activism in the United States, France, and the Netherlands. Wiley Blackwell, Hoboken (1970)
31. Cheung, S.Y., Phillmore, J.: Social networks, social capital and refugee integration. Research Report for Nuffield Foundation (2013)
32. Lamba, N.K., Krahn, H.: Social capital and refugee resettlement: the social networks of refugees in Canada. Migr. Integr. **4**(3), 335–360 (2003)
33. Granovetter, M.S.: The strength of weak ties. Am. J. Sociol. **78**(6), 1360–1380 (1973)
34. Pinelli, B.: I campi d'accoglienza per richiedenti asilo [Refugee camp]. In: Riccio, B. (ed.) Antropologia e migrazioni [Anthropology and migration], pp. 70–78, CISU, Rome (2014)
35. Agier, M.: Le camp comme limite et come espace politique [The camp as a limit and political space]. In: Makaremi, C., Kobelinsky, C. (eds.) Enformés dehors. Enquêtes sur le confinement des etrangers [Locked outside. Investigations into the containment of foreigners], Terra, Paris, pp. 27–40 (2009)

Networks in Education

From Connection to Community: A Medium-Term Contribution of a Mobile Teacher Training in Madagascar – The Genesis of a Social Network

Eilean von Lautz-Cauzanet(✉) and Eric Bruillard

Université Paris Saclay, Paris, France
e.lautzcauzanet@gmail.com, eric.bruillard@parisdescartes.fr

Abstract. This article presents a Social Network Analysis (SNA) of the medium-term contributions of a mobile phone supported teacher training project. Based on the case study of the IFADEM project in Madagascar, a social network analysis with a mixed-method approach was conducted, making use of qualitative and quantitative datasets covering both the training period and the aftermath (2 years) of the project. The results indicate that a mobile supported teacher network has been formed by former participants and sustains autonomously. The mobile phone plays a hybrid role, both as a tool for communication whilst also transforming relationships, leading to teachers identifying as a community which in turn generates professional practices such as collaboration and innovation. Finally, the SNA also allowed for the identification of network leaders whose skills, competences and activities contribute to the value and sustainability of the network.

Keywords: Teacher collaboration · Social Network Analysis · M-learning · Mobile learning · M4e · Teacher training · E-learning · Teacher leadership · Teacher networks

1 Introduction

During the last two decades, there has been increasing interest towards the use *of ICTs to improve the quality and access to teacher training.* The need to do so appeared particular urgent in developing countries, including in Sub-Saharan Africa. Here, around 6,4 million teachers must be recruited by 2030 in order to tackle amongst others the worrying student performance situation, according to the latest PASEC evaluation [1]. In this context, *mobile phones* have attracted the attention of research and development actors in the African region. With GSMA expecting that by 2025, half of the population will have a mobile subscription, the expectations towards mobiles' positive educational outcomes are high, especially when it comes to distance training [2]. Mobile phones are considered by Burns as the most promising technologies for teacher distance education in developing countries [3].

A. Antonyuk and N. Basov (Eds.): NetGloW 2020, LNNS 181, pp. 285–305, 2021.
https://doi.org/10.1007/978-3-030-64877-0_19

Mobile phones were considered as powerful tools for teacher training because of their ubiquity, cost efficiency and user friendliness [4]. In addition, research suggests that mobile phones could be used as a tool to spread and access knowledge and resources. Amongst others, research considers their potential to compensate for the lack of training institutions and enable personalized training as well as help teachers to develop new literacies [5, 6].

Furthermore, among the most cited benefits figures enhanced teacher *collaboration.* Mobiles would favor the creation of a collaborative community, as part of which teachers could construct collectively the knowledge they need [7]. This enthusiasm for mobile teacher training as an enabler for collaborative networks and processes appears comprehensible as research findings indicate a positive correlation between enhanced teacher collaboration and student performance [8, 9].

However, this overall enthusiasm contrast with a striking *lack of studies on medium-term contributions* of teacher training projects using mobile phones. Addressing this gap is important: knowing if, how, which and why contributions sustain over time appears indispensable for the development and evaluation of appropriate education policies and projects aiming for a long-term improvement of the education situation in developing countries.

In addition, given the higher familiarity of users with mobile phones, an appropriation process leading to *the development of usages* beyond those prescribed by the project can be expected – and its contributions observable only over time. Finally, the need for a medium-term perspective appears even clearer when looking at the lack of evidence on the extent to which mobile supported teacher training project do lead to the creation of *teacher networks* or *teacher communities* – a term that has been criticized for its vagueness and tendency to be used as buzzword:

> The word community is at risk of losing its meaning. From the prevalence of terms such as 'communities of learners', 'discourse communities' and 'learning communities' to 'school community', 'teacher community' or 'communities of practice' it is clear that community has become an obligatory appendage to every educational innovation. [10: 43]

In summary, there is a clear need for a medium-term perspective on the outcomes of mobile teacher training projects and for an analysis that sheds light on the question if and to which extent it is possible to identify sustainable collaboration mechanism taking place within a teacher network.

The here presented study builds on the results of a PhD research process that that took place from 2014 to 2017.

It analyzed medium-term contributions of mobile teacher training pilots in Sub-Saharan Africa. In this context, a Social Network Analysis of the IFADEM mobile training project Madagascar had been conducted [11]. As the same project served as case study for this article, the project and first conclusions made by organizers at the end of the project in 2012 will be exposed in detail at the beginning of this article. As concepts of theories like activity theory, appropriation and social network theory have played an important role in our methodological choices and interpretations made, these will be then presented before exposing then our research question, the methodological process and the used SNA measures and concepts. This shall allow for the understanding of our results we will discuss and, finally, put in context.

2 IFADEM – The Case Study

The IFADEM project in Madagascar is part of the IFADEM initiative that organized multiple training projects in various African countries. Initiated by the Association of Francophone Universities (Agence Universitaire de la Francophonie) it was implemented in Madagascar by the Ministry of Education, the French Development Agency (Agence française de développement) and Orange, a French telecommunication provider present in Madagascar.

2.1 Project Implementation

From 2012 to 2013, 456 teachers took part in the in-service and mobile supported teacher training project. The project aimed to improve teachers' pedagogical skills and knowledge of French, the language they must teach in but often do not master well. The training took place in a rural area consisting of four zones: Ambositra, Fandriana as well as Ambatofinandrahana and Manandriana, characterized by poor infrastructure and mobile coverage.

All trainees gathered every three months for three days, whilst also meeting every month in small groups in their nearby villages with an assigned tutor. During these meetings, they worked together on the workbooks they had received. They were also provided with mobile phones containing audio files and on that same phone, would receive a daily quiz question related to their training content. Most importantly, a teacher could use the phone for free during the training period in order to interact with their training peers and tutor. When the training ended, they could keep the phones but had to pay for communications.

2.2 Project Evaluation and First Findings

Project evaluators produced a total of six quantitative datasets, as well as an interview report summarizing interviews conducted with 18 training participants. In addition, two final evaluation reports were respectively elaborated by Orange and external evaluators. The quantitative datasets allowed evaluators to conclude that teachers had developed non-prescribed usages both for professional and private purposes during the training [12]. In addition, Orange conducted Social Network Analysis based on the Call Detail Records found that the mobile communications had a network structure [13, 14].

Availability and Potential of Call Detail Records. A presentation of a one specific dataset – the Call Detail Records (CDR) – is indispensable for the understanding for our research process and methodology. Used throughout our research, it is without any doubt a valuable dataset. Besides being difficult to retrieve because of data protection legislation, its added value is mainly related to its nature. Indeed, CDR – also known as Station Message Detail Recording (SMDR) – provide all information necessary for SNA algorithms. CDR provide details on all incoming and outgoing calls, such as date, time and duration of communication as well as the number of out- and incoming calls, the type of call partner (training related or other) and the type of communication (SMS or Voice). Available for each participant and retrieved from the telecom provider Orange, we analyzed the CDR during the training period (October and November 2012, as well as March, April, May, June 2013) and after the training (March, April, May, June, October, November and December 2014). The CDR of participants were not anonymized; it was hence possible to differentiate between training participant communications and communications towards non-participants. This enabled us to consider socio-economic characteristics and compare individual findings with statements made during interviews. It is important to underline that all participants gave their permission to access and use their CDR for research purposes.

3 Theoretical Framework

This research focuses on effects of technological tools like mobile phones on processes like collaboration for teachers working in a poor resourced environment. We consider that tools do not necessarily have an inherent positive outcome whilst assuming also that challenging environments affect individuals, usages and hence also the possible genesis of teacher networks. Therefore, we consider that activities with and not the tool alone inform about contributions. Therefore, willing to adopt a critical approach towards technological determinism, we consider here the social account of technology. For a better understanding of our perspective and process, we present in the following the theoretical background: a social constructive approach of analyzing technology use and appropriation, framed by theories like activity theory and social network as a theory.

3.1 Activity Theory Applied to Educational Technology Research: A Focus on the Development Process of Contributions and Appropriation

Activity Theory. Activity theory has been used to o study the design and implementation of learning supported by technology in various communities of practice [15]. It is a suitable framework for Social Network Analysis conducted in the study presented in this article, as it allows to focus on the learning and its outcomes involving technology introduction in three communities (teacher community, teacher – tutor community and trainees – non-trainee community) and considers the possibility that experiences and the reflection on experiences allows for the transformation of practices [16]. According to this approach, all human experience is shaped by the tools and sign systems that we use. In the case of technology, the unit of analysis within activity theory is not the individual, nor the technology alone, but the use of the technology within the context. Activity theory appears also to be a suitable frame for our research and the willingness to analyze contributions with a medium-term perspective. Activity theory takes into consideration the development of contributions over time, considering that 'tools, or as they are more commonly referred to today, technology, […] meet needs and serve purposes yet simultaneously create new needs and purposes. [Tools] develop over time, they are appropriated from other contexts' [17: 46].

Appropriation. This research adopts the appropriation of knowledge concept that draws on the developmental theories of Vygotsky, adopting both a cognitive and social-constructivist views of learning [18]. We assume hence that mobile usages developed at the individual level, outside the organizational environment and before entering the organization or a system, can influence organizational usage and the processes of appropriation at the organizational level. In sociology, researchers agree on defining appropriation as the development of new, altered usages [19, 20]. This concept implies that individuals are driven by intentionality's, willingness, desire and attribute a sense to the technology they use. Time, and hence duration, is an inherent part of appropriation process: the gap between the initial meaning and altered usages is created over time.

3.2 Social Network Analysis

As there are indications that a collaborative network has emerged as a medium-term contribution of the mobile teacher training, we study the possibility of this result being closely related to a collective appropriation process. Hence, social network theory was here an obvious framework for this research. Purposes of social network research. According to Stegbauer and Häussling even though social network research does not have a uniform theoretical orientation and methodology, 'it involves a "thinking in relations" mindset that ultimately leads to new insights into social processes' [21: 2].

It is commonly considered as an overarching paradigm used in the various branches of social science. The focus doesn't lie only on isolated persons and their characteristics, but on actors and their relationships. It is considered that perceptions, decisions, actions and social processes are shaped because and within social networks [22].

Social Networks and Social Capital. Scholars refer often to the concepts of social capital theory in order to explain why social network perspectives are helpful to understand human interactions. Both concepts are directly related. According to social capital theory,

> [...] social structure, or the web of relationships among individuals, offers opportunities and constraints for the exchange of resources. Individuals may tap into the resources that are available in the social structure in which they are embedded and leverage these resources to achieve individual or organizational goals. [23: 10]

According to Moolenaar, social capital theory focuses on social structures as means for acquiring resources, in contrast to social network theory which identifies regular patterns in social structure and the mechanisms linking social structure to social capital outcomes [23].

The importance of the mutual influence of individuals, tools and the environment and the importance over time had been in the scope of our research since 2014 and lead to the here presented study. In the following we will recall therefore the beginning of this research process, which had already looked at the question of collaboration as a contribution of mobile teacher training. The results in turn lead to our research question and methodological choices in order to analyze if the social network can be considered as a contribution of mobile teacher training.

4 Research Questions and Methodological Process

4.1 Findings from Our Previous Studies and Resulting Research Question for the SNA Study

The here presented social network analysis and its results build on the findings of two studies, which we conducted respectively in 2015 and 2016. The qualitative study involving interviews with former participants, a control group and class observations, found that the training had led to contributions in the professional and private sphere of teachers and former tutors, including practices in the classroom, professional sphere and for private purposes. These practices sustained after the end of the pilot project [13]. We found that the sustainability of these usages and contributions were a result of the collective training experience. The latter had increased their awareness and perception of usefulness of the phone and set the base of collaborative practices among former participants. We found also that the environment and role of each participant influences on the motivations and degree of sustainable mobile usage, and that that profiles such as school directors and individuals with administrative responsibilities differentiated themselves from other with diverse and innovative usages whilst appearing to push other participants to do the same. The study found also that the phone usages were overall more diverse, intense and diversified among former training participants in comparison to a control group, composed of teachers who also own mobile phones but did not take part in a training course. Hence, we concluded that the simple fact of owning a phone cannot explain the motivation to develop new usages or collaborate – the training experience acts as a key ‘kick off phase’.

Then, our quantitative study analyzed mobile communications through univariate descriptive statistics and analytical statistics (Chi-2 test followed by Multi Correspondence Analysis and inferential tests such as decision trees) showed that almost half of former training participants continue to communicate with former training participants, at their own cost and over two years after the end of the pilot project [13]. The already by evaluators observed significant ‘sustainability gaps’ according to trainees’ workstation could be confirmed and explained by infrastructural differences (access to electricity, quality of network coverage, previous phone experience) among which network coverage progressively appeared as key variable when predicting impact on trainees’ chance to use their phone on the medium-term. Most importantly, we found that network coverage was de facto a pre-requisite for a decisive appropriation process: participants concerned by a lack of satisfying network coverage were excluded from a process of apprenticeship and relatedness other training peers could be part of.

As a result, the research question for the SNA study was: How can we use social network analysis to analyze if there is effectively a teacher network, its contributions and the role of the mobile phone?

4.2 Methodology

In view of our research question, we combined three main SNA concepts and their methods: the structure of a network, its processes and dynamics (e.g. groups) and the role of individuals, for which we chose the ego-network analysis approach in form of interviews with network members. We cross-analyzed the CDR of individuals with their respective socio-economic characteristics, training characteristics and questionnaire results. After having extracted the CDR data from 185 individuals with intra-training communications with Python, we used Gephi, open source software for network visualization and analysis.

Structural Analysis of the Network. To identify if there was effectively a network, we used SNA methods that would allow us to qualify its structure.

Centrality. As the notion of degree centrality is a key indicator of the degree of importance and prominence of an individual within a social network, we used Gephi so that the graph highlights participants' centrality values. Groups were identified with the modularity class function and visualized with a common color code. We analyzed then the extent to which the groups and the network are interconnected and if there are disconnected or very isolated parts. Subsequently, we analyzed the structure, taking into consideration the characteristics of network members, looking at variables such as network satisfaction and training location. To know if some individuals would be more solicited than others, the out- and in-degree scores of each participant were calculated.

Diameter. We calculated the network diameter score, compared it to the single diameter score and average length path, and analyzed to which extent this length path would be lower than the diameter score of the network. Possible outliers were calculated through an analysis of the average eccentricity score and standard deviation.

Density. After visual density analysis – large or small distances between nodes – the overall density score was calculated as well as the clustering coefficient in order to identify the number of connections among all possible connections. As a second step, we compared the degree scores of individuals with their respective clustering coefficients. This allowed us to analyze if the structure of the network is composed of individuals evolving in cohesive or lose groups.

Closeness Centrality. Following a visual analysis of node sizes, we calculated eccentricity scores to identify gaps. We verified the range of eccentricity scores, knowing that a limited distribution of eccentricity among individuals would likely to be accompanied with a low closeness centrality score. The average closeness centrality, its mode and standard deviation were calculated: A score closer to 0 than 1 would signify a well-connected network, a value closer to 1 – a sparsely connected network [24]. At a later stage, we calculated the average closeness centrality score for each group in order to verify the homogeneity of the network, i.e. if some groups are more vulnerable than others in the sense of being less well connected and less well positioned for activities such as collaboration [25]. Finally, we also analyzed individuals according to their closeness centrality score and profession.

Network Dynamics – Homophily Analysis. An in-depth group analysis was conducted in order to understand their composition, activities and contribution to the sustainability of the network.

Characteristics and group homophily. Recent research findings indicate that teachers are more likely to interact with teachers who are similar to them with regards to gender, age, experience, ethnicity, grade level, subject matter, physical proximity, beliefs about teaching, and prior professional relationships [9, 26, 27]. Also, our previous studies had shown an uneven impact of gender, work station and infrastructure impact on the usage of the phone. Therefore, the group analysis focused on socio-economic homophily, a shared training experience and phone usage during the training as possible factors influencing on group composition.

Network Drivers and Leaders – Closeness, Betweenness Centrality and Perceived Leadership. In addition to an analysis of the networks' structure and group patterns, we also looked at the role of individuals within the network in order to detect possible teacher leaders. Four methodological and analytical approaches to leadership were applied in order to look at leadership in terms of

- popularity and influence – because of a high number of connections (centrality degree);
- the power to ensure rapid information transmission (betweenness centrality);
- a favorable position – close to powerful individuals and 'influencers';
- an individual being perceived as leaders by the environment.

Centrality Degree – Distribution of Leaders with High Number of Connections, their Location and Their Profession. First, we looked at the distribution of individuals with high centrality values within the network. To do so, we calculated the mode, average and maximum centrality value and defined leaders as those with >15 connections. Then, a visual analysis of their distribution within the graph was conducted. Finally, we identified the profession of each leader.

Betweenness centrality – leaders with rapid information transmission capacities and their location. While degree centrality looks at leadership from a quantitative perspective and the power to connect (to) people, betweenness centrality focuses on an individual's capability to enhance the efficiency of the network. We used the measure to count the paths between actors and in fine, analyze to which extent an actor will lie on paths between many other actors of the network – the more often, the higher the actor's capacity to influence the spread of information across the network. We computed a graph that would highlight those individuals with high transmission capacities and analyzed their geographic and socio-economic characteristics.

Eigenvector – A Favorable Position Thanks to Proximity to Leaders. Finally, we also looked at individuals who were in a favorable position because of their proximity to leaders as defined by centrality degree or betweenness centrality. For this purpose, we computed a graph representing individuals according to their eigenvector scores, analyzed their location in the network and finally, compared these 'leaders' with those who are leaders from a betweenness and centrality perspective – or both.

Ego-Network Analysis. An ego-network analysis based on targeted interviews with network members was conducted in order to provide in-depth explanations for the patterns and dynamics.

Qualification of Ties. Each participant of the ego network consulted a list with names of their respective group and were asked to qualify the ties by using three instruments inherent to ego-network analysis:

- name generators: Individuals were showed the list and asked if there were other training participants that should be on the list. If so, these were added;
- name interpreter: Individual were asked to indicate since when (prior to the training or since the training) they

  - knew the indicated individuals;
  - whom they would consider as colleague and whom as a friend;
  - whom they identify as having occupied the position of leaders during the training and what skills and competences the respective person has;

- density questions: Questions concerning the relationships in-between the contacts.

Interviewees for the ego-network analysis were selected among two pre-established lists of former participants, respectively listing

- individuals with a high degree centrality (high number of connections) and
- individuals who, according to the call detail record analysis, were not using the phone anymore.

We calculated scores for each type of relationship described in order to assess the proportion of individuals who would qualify group members as

- new colleagues made during the training;
- colleagues from before the training and;
- friends (independently from the training period).

Experiences and Perception of Contributions and Leaders. In addition, individuals were asked to talk in detail about their experience with other group members during the training sessions, the type of interactions in-between the indicated training participants and to which extent these relationships have evolved since the end of the training.

The questions were structured according to the following themes:

- teacher collaboration;
- leadership characteristics;
- perceived characteristics of 'network leaders';
- mostly used phone functions and
- perceived impacts of the training experience on pedagogical and professional practices.

5 Main Results of the Social Network Analysis

The methodological processes combining SNA with qualitative and quantitative methods allowed to ascertain that the mobile played a key role in the set-up and development of an autonomous teacher network. While this network drives collaborative and innovative practices, we also found that teacher network leaders play a key role in sustaining the existence of the network.

5.1 Existence of a Network and Role of the Mobile Phone

The centrality degree, closeness centrality degree and diameter analysis as well as the homophily analysis provided informing results to address the question whether the mobile teacher training has led to a teacher network.

Centrality Degree, Closeness Centrality and Diameter Confirm the Existence of Mobile Communication in a Single Network Structure. The Fig. 1 presented below represents teachers according to their centrality scores and colors them according to the group they belong to as calculated by the modularity class function, a SNA algorithm for group detection. A single, connected network composed of two major clusters and sub-groups is clearly visible, as well as the presence of individuals from 3 training districts; former trainees from Manandriana are absent. The analysis of participants' out-degree centrality and in-degree indicates no extreme difference among network members, a few are solicited more than others.

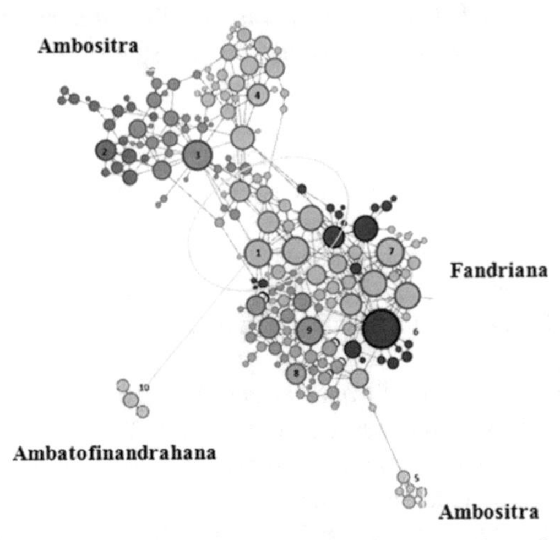

Fig. 1. Mobile teacher network – Node sizes correspond to centrality values of former trainees located in cluster areas in three training areas (Ambositra, Fandriana, Ambatofinandrahana). Colors refer to a common group.

Closeness Centrality – (Only) Two Vulnerable Spots. Figure 2 highlights the nodes according to their closeness centrality scores shows and shows beside a few exceptions a homogenous network with nodes of same size. Knowing that the distribution of eccentricity is limited, a high distribution of closeness centrality was not expected, and closeness centrality scores confirm this intuition: With an average closeness centrality of 0.23 and a mode of 0.0 (standard deviation = 0.17) the visual homogeneity impression can be confirmed. Furthermore, with the average value per individual being closer to 0 than 1, the presence of a single and complete network can be confirmed.

Finally, the actual diameter score (14), much lower than the diameter score of the network (185), confirms not only the network structure but also indicates that the networks nature is reachable by all, which can be interpreted as a network capable of regular and efficient information processes. Only two individuals from Ambatofinandrahana, a very isolated district, have high closeness rates and appear as particularly vulnerable i.e. dependent from single actors. They have the lowest score of the network (0.6).

Geographic Homophily – Effect of Training Location. The analysis shows that the upper area is composed of individuals living mostly in Ambositra, whilst the lower

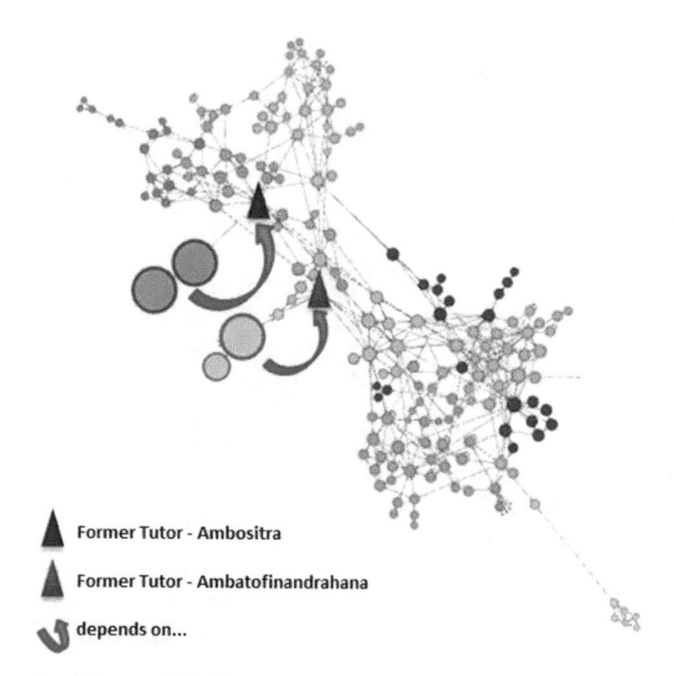

Fig. 2. Network of former IFADEM trainees: Detection of individuals with high closeness centrality scores vs individuals with high scores. Colors refer to common groups.

part is dominated by individuals from Fandriana. Each group is mostly composed of members from the same school district (Group 2, 3, 4 and 5: Ambositra, Group 6, 7, 8, and 9: Fandriana; Group 10: Ambatofinandrahana).

This geographical homogeneity of these areas contrasts with group 1. This group is composed mostly of former tutors (pedagogic councilors), they are hence naturally from multiple assigned school districts as they work in different districts. This finding, and a closer analysis of all groups, shows that geographic homophily reflects the training experience. Indeed, most of the groups in the upper area have been trained in wave 1 (organized for Ambositra members) and in the bottom part in training wave 2 (organized for those from Fandriana) (Fig. 3).

Effect of Network Satisfaction. The upper area of the graph is composed of people with an average network satisfaction score of 3.8 out of 4, which is slightly higher than the network satisfaction of the bottom part (3.2 out of 4). The rather distanced and small group 10 belongs to an area already known for poor network coverage and their vulnerability (see closeness centrality analysis).

5.2 Contributions of a Mobile Teacher Network – Are There Signs of Collaboration and Innovation?

The question whether the phone has led to contributions such as collaboration and innovative practices could be addressed through the analysis of measures such as density,

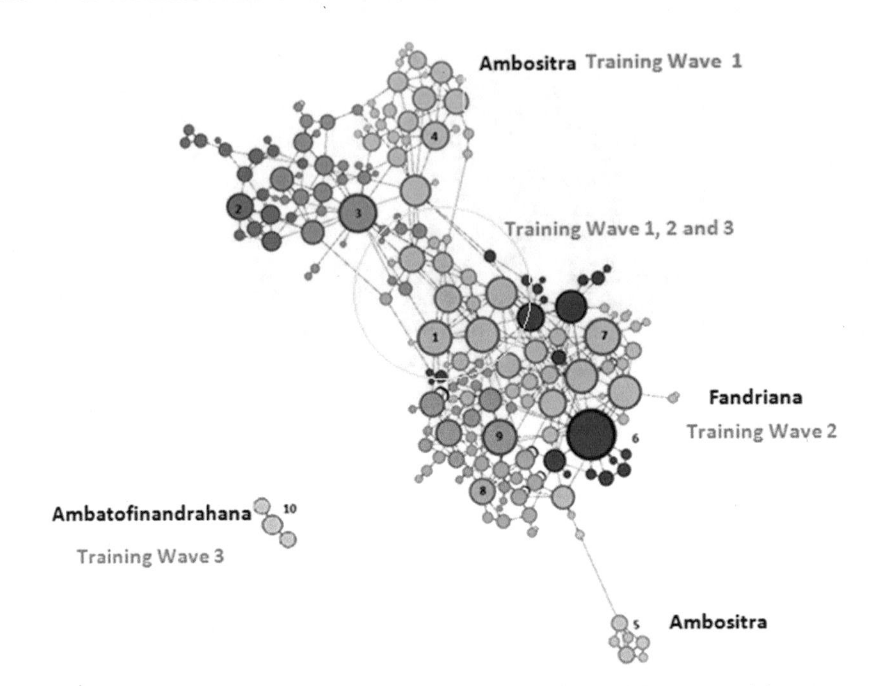

Fig. 3. Mobile teacher network – Distribution of individuals of the same training wave

cluster coefficient scores, and their respective concepts of cohesion. The ego-network analysis provided further insights on the nature of contributions.

Density – Group Cohesion Drives Collaboration. There is a higher density in the bottom part of the network, mostly composed of members from the same geographic location (Fandriana). This higher density is accompanied by an overall higher number of individuals. There are also more sub-groups in that bottom area (5 versus 3 in the lower area). The calculated density score is 0.015 and hence closer to 0 than to 1 – meaning that the average number of connections to network members is much lower than the theoretically possible number of 201 connection. A clustering coefficient of 0.208 was calculated, indicating that 20% of all possible connections within the graph are established, 100% would mean that all individuals are connected directly to all network members. This is a sign for cohesion within groups. In these areas, reported collaborative practices were frequent. A teacher reflected: "My former IFADEM tutor… she's also my pedagogic councilor – but I speak to her as my former tutor, as I continue to ask questions that concern the workbooks […] we [former trainees] are geographically distant from each other so we use the phone."

The Phone Nourishes Instrumental and Expressive Relationships. The ego-network analysis shows that in each group of the network, there are individuals who qualify their respective group members as either former colleagues or new connections they made during the training (Fig. 4). The observed group density is not a result of 'friends who just catch up'. It confirms also the qualitative study results: users developed professional and private phone communication towards former IFADEM trainees.

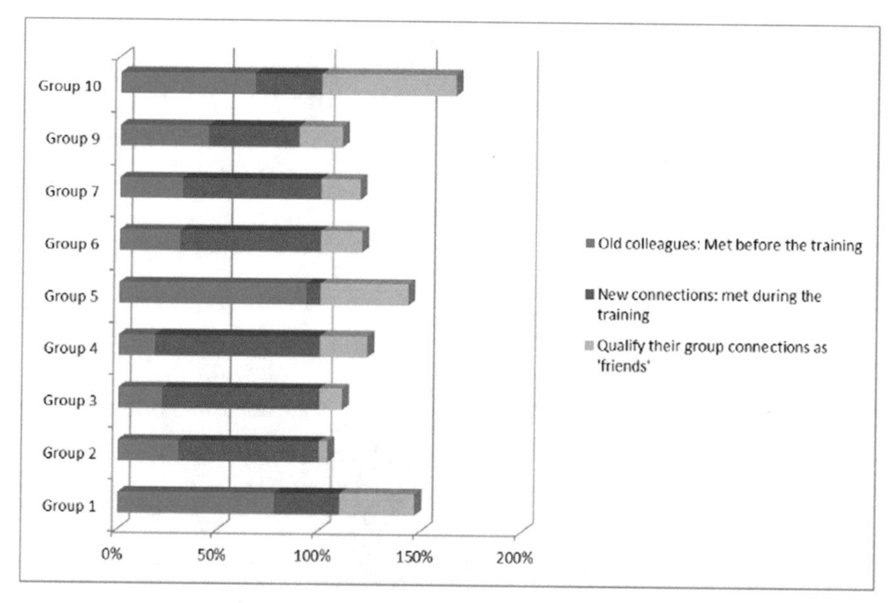

Fig. 4. Nature of ties ('old colleagues', 'new connections', 'friends') in each group

High Group Density Comes with More Frequent Emergence of Innovative Practices. When comparing the statements about collaboration and innovative school practice descriptions made by interviewees from the upper graph level (lower density/cohesion) and the lower graph level (higher density/cohesion), we found that interviewees in high density zones reported more often and in more detailed manner innovative practices they developed and applied within the group and school, as illustrates the following example. A Teacher (Ambositra) reflects:

> With those from the same school [we are in contact] on a daily basis, and face to face. And the others, when they come to Ambositra to receive their salary, they call me first and we meet up… asking for advice and exchange… [But they also call for advice and exchange] in general, not only when they come to Ambositra. We've become friends… we also discuss personal things. With regards to the phone, I take pictures during a lesson… but I also use the voice recorder. I record my students and then I make them listen so they can improve.

5.3 Teacher Leaders – Are There Individuals Who Contribute to the Networks' Sustainability?

In order to understand the role and contributions of teacher leaders for the network, a leadership analysis based on degree centrality, and betweenness centrality was conducted. This was followed by an ego-network analysis to comprehend the key characteristics of their skills and personality.

Degree Centrality – Popularity and Influence. When looking at a number of connections per individual, the resulting mode value was 2, the average 6 and the maximum value 31 (Fig. 5). A minimum threshold of 15 connection was determined to identify network leaders; the graph was computed so that this threshold would be taken into consideration. As expected – considering the results of the density analysis – there are more individuals with >15 connections in Fandriana (13) than in Ambositra (5); they are easily identifiable because of their size.

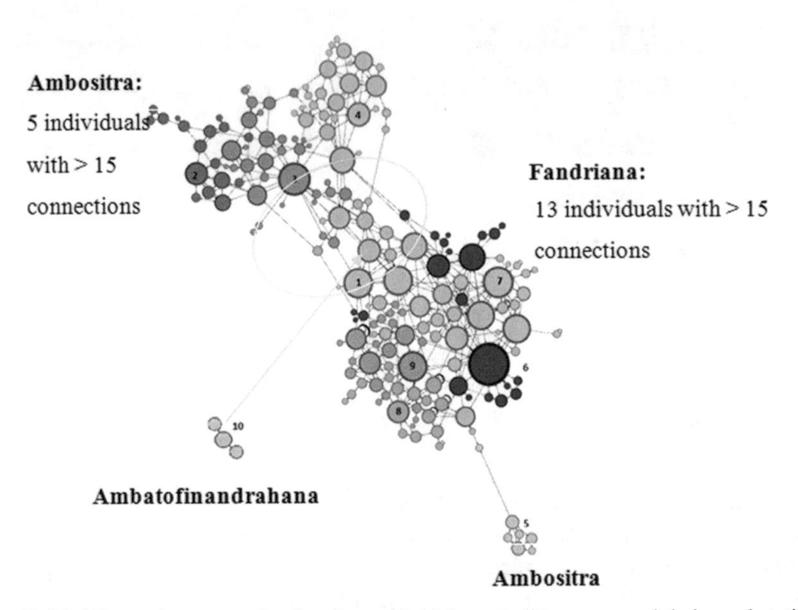

Fig. 5. Mobile teacher network – Leaders with high centrality scores and their workstation

Network leaders are well distributed within the districts. A majority of network leaders are located in the district of Fandriana. Also, every group hosts a network leader. Besides three individuals, top leaders are either former tutors, working as *conseillers pédagogiques* (regional supervision of teachers) or *Chef ZAP*[1]. All top leaders have already higher phone familiarity and have been categorized as very active phone users during the training.

Betweenness Centrality – Rapid Information Transmission. The graph highlighting individuals with high degree centrality scores shows that only a few network leaders of this category exist in the network, in particular in the upper part of the network (Fig. 6). A majority of these leaders are again former tutors or *Chef ZAP*.

Eigenvector Centrality – Well Positioned Individuals. Figure 7 highlighting individuals according to their eigenvector centrality scores shows a domination of these profiles in the bottom part, mostly composed of former trainees from Fandriana. The two leaders who have been important for information transmission in the Ambositra (top area) 'disappear', indicating that they are not well connected to what can be considered as 'hot spots' of the teacher network, all located in another area, Fandriana. It can also be

[1] *Zone d'Administration Pédagogique* ('Zone of Pedagogic Administration').

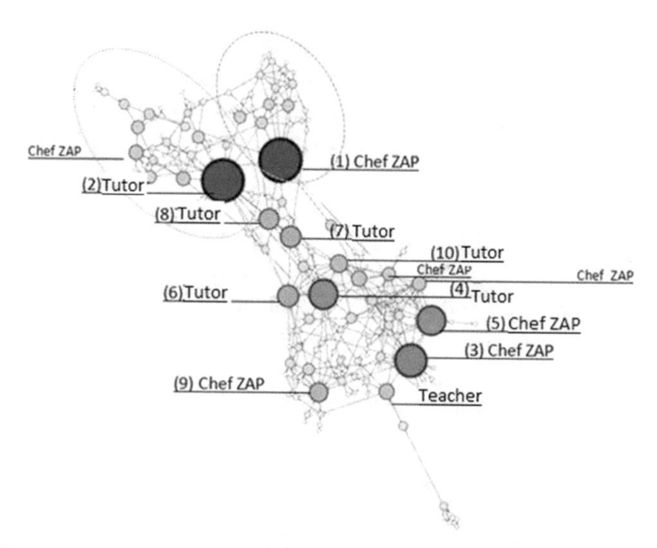

Fig. 6. Mobile teacher network – Detection and ranking of individuals with high betweeness-centrality score

observed that 'betweenness centrality leaders' of group 6, 7 and 8 remain also 'Eigenvector centrality leaders'. They are leaders who have simultaneously many connections and are connected to well-connected individuals.

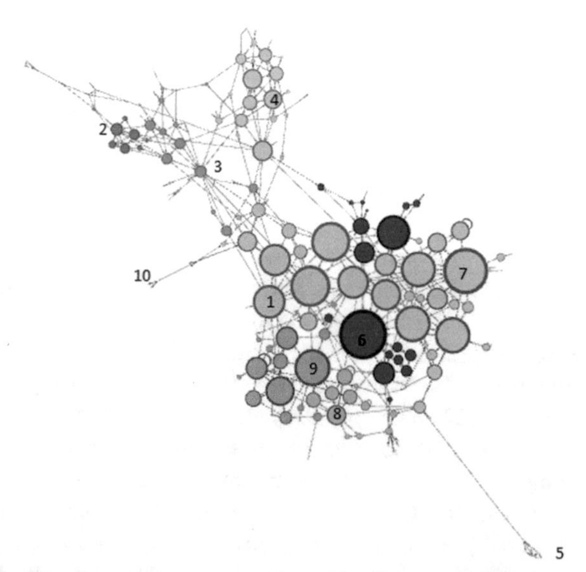

Fig. 7. Mobile teacher network – Detection and ranking of individuals with high eigenvector centrality

Ego-Network Analysis – Competences, Personality and Contributions of Leaders. All network leaders distinguished themselves from the group and the network as they had advanced French and pedagogical skills, a high sense of professionality, commitment and proactivity. Among the most often cited social skills, we found that network leaders have

- a personal, inherent capacity to influence their environment;
- a high sense of generosity, availability and flexibility;
- a dynamic, audacious and self-confident character;
- great listening skills, are able to resolve conflicts and open-minded.

Among frequently described contributions figured

- proactivity, e.g. decision to organize meetings with former trainees since the end of the pilot;
- active encouragement of collaboration among peers;
- advocating for the use of mobile phones in class and student-centered teaching practices (e.g. the use of the audio files in classroom, in order to encourage students to 'come out of their shell');
- ensuring information flow *within the network* (administrative and pedagogical information and resources) via the phone and face to face;
- …and between the *network and individuals 'outside'*.

6 Discussion

6.1 SNA Shows the Presence of a Network – Supported by the Mobile Phone as a *Tool*

The SNA found that mobile communications of former training participants are structured in the form of a network that sustains over two years after the end of the pilot. The structure indicates further shared power mechanisms – no network member has particular more access to resources than other [28]. This characterizes the network as stable and possibly innovative [9]. Furthermore, the diameter score allows us to conclude that the present network structure is capable of efficient information transmission. There is barely any isolation that could, for example, lead to information withhold or isolation at some part of the network [22]. The low closeness centrality is another sign for stability and the use of the tool for collaborative communications, allowing former trainees to access multiple resources from multiple other actors of their network.

The analysis of geographical characteristics confirms the importance of geographical homophily and indicates that the phone is a particular important tool when it comes to communicating with former training members living nearby; as observed by studies which found that phones are often used to intensify the contacts of those who live in the same locations, this allows for the creation and reinforcement of communities living near to each other [29]. However, the network structure reflects even more the importance of a shared training experience. Indeed, interviews showing that individuals continued to communicate once they had been appointed to different indicates that phone is not only a tool to sustain a relationship which has emerged thanks to geographical proximity. They also use it in case this relationship later faces geographic distancing.

The importance of a common experience as a starting point for a network genesis is illustrated by the distribution of tutors within the network. It could be expected that they are present in the group they have tutored during the training. However, they form a group of their own. This homophily reflects the principle of 'similarity breeds connection'. Tutors have made their very own experience as 'Ifademiens' and have developed a strong cohesion with their peers.

Finally, the absence of satisfying network coverage has certainly a negative impact on the likelihood to generate or join a mobile supported network – the absence for trainees from Manandriana testifies this well. The presence of individuals with poor network satisfaction reveals however that individuals can be motivated enough to overcome infrastructural challenges. Teachers reported problem-solving attitudes, e.g. climbing on a hill or typing questions and waiting to send them once the individual is in a better covered zone. This can be considered as a strong sign of appropriation [30]. It reveals a teacher network whose contributions are influenced but not determined by the quality of mobile network coverage.

6.2 SNA Reveals a Mobile Network Which Enhances and Transforms Relationships, Favoring Contribution Such as Collaboration and Innovation

The analysis also revealed a strong cohesion within groups; this cohesion sustains over time and is supported by the usage of the phone. The visual graph analysis and interviews confirm this interpretation: individuals feel close to those individuals of their respective group and report regular collaboration with group members. This is an illustrative example of interpreting density. With regards to the overall low network density, low cohesion could be expected. However, in contrast to a network based on simple interaction (e.g. Facebook), the process of collaboration requires that individuals actively work together [28]. In the IFADEM case, this means using the phone to ask, answer a question or discuss a problem – a more demanding task than those observable in simple interaction networks. Strong cohesion with all network members is not realistic, it is hence normal that network density is overall low. However, group density is high and so is the cohesion in these groups.

The Ego-Network interview indicated the presence of instrumental and expressive ties. This confirms the results of our previous qualitative study as they show the hybrid role of the mobile phone. Teacher have the possibility to combine professional and personal communications in a single connection. This is a key added value of the network groups and a factor for motivation to continue using the phone. It appears that the appropriation of the phone and transformation of relationships led to the genesis of a network that serves "the circulation of information and resources that pertain to organizational goals", while also reflecting "patterns of more affect-laden relationships [...] that transport and diffuse resources such as social support, trust, and values" [31: 635]. SNA identifies network leaders whose skills, competences and activities contribute to the sustainability of the network.

The SNA analysis had led to the identification of individuals who contribute to the sustainability of the network because of their role and competences: while some are capable to share resources and connect with a large number of individuals (high degree centrality) others contribute to an efficient information flow (high betweenness centrality) – a key contribution when it comes to distance, mobile supported communications. Furthermore, one part of the network (Fandriana) benefit even from a few 'top' network leaders who fulfill both roles (eigenvector analysis).

Furthermore, we found that former tutors and Chef ZAP are key network leaders because they are also leaders on a daily life in their work. These results are in line with the conceptual model of teacher leadership developed by York-Barr and Duke [32]. These leaders are seen as occupying a respected position, they act in a continuous learning environment and their work is valued by their hierarchy as well as visible and shared. In addition to intense mobile communication with teachers, a Chef ZAP is endowed with a motorcycle and can also meet teachers more often face to face. This mobility is also a typical characteristic of teacher leadership, particularly when it comes to leadership based on the use of technology [33].

It appears that being used to leadership comes with professional and social skills that in turn favor their position as leaders. 'Leadership familiarity' also comes with the capacity to push professional activities within the network and ultimately, increase value of the network in the eyes of teachers. Teachers evolve and benefit from a network that is driven by network leaders pushing innovative practices and collaboration.

Outstanding leaders are often Chef ZAPs. Their high betweenness centrality score is in line with studies, which found that individuals perceived as instructors and experts were those with the highest betweenness centrality scores. They are also the most active when it comes to initiating collaboration between network members in distance learning environments [27]. The importance of both instrumental and expressive ties for teacher networks can again be confirmed, as we found a direct correlation between close relationships and collaborative practices within groups, especially in those 'hosting' Chef ZAP leaders.

Finally, leaders also contribute to the network's quality as they connect the network to the regional administration and to other teachers. They continuously 'feed' external resources to the network and prevent its isolation. Overall, the presence of the leaders within the network contributes to its sustainability as they enhance contributions valuable for teachers whilst also making sure that the network can evolve with its immediate environment.

7 Conclusion

This study illustrates how the introduction of mobile phones in hybrid teacher training schemes can lead to the genesis of collaborative teacher networks – even in challenging environments such as Madagascar. Given the multiple constraints teachers face in developing countries, this positive educational contribution is a promising argument for the use of mobile phones in teacher training projects. The potential of mobile technologies in Sub-Saharan Africa goes beyond their ubiquity, it is related to a progressive appropriation that comes with usages which can sustain over time, without the support a development project. This is also a substantive result for development actors looking for the long-term contributions that public policies aim for but constrained by multiple factors making medium-term evaluations often difficult.

Moreover, the study shows that SNA can be a powerful approach to comprehend how teacher networks are structured and sustain over a longer period. SNA concepts, graphs and measures allow for both a visual representation of structures and processes, whilst also allowing to adopt multiple perspectives on the effect of ties and group dynamics. It also confirms the need for more mixed-method studies in that field, allowing for an in-depth understanding of context, processes and individuals.

References

1. UNESCO UIS report (2019). http://uis.unesco.org/sites/default/files/documents/sdg4-data-digest-data-nurture-learning-2018-en.pdf
2. GSMA Intelligent. The mobile economy Africa (2016). http://www.gsma.com/mobileeconomy/africa/. Accessed 30 Aug 2020
3. Burns, M.: Distance education for teacher training: modes, models and methods. Education Development Center Inc., Washington (2011). http://www.ltd.edc.org/sites/ltd.edc.org/files/DE%20Book-final.pdf
4. Adam, L., Butcher, N., Tusubira, F., Sibthorpe, C., Souter, D.: Transformation-ready: the strategic application of information and communication technologies in Africa. Education Sector Study Final Report, 142 (2011)
5. Kularbphettong, K., Putglan, R., Tachpetpaiboon, N., Tongsiri, C., Roonrakwit, P.: Developing of mLearning for discrete mathematics based on Android platform. Proc.-Soc. Behav. Sci. **197**, 793–796 (2015)
6. Husbye, N., Elsener, A.: To move forward, we must be mobile: practical uses of mobile technology in literacy education courses. J. Digit. Learn. Teach. Educ. **30**(2), 46–51 (2013)
7. Baran, E.: A review of research on mobile learning in teacher education. J. Educ. Technol. Soc. **17**(4), 17–32 (2014)
8. Goddard, Y., Goddard, R., Tschannen-Moran, M.: A theoretical and empirical investigation of teacher collaboration for school improvement and student achievement in public elementary schools. Teach. Coll. Rec. **109**(4), 877–896 (2007)
9. Moolenaar, N., Sleegers, P., Daly, A.: Teaming up: linking collaboration networks, collective efficacy, and student achievement. Teach. Teach. Educ. **28**(2), 251–262 (2012)
10. Grossman, P., Wineburg, S., Woolworth, S.: What makes teacher community different from a gathering of teachers. Center for the Study of Teaching and Policy, pp. 5–56 (2000)
11. von Lautz-Cauzanet, E., et al.: Mobile supported teacher training in Sub-Saharan Africa. Which contributions and how to analyze them? Ph.D. thesis, Paris Saclay (2018)

12. Gaboton, J., Despover, C., Jarousse, J.: Rapport de mission en république d'haiti, suiviévaluation de l'initiative francophone pour la formation á distance des maitres (IFADEM) [Report on the mission to the Republic of Haiti, Monitoring and Evaluation of the francophone initiative for distance learning for teachers]. Technical report (2013)
13. von Lautz-Cauzanet, E., Bruillard, E., Le Quentrec, E.: Unexpected benefits of a mobile supported distance training initiative in Madagascar. In: Society for Information Technology & Teacher Education International Conference, pp. 2413–2420. Association for the Advancement of Computing in Education (AACE) (2016)
14. Gire, F., Le Quentrec, E., Loiret, P., Razafindrakoto, T., Rakotovao, L., Hanitriniala Ratompomalala, H., Ben Abid-Zarrouk, S., Pourcelot, C.: Retour d'expérience de l'utilisation du mobile dans le dispositif de formation continue des enseignants du primaire a Madagascar [Feedback on the use of mobile phones in in-service training for primary school teachers in Madagascar] (2013)
15. Cobb, P., McClain, K., de Silva Lamberg, T., Dean, C.: Situating teachers' instructional practices in the institutional setting of the school and district. Educ. Res. **32**(6), 13–24 (2003)
16. Engeström, Y.: Learning by Expanding. Cambridge University Press, Cambridge (1987)
17. Murphy, E.: Activity Theory Perspectives on Technology in Higher Education. IGI Global, Philadelphia (2013)
18. Vygotsky, L.: Socio-Cultural Theory. Harvard University Press, Cambridge (1978)
19. Bachelet, C.: Usages des TIC dans les organisations, une notion à revisiter? [ICT uses in organisations, a concept that needs to be revisited?]. Ph.D. thesis, IREGE, Université de Savoie (2003)
20. Méadel, C., Proulx, S.: L'usager en chiffres, l'usager en actes [The user in figures, the user in deeds]. Presses de l'Université Laval (1998)
21. Stegbauer, C., Häußling, R.: Einleitung in das Handbuch Netzwerkforschung [Introduction to the Network Research Handbook]. In: Stegbauer, C., Häußling, R. (eds.) Handbuch Netzwerkforschung [Network Research Handbook], pp. 13–16. VS Verlag für Sozialwissenschaften, Wiesbaden (2010)
22. Adler, S.: Konzepte der Netzwerkanalyse [Concepts of Network Analysis]. Universität Wien (2010)
23. Moolenaar, N.: A social network perspective on teacher collaboration in schools: theory, methodology, and applications. Am. J. Educ. **119**(1), 7–39 (2012)
24. Mutschke, P.: Zentralitäts-und Prestigemaße [Measures of centralisation and prestige]. Springer (2010)
25. Wu, Y., Duan, Z.: Social network analysis of international scientific collaboration on psychiatry research. Int. J. Ment. Health Syst. **9**(1), 2 (2015)
26. Bidwell, C., Yasumoto, J.: The collegial focus: teaching fields, collegial relationships, and instructional practice in American high schools. Sociol. Educ. **72**, 234–256 (1999)
27. Penuel, W., Fishman, B., Yamaguchi, R., Gallagher, L.: What makes professional development effective? Strategies that foster curriculum implementation. Am. Educ. Res. J. **44**(4), 921–958 (2007)
28. Cherven, K.: Mastering Gephi Network Visualization. Packt Publishing Ltd., Birmingham (2015)
29. Riviére, C.-A.: Le téléphone, un facteur d'intégration sociale [The telephone, a factor for social integration]. Insee (2001)
30. Hahn, H., Kibora, L.: The domestication of the mobile phone: oral society and new ICT in Burkina Faso. J. Mod. Afr. Stud. **46**, 87–109 (2008)
31. Moolenaar, N.M., Daly, A.J., Sleegers, P.J.C.: Occupying the principal position: examining relationships between transformational leadership, social network position, and schools' innovative climate. Educ. Adm. Q. **46**, 623–670 (2010). https://doi.org/10.1177/0013161X10378689

32. York-Barr, J., Duke, K.: What do we know about teacher leadership? Findings from two decades of scholarship. Rev. Educ. Res. **74**(3), 255–316 (2004)
33. Riel, M., Becker, H.: Characteristics of teacher leaders for information and communication technology. Springer (2008)

Good Intentions or Good Relations? A Longitudinal Analysis of the Interrelation Between Network Embeddedness and the Perception of Obstacles to Cooperation

Dumitru Malai(✉) and Frank Lipowsky

University of Kassel, Kassel, Germany
{malai,lipowsky}@uni-kassel.de

Abstract. The longitudinal analysis of the development of 35 PRONET sub-projects in teacher education at the University of Kassel has its focus on the explanation of factors that are either inhibitory or stimulating for the cooperation between the sub-projects. The members of sub-projects were interviewed using an online questionnaire in order to distinguish the perceived impediments when cooperating with other sub-projects. Three kinds of perceived obstacles (internal, external, and cooperation intention) were analyzed. Indicators of cooperative activity in deliberate cooperation and embeddedness in structural/organizational linkage were obtained from the network data of the sub-projects' members and correlated with the data on perceived obstacles to cooperation. The analysis of the correlations revealed the effect of the embeddedness in deliberate cooperation as well as in the structural linkage on the future perception of obstacles to cooperation.

Keywords: Cooperation networks · Organizational cooperation · Perception of obstacles · Interlock directorates · Teacher education

1 Introduction

The cooperation of academic staff in higher education, for instance, in research and teaching, is known to be beneficial to a broad spectrum of actors [1]. In this respect, enhancing the cooperation in the higher field of education has become a goal at many universities. Worldwide programs were launched that aim at fostering the cooperation between academic staff [2, 3].

However, monitoring the implementation of cooperation although lacks concrete indicators of the output and development of cooperation. Factors that are either inhibiting or stimulating for cooperation can serve as (such) indicators. Understanding these factors can be useful for the management of higher education institutions because of the adjustments that can be made, therefore reducing the costs, and demanded resources of the research and learning programs [4].

A. Antonyuk and N. Basov (Eds.): NetGloW 2020, LNNS 181, pp. 306–321, 2021.
https://doi.org/10.1007/978-3-030-64877-0_20

One key factor in implementing cooperation is the understanding of possible obstacles that prevent actors to collaborate. Perceptions and attitudes are factors that influence the behavior, including also cooperative behavior. In this respect, the focus should be put on the perception of obstacles by the actors of cooperation. In order to evaluate and to explain their impact, obstacles to cooperation need to be distinguished and measured.

The depiction of the obstacles to cooperation is also a goal the evaluation of the project PRONET includes. As part of a Germany-wide campaign, which has the aim to improve the quality of teacher training (*Qualitätsoffensive Lehrerbildung*), the project PRONET[1] (professionalization through networking – in German: ***Pro****fessionalisierung* *durch* ***Vernet****zung*) was launched in 2015 at the University of Kassel. PRONET consists of 35 sub-projects aimed at jointly developing innovative course concepts and at offering them to teacher students in order to improve teacher education at the university. In the development of innovative concepts, the collaboration between sub-projects is a crucial factor which is intended to influence the improvement of teacher training. The cooperation of the PRONET actors and the significant intellectual capital that emerges as its result can reduce the redundancy and the excessive fragmentation of the university teacher training. Also, the sub-projects are composed of experts of different branches of science (e.g. German philology, biology, theology, mathematics) and disciplines of teacher training (subject, didactics, pedagogy) which facilitate an interdisciplinary discourse and can, for this reason, be beneficial for involved actors.

The sub-projects are oriented towards research and teaching and are assigned to three so-called fields of action (*Handlungsfelder*). In the first field of action, sub-projects are collaborating to further develop the 'practical studies' at the University of Kassel that focuses on enhancing the reflexive skills of teacher students. In the second field of action, the sub-projects work on advancing pedagogical concepts towards inclusion, sensitize students of teacher education to topics on heterogeneity and diversity, and equip them with tools for inclusive teaching. Finally, the third field of action includes sub-projects that aim at interconnecting disciplines, didactical, and pedagogical aspects of teacher education.

Besides their activity in fields of action, the sub-projects have the opportunity to get in touch at the obligatory face-to-face meetings, where the management of PRONET and the sub-projects meet to discuss the current challenges and exchange information. The sub-project PRONET Meta-Evaluation is in charge of evaluating the development of the other PRONET sub-projects. As part of the evaluation, PRONET Meta-Evaluation investigates the emerged connections between sub-projects and visualizes them using organizational network analysis. Moreover, it investigates opinions towards the cooperation of the academic members in projects as well as the perceived obstacles to cooperation.

[1] This project is part of the "*Qualitätsoffensive Lehrerbildung*", a joint initiative of the Federal Government and the *Länder* which aims to improve the quality of teacher training. The programme is funded by the Federal Ministry of Education and Research. The authors are responsible for the content of this publication.

First, the theoretical and empirical background on the formation of attitudes and its relation to cooperative behavior will be introduced. This includes psychological, network, and cooperation theories that were considered to formulate several hypotheses. In the following chapter, the sample, investigation design, measurement instruments, and the empirical approach will be presented. Finally, the results and the conclusion will be outlined.

2 Theory and Hypotheses

The relation between attitude and subsequent behavior is described in social psychology as situational [5]. The behavior depends on the interpretation or the perception of a certain situation and is proved not to be solely an expression of the attitude one person has [6–8].

The formation of attitudes is, on the one hand, habitual and depends on past experiences but, on the other hand, situational, and rely on the interpretation of the current situation. Also, the attitudes can be revisited over time or change due to an event [9]. This introduces the concept of attitude as a process in which time is playing a crucial role.

The theory of *planned action* joins the concepts of perception, attitude and behavior intention and postulates that "people act in accordance with their intentions and perceptions of control over the behavior, while intentions, in turn, are influenced by attitudes toward the behavior, subjective norms, and perceptions of behavioral control" [10]. According to this theory, the perception of control corresponds with the perception of obstacles and the belief of the individual's ability to overcome them [11].

When speaking about public goods, cooperation is one form of planned action, intended to maximize collective benefits. In other words, the idea of planned action shares several concepts with the public good theory [12]. Both of the theories imply a purposeful action by the actors. Actors that pursuit the same collective goal will be intrinsically more motivated to cooperate and will also share the intentions to join their efforts. The attitudes of the actors towards cooperation and its obstacles will also be shaped by the initial common embeddedness in a collective work. However, the aspect of the interest is also present in the public good approach [12].

Besides that, the aspect of the expectancy towards the output of behavior shapes the formation of an attitude. This idea represents the corpus of the *expectancy-value* model: "Each belief associates the object with a certain attribute, and a person's overall attitude toward an object is determined by the subjective values of the object's attributes in interaction with the strength of the associations" [10]. Many cooperation theories (e.g. the social exchange theory [13]) contain this aspect of the expectancy-value theory and define the cooperation actors as driven by their interest. The actors with related interests are thus more likely to cooperate as they have similar value orientations and expect a mutual meeting of their interests [14]. The value of cooperation become visible insofar as the goals that are not feasible for the individual actors themselves can be achieved through cooperation. Moreover, cooperation leads to new ideas and solutions to all members involved. This reasoning speaks in favor of the actors' external motivation to cooperate. However, the interest also implies time and resources spend on cooperation. As well as the aspects of the trustfulness and the output of the previous cooperation, which can influence the perception of external obstacles. The assumption of the planned action and expectancy-value theory were operationalized and empirically tested for diverse disciplines [15]. Moreover, there is empirical evidence from both theoretical perspectives for the special case of cooperative behavior. For instance, the investigation of Roos and Hahn [16] found that collaboration is determined by personal/economic and normative beliefs. Another study of teacher cooperation [17] has similarly shown that the subjective value of cooperation determines cooperation. Nevertheless, limitations of the theories of planned behavior and expectancy-value model are discussed in empirical studies and evaluated as not entirely predictive for cooperative behavior [18].

Other important and recent insights into the conceptualization of attitudes were introduced by the network theory. It defines attitudes as interconnected entities in a broader network of attitudes, which constitutes an organizational culture or an ideology [19]. The major contribution of the network perspective is its view of the networks as dynamic structures that change over time. This change in attitudes and perceptions happens not just on the individual level but is also related to the changes in the attitudes and perceptions of other actors in the network [20]. Thus, the experience of past cooperation impacts future cooperation, or, in other words, the establishment of new connections or ties in a cooperative network. Concerning inter-organizational cooperation, new ties tend to be integrated into the existing structure of relation [21].

A variety of studies support the theoretical view of the dominance of the structures actors are embedded in. The position inside a network or group can be beneficial or disadvantageous for an actor. It also subscribes a role to an actor, according to which he acts, and in the long term even forms its behavior. The experiments of Leavitt [22] have shown that the centrality of an actor's position in a group influences communication behavior. Thus, the actors in a central network position are more influential in the communication process. Also, they are central to the formation of attitudes, playing an important role in the spread and circulation of attitudes. With the development of the methods of network analysis, the relationship between networks and behavior was analyzed more broadly supporting the previous findings [23].

In networks where actors are free to cooperate with any other actors, structures of cooperation still have a great impact on their behavior. A free i.e. active engagement in cooperation determines the actors in adopting a set of rules of behavior, in order to make the collaboration of both parties convenient [24]. This process is illustrated by the theory of negotiated order [14], which explains how the process of the academic or faculty cooperation is taking place. A negotiation order emerges when the cooperation actors share their perceptions of the issue or problem and make agreements about the dimensions of their work together: "As they work together, collaborators each contribute and identify their interests as feasible, reframe the problem, and search for approaches" [25].

Also, organizational networks have an impact on the behavior and attitude of their members. This is the example of the institutions, which are organized in a certain way and bring actors in different positions together. One example of this kind of structure are the networks created through multiple memberships of the actors in central positions, like the co-membership in multiple organizations, known in the literature as interlock directorates [26]. Embeddedness, in this respect, denominates an affiliation to a structure that is rather rigid and stable. The findings regarding active cooperation and embeddedness in organizational structures consider the economic behavior or the collaboration between actors who try to fulfill their interests [27] and access needed resources [28]. However, in organizational networks oriented towards the development of collective goods like teaching concepts in teacher education, the idea of competition is not as prominent as in the economic organizations.

Criticism and the limits of cooperation are, nevertheless, also present. For example, there are studies [27] that mention a threshold after which cooperation can derail the performance of organizations. The overembeddedness can lead to dependence, resource loss, and inflexibility in the actors' action. This is an important point to consider; however, the findings refer to the economic action and not for public goods cooperation.

The three perspectives of planned behavior, expectancy-value, and the network theories can be applied to the organization of the PRONET sub-projects and will be further outlined below.

The cooperation between sub-projects has the characteristics of planned action. It can be argued that the idea of cooperation is intrinsic to the underlying concept of PRONET, since teacher training is a public good that is used by many members of the community. Sub-projects, which follow this underlying concept, also embrace the idea of cooperation. The active cooperation of the sub-projects is expected to be a consequence of their positive intrinsic motivation. Concerning the organizational structures, the earlier presented concept of fields of action is an example of the cooperation orientation of PRONET, and represents places for exchanging information and experiences. The assignment of sub-projects to the fields of action was intended to enhance cooperation and is an essential part of the PRONET underlying concept. From this point of view, sub-projects are expected to have a positive attitude to engage in cooperation with other sub-projects and to perceive fewer obstacles to cooperation position. For instance, the amount of perceived internal obstacles (i.e. intrinsic to the actors) is supposed to be low. The assumption is that network cooperation activity correlates with the subsequent formation of attitudes towards cooperation, respectively the perception of obstacles to cooperation. According to this, the first hypothesis of the effect of prior cooperation on the subsequent perception of internal obstacles is formulated (H1): The higher the past cooperative activity and the embeddedness in cooperation, the smaller are the perceived internal obstacles to cooperation.

Moreover, cooperation is proved to be significant and beneficial to achieve organizational goals which cannot be achieved by individual actors [29], because emergent attributes of the organization allow the achievement of more elaborate goals. Thus, the achievement of PRONET goals, such as the development of 'practical studies' or the development of inclusive teaching concepts with a focus on diversity, cannot be done by isolated sub-projects and require joint working. The value of cooperation is determined by the necessity of the sub-projects to be informed about the work of other sub-projects of the same field of action and to communicate in order to contribute to the general goal. Also, the cooperation of PRONET sub-projects is expected to be beneficial because it connects many experts from different academic fields, creating a favorable interdisciplinary environment. The embeddedness in cooperation structures is expected to increase the benefits of the sub-projects, and to improve the perceived i.e. subjective value of cooperation. Thus, the value of cooperation perceived (i.e. expected) by the actors relates to extrinsic factors. The perception of *external* obstacles to cooperation is expected to be smaller in a favorable cooperative environment. Following the assumption that the embeddedness in organizational networks influences the formation of attitudes towards cooperation, respectively the perception of obstacles to cooperation, the second hypothesis of the effect of cooperation on the subsequent perception of external obstacles (H2) was formulated: The higher the past cooperation activity and the embeddedness in cooperation, the smaller are the perceived external obstacles to cooperation.

From the network perspective on attitudes is expected that past cooperation activity and the former embeddedness in cooperation structures will determine the subsequent perception of obstacles to cooperation and the subsequent cooperative behavior. Thus, the embeddedness and the activity in cooperative structures as well as the position inside the network are expected to influence the perception of obstacles to initiate new cooperation. The past experiences of cooperation and the changes in the structures of the cooperation network might influence therefore the formation of the future perception of obstacles to start/initiate cooperation. Following these assumptions, the third hypothesis of the effect of cooperation on the subsequent perception of taking initiative in cooperation (H3) was formulated: The higher the past cooperation activity and the embeddedness in cooperation, the smaller are the perceived obstacles to initiate cooperation.

3 Method

3.1 Sample and Dataset

The PRONET Meta-Evaluation study is designed longitudinally. Network data were collected at three time points by now: the first measurement was carried out in December 2015, shortly after the project had started, the second one in summer 2017, and the last one at the end of the first project period in December 2018.

The first measurement consisted of data of 59 members of the academic staff out of all departments at the University of Kassel engaged in teacher education. The second measurement delivered data of 61 members, and the third one of 44 members.

The respondents represented researchers and teachers from different branches of science and disciplines of teacher education, for example, German philology, natural sciences, or foreign languages. The sample included professors, assistant researchers, and educational staff members. The respondents were interviewed via an online questionnaire. The data were aggregated on the level of the PRONET sub-projects. Therefore, the sample of each measurement point consisted equally of the 35 PRONET sub-projects.

3.2 Measures and Variables

Two kinds of cooperation were distinguished: the structural linkage and deliberate cooperation. The structural linkage emerges as a result of the fact that some staff members work in more than one sub-project, thereby creating connections between sub-projects (interlock directorates). The deliberate cooperation emerges voluntarily between sub-projects in order to come together and collectively elaborate new products such as teaching concepts and materials, seminars, and workshops. Both of these connecting activities were collected as network data. For the data on deliberate cooperation, the directions of the links/nominations between sub-projects were recorded. The links of the structural linkage network (interlock directorates) are undirected, the relation between sub-projects being the reflection of co-membership of certain actors in multiple sub-projects.

For both networks at each point of time, centrality measures were calculated and exported as separate variables for each node (sub-project). Outdegree and transitivity were the main extracted centrality measures. The outdegree values are available for the network of deliberate cooperation, which is recorded as a directed graph. Outdegree measures the outgoing cooperation, i.e. the number of cooperation partners that are nominated by a sub-project, being this way an indicator of the cooperative effort of the sub-projects. In other words, the use of outdegree intends to capture the cooperation activity in the network of deliberate cooperation. The estimates of the density and reciprocity of the network of deliberate cooperation (formed between 35 sub-projects) were: $d_{\text{tp1}} = .18$, $d_{\text{tp2}} = .15$, $d_{\text{tp3}} = .10$, $rc_{\text{tp1}} = .42$, $rc_{\text{tp2}} = .52$, $rc_{\text{tp3}} = .45$. The estimates of the density of the network of structural linkage were: $d_{\text{tp1}} = .05$, $d_{\text{tp2}} = .05$, $d_{\text{tp3}} = .04$. The network of structural connection is less dense than the network of deliberate cooperation since organizational structures cover only a small amount of possible cooperation combinations. The sub-projects had on average 6.03 cooperation contacts to the first, 5.11 to the second, and 3.14 to the third measurement point. The range of the outdegree varies between 1 and 22 (t1), between 0 and 9 (t2), and between 0 and 8 (t3). For the network of structural linkage, the measure of transitivity was used to represent the cooperation. Transitivity denominates the embeddedness in organizational structures or the belonging to high connected regions of the structural network. The cooperation between sub-projects is here created by actors in a central position who establish a connection between them. This value suggests a greater influence of the respective actors. General transitivity of the network of structural linkage is represented as follows: $t_{\text{tp1}} = .79$, $t_{\text{tp2}} = .75$, $t_{\text{tp3}} = .82$.

The scale *perceived obstacles to cooperation* was developed by the PRONET Meta-Evaluation and implemented in order to distinguish the sub-projects' impediments to cooperation with other sub-projects. The three facets of the scale are external obstacles, internal obstacles, and the cooperation initiative. The responses were recorded on a six-point Likert scale (1 = "completely disagree" to 6 = "completely agree").

The subscale *internal obstacles to cooperation* measures perceived impediments to cooperation due to person/team-related motives. The subscale consists of 6 items (e.g. "I don't believe our project would benefit from cooperation.") with the following mean, standard deviation and reliability (Cronbach's alpha) values to each of the three measurement points ($M_{tp1} = 2.02$, $M_{tp2} = 2.48$, $M_{tp3} = 2.50$, $SD_{tp1} = 1.29$, $SD_{tp2} = 1.04$, $SD_{tp3} = 1.28$, $\alpha_{tp1} = .76$, $\alpha_{tp2} = .63$, $\alpha_{tp3} = .78$). *External cooperation* refers to the perceived obstacles related to the factors which are not in the personal or team sphere of action. This subscale consists of 4 items (e.g. "Contacted PRONET sub-projects don't have enough time for cooperation.") with the following values ($M_{tp1} = 1.55$, $M_{tp2} = 1.97$, $M_{tp3} = 1.95$, $SD_{tp1} = 1.02$, $SD_{tp2} = 1.03$, $SD_{tp3} = 1.08$, $\alpha_{tp1} = .75$, $\alpha_{tp2} = .57$, $\alpha_{tp3} = .75$). Finally, the subscale *cooperation initiative* measures the perceived personal/team deliberate and purposive networking activity. It consists of three items (e.g.; "I don't have enough personal contacts with PRONET colleagues to initiate cooperation.") with the following values of the subscale ($M_{tp1} = 2.76$, $M_{tp2} = 2.51$, $M_{tp3} = 2.38$, $SD_{tp1} = 1.56$, $SD_{tp2} = 1.18$, $SD_{tp3} = 1.28$, $\alpha_{tp1} = .82$, $\alpha_{tp2} = .73$, $\alpha_{tp3} = .78$).

3.3 Empirical Approach

For the current analysis, the relation between the values of the above-mentioned scales on the perception of obstacles to cooperation and the centrality measures (outdegree, transitivity) was examined by the means of Pearson (bivariate) correlation [30]. The examination was done for both, the network of deliberate cooperation and structural linkage (i.e. interlock directorates). Repeated (i.e. systematic) time dependencies between pairs of variables based on psychometric and network data were recorded for further analysis.

The specific characteristic of this analysis consists of a mix of network and psychometric data.

Because of the small sample, the statistical technique of bootstrapping was used to estimate the confidence intervals of the bivariate correlations [31]. The bootstrap sampling was based on 1000 iterations.

4 Results

All the analyzed relations between variables are presented in the following correlations table (Table 1). The table with confidence intervals (Table 2) for the used bivariate Pearson correlations can be found in the Appendix.

Table 1. Correlation table between outdegree and transitivity values and the subscales of perceived obstacles. outdeg: outdegree in the network of deliberate cooperation, t_p: transitivity in the structural linkage network, IO: internal obstacles; EO: external obstacles, CI: cooperation initiative; $*p < .05$; $**p < .01$.

	outdeg_t1	outdeg_t2	outdeg_t3	t_p_t1	t_p_t2	t_p_t3	IO_t1	IO_t2	IO_t3	EO_t1	EO_t2	EO_t3	CI_t1	CI_t2
outdeg_t2	.77**													
outdeg_t3	.63**	.55**												
t_p_t1	.88**	.64**	.68**											
t_p_t2	.61**	.38**	.38**	.77**										
t_p_t3	.72**	.52**	.59**	.90**	.80**									
IO_t1	−.25	−.24	−.11	−.26	−.26	−.18								
IO_t2	−.20	−.28	−.17	−.14	−.14	−.14	.63**							
IO_t3	.15	.17	−.11	.11	.14	.13	.24	.09						
EO_t1	.28	.29	.27	.17	.14	.21	.53**	.08	.20					
EO_t2	.01	−.04	−.20	.10	.10	.10	.20	.60**	.08	.09				
EO_t3	.44*	.39*	.43*	.54**	.50**	.58**	.09	.15	.51**	.37	.28			
CI_t1	−.22	−.23	.06	−.29	−.30	−.24	.51**	.39*	.23	.37	−.03	.10		
CI_t2	−.29	−.31	−.04	−.33	−.25	−.31	.53**	.49**	.09	.27	.13	.17	.90**	
CI_t3	−.44*	−.55**	−.45*	−.50**	−.48**	−.49**	.09	.04	.17	−.10	−.07	−.08	.51*	.53**

To prove the first hypothesis (H1) time interdependencies between the variables perceived internal obstacles to cooperation, and the embeddedness in structural and active cooperation (outdegree, transitivity) were tested. Neither in the network of deliberate cooperation nor in the network of structural connection the relation between the perception of internal obstacles and network variables yields significant results ($r_{\text{iooT1}} = .15$, $p > .05$; $r_{\text{iooT2}} = .17, p > .05$; $r_{\text{iotT1}} = .11, p > .05$; $r_{\text{iotT2}} = .14, p > .05$). No repeated time-delayed dependencies between perceived internal obstacles and the centrality measures of structural linkage and deliberate cooperation were found.

Therefore, the first hypothesis addressing the influence of past cooperation on perceived internal obstacles in the future (H1) cannot be proved.

Next, in the correlation table, we can observe that outdegree activity has a delayed time effect on the perception of external obstacles. The correlation between the outdegree activity to the first and second time points and the perception of external obstacles to cooperation is in both cases positive, moderate, and significant ($r_{\text{oeoT1}} = .44$, $p < .05$; $r_{\text{oeoT2}} = .39, p < .05$). Conversely, there is no time-delayed significant correlation between the perception of external obstacles (t1 and t2) to cooperation and the outdegree activity to the third time point. The significance of these results can be described as that the cooperative activity in the past goes along with the perceived external obstacles in the future. This result reveals the influence of the perception of external obstacles on the outgoing cooperative activity. Sub-projects which were active in cooperation in the past tend to perceive more obstacles to cooperation in the future.

The transitivity of the structural linkage network affects the future perception of external obstacles to cooperation. The structural linkage (like interlock directorates) are formed by the academics, who work in several projects and thus create the connection between them. There is also a time-delayed correlation between the transitivity of the structural linkage in both first and second time points and the perception of obstacles to cooperation ($r_{\text{teoT1}} = .54$, $p < .01$; $r_{\text{teoT2}} = .50$, $p < .01$). Both correlations are positive, moderate, and significant. Thus, the position in the structural network in the past correlates with the formation of the perception of external obstacles to cooperation in the future. These results reveal the empirical relationship between the embeddedness in organizational structures and the future perception of external obstacles to cooperation. The delayed effects in time of the perceived obstacle to cooperation on the transitivity are not significant. Therefore, the second hypothesis (H2) can be confirmed. The past cooperation measured by the network embeddedness correlates with the future perception of external obstacles to cooperation.

In the correlation table, we can also observe the relationship between the initial outdegree activity and the future cooperation initiative. The correlations of the degree activity of the first and second time points are moderately negative and significant ($r_{\text{ociT1}} = -.44, p < .05$; $r_{\text{ociT2}} = -.55, p < .01$). That means that sub-projects which cooperated actively in the past encounter fewer impediments to start a new cooperation. Thus, the experience of cooperation activity is associated with the formation of a favorable attitude to initiate cooperation. A long-term relation could be shown between cooperation activity (measured by outdegree) in the past and the perception of fewer obstacles to initiate cooperation in the future. The time-delayed effect of the cooperation initiative on outdegree activity is here also not significant ($r_{\text{cioT1}} = -.23$, $p > .05$; $r_{\text{cioT2}} = .06$, $p > .05$).

Finally, we observe the correlation of the transitivity of the structural linkage network on the perception of obstacles to cooperation initiative. This correlation is for the first and the second time points moderately negative and significant ($r_{\text{tciT1}} = -.50$, $p < .01$; $r_{\text{tciT2}} = -.48$, $p < .01$). This means that sub-projects that are in a tight organizational structure from the beginning are predisposed to develop a reserved attitude to initiate cooperation. The past embeddedness in the networks of the deliberate cooperation and the structural linkage network (interlock directorates) is in negative interdependence with the future perception of obstacles to initiate cooperation. Thus, the embeddedness in the structural network is in an empirical relationship with the perception of more impediments to starting new cooperation. This result reveals the time interdependence of the perceived obstacles by the transitivity in the network of structural connectivity in the long term. The significant time-delayed effect of the cooperation initiative on transitivity was found. The third hypothesis (H3) can be confirmed as well.

5 Conclusion

Based on the analysis through bivariate Pearson correlations, the effects of the former cooperation activity in the deliberate cooperation and the embeddedness in organizational structures on the future perception of obstacles to cooperation were illustrated. Sub-projects which actively connected with other projects early in their project phase have a higher perception of external obstacles to cooperation in the future. This can be caused by the limit of connection they reach early at the time of the first measurement. It was also shown that a lower perception of obstacles to cooperation not necessarily leads to more cooperation effort (outdegree activity) in the future. Hence, there is a time interdependence between the prior cooperative behavioral variable and the future perception of obstacles.

Furthermore, projects which are connected structurally to each other, i.e. due to a given organizational structure, e.g. staff members who work in several projects (analog to interlock directorates), perceive more obstacles depending on the level of their structural connection (transitivity). Given the structural connection from the beginning, they tend to form a more pronounced perception of external obstacles to cooperation in the future. The tighter (or more transitive) the prior linkage between sub-projects, the higher is the value of their subsequent perception of the obstacles to cooperation.

Concerning the cooperation initiative, it was shown that the cooperation activity (outdegree) at the sub-project beginning goes along with the future perception of obstacles of cooperation initiative. Furthermore, the relation goes from the network variable of cooperation activity to the subsequent attitude in the future. The reason for this effect may be as well the limit of connections reached, which determines the sub-projects to be rather reserved for starting new cooperation.

As in the previous findings, the level of embeddedness (transitivity) in the structural linkage goes along with the perception of obstacles to initiate cooperation. Embeddedness in a tight organizational structure may be perceived as sufficient by the sub-projects and see no necessity in initiating new cooperation. These results are coherent with the previously mentioned findings that point out the limits of the cooperation up to a certain performance threshold [27].

Finally, neither the effect of cooperation activity (outdegree) nor the embeddedness in organizational structures (transitivity) could predict the formation of perceptions towards internal obstacles to cooperation. One reason for this can be the tendency to ascribe perceived obstacles to external factors. This phenomenon is known in the literature as *external causal attribution* [32]. Because obstacles have a negative connotation, the sub-projects will rather ascribe them to factors that are out of their control and preserve their self-efficacy beliefs. Also, since the cooperation demands time and resources and there are no prescribed criteria of cooperation amount sub-projects are not expected to question their effort and cooperation activity. The existing interdependencies between the perception of external obstacles and embeddedness in cooperation and the missing relationship between the perception of internal obstacles and the embeddedness in cooperation speak for the assumption of the external attribution by the sub-projects. Another reason for these results can be the limitations of the rather small sample that cannot cover small or medium effect sizes.

Two general conclusions can be drawn from the presented results. First, the results show the time interdependence of the social and organizational structures/networks with the perception of obstacles to cooperation. The embeddedness in highly cooperative structures goes along with the development of attitudes over time as a reaction to the current situation which indicates an adaptive behavior of the actors.

The second conclusion can be drawn concerning the limits of cooperation. Given the fact that cooperation demands resources and time, it appears plausible that actors reach saturation in cooperative contacts, and as a result, perceive more obstacles to cooperation and cooperation initiative. Further analysis, for instance, using repeated-measures analysis of variance can reveal new information about the development of attitudes and perception.

The limitations should also be pointed out. The use of actor-oriented models or exponential random graphs would have been more suitable for analyzing network data. Further analysis using models that account for the network structure will be used to validate the results. Moreover, the data analysis was conducted based on a relatively small sample; consequently, only very strong test effects could be identified. The identification of moderate or small effects will require a replication based on a larger sample.

The presented findings are relevant insofar as they can be used in the organizational planning for enhancing cooperation. The emphasis should be placed either on creating opportunities for direct cooperation, which were not too demanding in terms of resources and time or according to the sub-projects' extra resources only for cooperation purposes, which can help sub-projects to develop a better attitude towards cooperation at the same time being highly active in cooperation.

Appendix

Table 2. Confidence intervals. Bias-corrected and accelerated bootstrap. outdeg: outdegree in the network of deliberate cooperation, t_p: transitivity in the structural linkage network, IO: internal obstacles; EO: external obstacles, CI: cooperation initiative.

	CI 95% bound	outdeg_t1	outdeg_t2	outdeg_t3	t_p_t1	t_p_t2	t_p_t3	IO_t1	IO_t2	IO_t3	EO_t1	EO_t2	EO_t3	CI_t1	CI_t2
outdeg_t2	Lower	.72													
	Upper	.93													
outdeg_t3	Lower	.17	.24												
	Upper	.77	.80												
t_p_t1	Lower	.73	.52	.31											
	Upper	.96	.86	.81											
t_p_t2	Lower	.44	.36	.20	.64										
	Upper	.94	.85	.73	.98										
t_p_t3	Lower	.33	.28	.19	.72	.78									
	Upper	.88	.77	.74	.96	1.00									
IO_t1	Lower	−.60	−.57	−.42	−.55	−.55	−.50								
	Upper	.15	.19	.21	.08	.11	.18								
IO_t2	Lower	−.49	−.68	−.51	−.37	−.39	−.36	.32							
	Upper	.25	.16	.42	.21	.07	.15	.85							
IO_t3	Lower	−.08	−.14	−.52	−.	−.10	−.13	−.23	−.50						
	Upper	.51	.60	.36	.44	.41	.40	.63	.65						

(continued)

Table 2. *(continued)*

	CI 95% bound	outdeg_t1	outdeg_t2	outdeg_t3	t_p_t1	t_p_t2	t_p_t3	IO_t1	IO_t2	IO_t3	EO_t1	EO_t2	EO_t3	CI_t1	CI_t2
EO_t1	Lower	−.26	−.33	−.20	−.31	−.31	−.23	.24	−.34	−.15					
	Upper	.69	.76	.62	.56	.59	.55	.83	.55	.52					
EO_t2	Lower	−.22	−.45	−.42	−.14	−.15	−.24	−.08	.35	−.32	−.26				
	Upper	.51	.28	.46	.58	.54	.53	.50	.74	.44	.62				
EO_t3	Lower	.14	.03	.05	.30	.24	.37	−.26	−.45	.30	−.12	−.22			
	Upper	.73	.72	.77	.77	.72	.80	.44	.71	.80	.72	.63			
CI_t1	Lower	−.70	−.64	−.54	−.71	−.70	−.64	.19	.14	−.34	−.03	−.27	−.43		
	Upper	.10	.18	.23	−.09	−.08	−.07	.77	.79	.54	.65	.50	.53		
CI_t2	Lower	−.70	−.59	−.46	−.71	−.73	−.68	−.06	−.02	−.47	−.11	−.34	−.43	.81	
	Upper	.06	.14	.39	−.04	−.01	.05	.80	.81	.46	.62	.42	.49	.97	
CI_t3	Lower	−.68	−.85	−.69	−.75	−.70	−.77	−.22	−.36	−.39	−.48	−.54	−.70	.10	.13
	Upper	−.07	−.02	.12	−.15	−.13	−.04	.58	.47	.69	.30	.29	.54	.85	.78

References

1. van Santen, E., Seckinger, M.: Kooperation: Mythos und Realität einer Praxis. Eine empirische Studie zur interinstitutionellen Zusammenarbeit am Beispiel der Kinder- und Jugendhilfe [Cooperation: Myth and Reality of a Practice. An Empirical Study on Inter-Institutional Cooperation Using the Example of Child and Youth Welfare]. Doctoral dissertation, Free University of Berlin. Leske und Budrich, München, Opladen (2003)
2. Rippner, J.A.: State P-20 councils and collaboration between K-12 and higher education. Educ. Policy **31**, 3–38 (2017). https://doi.org/10.1177/0895904814558008
3. Guri-Rozenblit, S.: Collaboration between teacher training colleges and the open university of Israel. Teach. Educ. **7**, 59–67 (1995). https://doi.org/10.1080/1047621950070209
4. Eddy, P.L.: Partnerships and collaborations in higher education. ASHE High. Educ. Rept. **36**, 1–115 (2010). https://doi.org/10.1002/aehe.3602
5. Crano, W.D., Gardikiotis, A.: Attitude formation and change. In: Wright, J.D. (ed.) International Encyclopedia of the Social & Behavioral Sciences, pp. 169–174. Elsevier, Oxford (2015)
6. Rokeach, M., Norrell, G.: The Nature of Analysis and Synthesis and Some Conditions in the Classroom which Facilitate or Retard These Cognitive Processes. Michigan State University, East Lansing (1966)
7. Rokeach, M.: Attitude change and behavioral change. Publ. Opin. Q. **30**(4), 529–550 (1966)
8. Fendrich, J.M.: Perceived reference group support. Racial attitudes and overt behavior. Am. Sociol. Rev. **32**(6), 960–970 (1967)
9. Bamberg, S., Ajzen, I., Schmidt, P.: Choice of travel mode in the theory of planned behavior. The roles of past behavior, habit, and reasoned action. Basic Appl. Soc. Psychol. **25**(3), 175–187 (2010). https://doi.org/10.1207/S15324834BASP2503_01
10. Ajzen, I.: Nature and operation of attitudes. Annu. Rev. Psychol. **52**(1), 27–58 (2001). https://doi.org/10.1146/annurev.psych.52.1.27
11. Ajzen, I.: Perceived behavioral control, self-efficacy, locus of control, and the theory of planned behavior. J. Appl. Soc. Psychol. **32**(4), 665–683 (2002). https://doi.org/10.1111/j.1559-1816.2002.tb00236.x
12. Marwell, G., Oliver, P.: The Critical Mass in Collective Action. A Micro-Social Theory. Studies in Rationality and Social Change. Cambridge University Press, Cambridge (1993)
13. Homans, G.C.: The Human Group. Harcourt Brace, New York (1950)
14. Gray, B.L.: Collaborating: Finding Common Ground for Multiparty Problems. Jossey-Bass, London (1989)
15. Wigfield, A., Eccles, J.S.: Expectancy-value theory of achievement motivation. Contemp. Educ. Psychol. **25**(1), 68–81 (2000). https://doi.org/10.1006/ceps.1999.1015
16. Roos, D., Hahn, R.: Understanding collaborative consumption. An extension of the theory of planned behavior with value-based personal norms. J. Bus. Ethics **158**(3), 679–697 (2017). https://doi.org/10.1007/s10551-017-3675-3
17. Drossel, K., Eickelmann, B., van Ophuysen, S., Bos, W.: Why teachers cooperate. An expectancy-value model of teacher cooperation. Eur. J. Psychol. Educ. **34**(1), 187–208 (2018). https://doi.org/10.1007/s10212-018-0368-y
18. Kashima, Y., Gallois, C., McCamish, M.: The theory of reasoned action and cooperative behaviour. It takes two to use a condom. Br. J. Soc. Psychol. **32**(3), 227–239 (1993). https://doi.org/10.1111/j.2044-8309.1993.tb00997.x
19. Schlicht-Schmälzle, R., Chykina, V., Schmälzle, R.: An attitude network analysis of post-national citizenship identities. PloS One **13**(12) (2018). https://doi.org/10.1371/journal.pone.0208241

20. Fowler, J.H., Christakis, N.A.: Dynamic spread of happiness in a large social network. Longitudinal analysis over 20 years in the Framingham heart study. Br. Med. J. **337** (2008). https://doi.org/10.1136/bmj.a2338
21. Gulati, R., Gargiulo, M.: Where do inter-organizational networks come from? Working papers, 97/69/OB. INSEAD, Fontainebleau (1997)
22. Leavitt, H.J.: Some effects of certain communication patterns on group performance. J. Abnorm. Psychol. **46**(1), 38–50 (1951)
23. Mercken, L., Snijders, T.A.B., Steglich, C., Vartiainen, E., de Vries, H.: Dynamics of adolescent friendship networks and smoking behavior. Soc. Netw. **32**(1), 72–81 (2010). https://doi.org/10.1016/j.socnet.2009.02.005
24. Giovannetti, E., Piga, C.A.: The contrasting effects of active and passive cooperation on innovation and productivity. Evidence from British local innovation networks. Int. J. Prod. Econ. **187**, 102–112 (2017). https://doi.org/10.1016/j.ijpe.2017.02.013
25. Austin, A.E., Baldwin, R.G.: Faculty Collaboration. Enhancing the Quality of Scholarship and Teaching. ASHE-ERIC higher education report No.7. School of Education and Human Development, Washington DC, p. 48 (1991)
26. Mizruchi, M.S.: What do interlocks do? An analysis, critique, and assessment of research on interlocking directorates. Annu. Rev. Sociol. **22**, 271–298 (1996)
27. Uzzi, B.: Social structure and competition in interfirm networks: the paradox of embeddedness. Adm. Sci. Q. **42**(1), 35–67 (1997). https://doi.org/10.2307/2393808
28. Granovetter, M.: Economic action and social structure: the problem of embeddedness. Am. J. Sociol. **91**, 481–510 (1985). https://doi.org/10.1086/228311
29. Håkonsson, D.D., Obel, B., Eskildsen, J.K., Burton, R.M.: On cooperative behavior in distributed teams. The influence of organizational design, media richness, social interaction, and interaction adaptation. Front. Psychol. **7**, 692 (2016). https://doi.org/10.3389/fpsyg.2016.00692
30. Bühner, M., Ziegler, M.: Statistik für Psychologen und Sozialwissenschaftler [Statistics for Psychologists and Social Scientists]. Pearson-Higher Education, München (2009)
31. Efron, B., Tibshirani, R.: An Introduction to the Bootstrap. Chapman & Hall, New York (1993)
32. Kelley, H.H.: The processes of causal attribution. Am. Psychol. **28**(2), 107–128 (1973). https://doi.org/10.1037/h0034225

Influence of University Groups Formation on Academic Performance

Deniza Alieva(✉), Akmal Alikhodjaev, and Otabek Abdubositov

Management Development Institute of Singapore in Tashkent, Tashkent, Uzbekistan
deniza.alieva@gmail.com, alikhodjaev.akmal@gmail.com,
abdubasitov@gmail.com

Abstract. Present research is conducted at the management and business faculties of two universities in Tashkent (Uzbekistan): one international and one local. The data was collected from 136 participants from two faculties in the first case and 128 participants from one faculty in the second one. According to the system of studies used in the international university, each academic year student groups are mixed, and students have to build new communications and relationships with fellow group members. To analyze the influence of group changing on academic performance, we conducted social network analysis among bachelor students of the last year of studies. The results show that the biggest share of relationships (62.3%) is built during the first year at the university. Even when friends move to different groups after completion of the first year of studies, 78.2% of such friendships remain till their last year in the bachelor program. We have found positive correlation between changes in composition of social networks in classroom and academic performance: the more drastic is the change, the better students are performing in class. The performance showed by students from the local university who do not experience changes in the composition of their social networks during four years of study is lower in comparison with international university.

Keywords: Social networks analysis · Academic performance · University studies

1 Introduction

1.1 Educational System in Uzbekistan

The Republic of Uzbekistan, a country in Central Asia, is located between Amu Darya and Syr Darya rivers, bordering on all the states of the regions. The list of main cities of the country includes Tashkent, the capital, Samarkand, Bukhara and Khiva. The country is famous because of its cultural and historical heritage, as it was a center of Great Silk Road, and nowadays it attracts people all over the world willing to learn more about the region.

Uzbekistan has a high literacy rate among its population – almost 100%, according to the National Committee of Statistics [1]. The annual budget spending on education

A. Antonyuk and N. Basov (Eds.): NetGloW 2020, LNNS 181, pp. 322–334, 2021.
https://doi.org/10.1007/978-3-030-64877-0_21

in Uzbekistan is around 10–12% of GDP, while its share in state budget expenditures exceeds 35% [2].

The government aims to guarantee continuing education to the country's residents. To comply with that objective, a National Program on education was created, stating as the main goal the necessity to educate generations with modern knowledge, active civic position and abilities to think independently and make informed decisions. The authorities try to ensure that the professionals graduating from local and international universities located in Uzbekistan possess not only hard but also soft skills that can guarantee them success in their professional sphere.

The country created the system of education based on foreign and own experience, modifying it in the process. The critical analysis and application of its results include a cultural component as well, as Uzbekistan is a country with the Soviet past, where a mixture of different cultures, ideas and religions can be found. The state creates programs of continuing education that start from pre-school and finish with postgraduate courses, including on-the-way trainings and out-of-school education.

If we focus on university studies, we can discover that recently the situation has changed drastically. Until recent years the country was following state education system, inherited from its Soviet past. The students had the possibility to pass entrance exams only for one specialty at one university per year that was influencing the quality of education, increasing the competency between candidates and universities. However, at the moment, the market is opening for foreign private universities and the offer is increasing each year. In Uzbekistan, universities from Europe and Asia and the Webster University from the United States that opened its Tashkent branch in 2019 are operating successfully. On the other hand, we have local universities, which are supported by state, and are providing all types of university education.

The studies in local universities are held in Uzbek and Russian. An exception is the Islamic Academy, where studies are partially conducted in Arabic. The classes in international universities in the majority of cases are held in English; Russian Universities that are present in the market prefer to provide them in Russian.

It has to be mentioned that not every university is giving the opportunity to obtain Bachelor, Master and Doctorate degree. In the local ones, it is easier to find all three levels, while in the international ones, Bachelor degree is more popular. At the same time, there are joint programs organized between the local and international universities at all three levels of studies.

The international universities are operating autonomously, while the local ones are working based on governmental support. Therefore, in many cases the tuition fee of students can be several times higher in international universities in order to cover operational expenses and make payments to their partner universities overseas. The other particular characteristic of international universities is that they are working using their own study programs that should be confirmed by partner university and not local authorities. Usually those programs are copies of those used overseas and they are adapted to Uzbekistani market by usage of relevant examples and terms.

The specialties offered by universities differ and cover a wide range of matters: from computer engineering and medicine to economics, law or psychology. Overall, the universities are covering 174 areas of education [3].

The duration of undergraduate studies at most universities is 4 years. During this period, students obtain fundamental knowledge in an area of their choice, combining theoretical lectures with practical tutorials and work in the field. However, several international universities had to adapt their study programs to the local market to comply with the requisite of 4-year bachelor degree studies. Those who are collaborating with British universities faced a problem of non-matching in terms of number of years in undergraduate studies. Therefore, they are offering a Foundation year course, where students develop their English, mathematics and logical skills receiving basic knowledge of economy, IT, etc. After finishing the Foundation year course, students can choose the specialty they would like to study and cover all its modules in next three years.

Lately, Uzbek government decided that the higher education system of the country is ready to transit to the concept of "University 3.0", where not only academic but also scientific work should be carried out at the universities. This idea uncovered the necessity of creation of laboratories, incubators, techno-parks and even scientific museums within higher educational institutions of Uzbekistan. It led universities to seek talented students, who are showing high academic performance levels, and to create conditions needed to preserve those levels.

Nowadays the universities of the country are making a transition to a credit-based system to give students more opportunities for participation in international exchange programs. However, even if the country is opening its borders, Uzbekistan culture is still influencing the society and education in particular. According to the research [4], depending on a type of culture a student is coming from, the aim of receiving higher education can differ. Importance is given by students not only to educational outcome (well-payed job, prestige, etc.), but also to opportunity to create networks of good connections. On the one hand, here we can feel the influence of Central Asian culture and Uzbek one in particular. The value of personal relationships is quite high in the region [5–7] and people use them to make their lives easier and more comfortable. On the other hand, the Soviet heredity in a form of *blat* is still present in the country. It means that if one has connections, one has the power: knowing someone in a good restaurant can help you to enter without queue, while contacts in different organizations can provide you job opportunities that are not given to other candidates. These particularities influence the perception of networks of contacts that people establish during their lives. And, as four university years give opportunity to create a wide network, there can be a percentage of students who choose with whom they should form a link on the basis of their utility for the future.

In the next part we describe the processes of entrance and academic groups formation in Uzbek local and international universities.

1.2 Educational Process

Currently, higher educational system in Uzbekistan uses test and exams system for those who would like to enter a university. Future students can apply to several universities at once: three in the case of local and up to all in the case of international universities. The universities are accepting international language certificates as proof of language proficiency and are providing different benefits to those who have won republican Olympiads on the subjects they have to get tested in order to enter a university.

In the majority of cases, the applicant for bachelor studies in an international university has to pass English test or present the results of IELTS that have to comply with a minimal entrance bar. An international university usually is a branch of the one from another country (e.g., a Branch of Lomonosov Moscow State University), franchise (e.g., Westminster International University in Tashkent), joint venture (e.g., Management Development Institute of Singapore in Tashkent) or part of the university that has some similarities with a branch, but is called a campus (e.g., Webster University). The majority of students in international universities are local residents who obtain an international diploma after graduation.

After entering a university, a student forms part of his/her academic group. In local universities, the groups are created randomly, the only criterion is to preserve the equality in terms of numbers. In the international ones, especially in those where the classes are held in English, as the first year is traditionally the foundation or freshmen year, the groups are formed according to students' level of English.

In the following years, students from local universities can change their groups by writing a request to a faculty dean. At the same time, at an international university the change in groups formation occurs in the second year, when students choose their major and, following their choice, the groups are created by administration. After that students have the possibility to ask for group change by writing an official request and explaining the reasons.

There are no studies that analyze positive and negative aspects of such policies used for group formation in universities of Uzbekistan. Therefore, the current study can provide a basis for future analyses in the field.

The next part discusses academic performance and its relationship with different factors using the results of previous research. In addition, we determine the possibility of application of those results in Uzbekistani context.

1.3 Academic Performance

A wide range of studies has been performed recently to evaluate academic performance and the influence of different factors on it. Among those factors researchers and practitioners studied demographic, cultural, economic ones, extracurricular activities, peer and family influence, student attendance, etc. The analysis shows that each of those factors individually or in group can affect students' behavior and influence positively or negatively his or her academic performance.

For instance, demographic variables and cultural background have shown influence on grades in cases of mixed groups. Non-nationals in the study conducted showed better performance on average and more competency in English in comparison to local ones. The influence of gender and nationality on grades level in general was proven by other researchers [2, 8, 9].

The other group of demographic variables analyzed by scientists focuses on economic factors, such as family income, student's income, or student's debt. A positive correlation was found between high-income in students' families and their academic performance, while in the majority of cases provenience from low-income families was causing a low level of academic performance [10].

Psychological factors such as self-esteem [11] seem not to be an important characteristic that can impact positively or negatively academic performance of University students. The students tend to report on having good or high self-esteem, even if they are obtaining low or middle-level grades. Emotional and extrinsic motivations, emotional intelligence level also constitute important tools that can be used to improve grades of university students [12–14].

As it can be seen, income constitutes an important factor while analyzing academic performance. However, Beblo & Lauer [15] have determined that when combined with parent's status in the labor market, high income can lead to lower academic performance, as a student thinks that he or she has a workplace guaranteed by his or her parents.

Significant relationship was found between parents and child educational levels [10, 16]. It is worth notice that mother's education becomes more relevant in a majority of cases and affects academic performance of a child more. However, father's educational level does not demonstrate such strong influence.

Another important factor whose effect on academic performance was studied is the level of absenteeism. The measurement of correlation between those two indices [17–19] has found a small, but statistically significant number. It was confirmed that those who missed class on a specific topic or a given day were making more errors in tests regarding materials of the missed class in comparison with the student who were present. Moore [20] determined that the learning process itself can be improved by attending classes, which, as a consequence, improves students' grades. From psychological and motivational perspective, academic performance also can be related with attendance [21], as the lower is the number of classes attended, the lower is students' productivity.

Reinforcement of academic performance can be also reached by different extracurricular activities, even if those are not directly or indirectly related with academic modules [22–24]. An analysis showed that those who take part in these activities see more positively the school, college or university and it leads to an increase in academic performance level [25]. Victory produces more positive effect on students, however even the participation in this informal aspect of education itself can provoke more interest in studies and improve grades. On the other hand, it is worth noticing that not all activities can have a positive effect on students' academic performance. For example, it was proven that sports impact negatively due to problems caused with personal development [26], higher levels of alcoholism [27], physiological injuries [28], etc.

The negative impact discussed can be leveraged by positive influence of the stability of social networks formed in academic process or even in a neighborhood. This factor also has influence on academic performance of students at all educational levels [29–32]. Creation of sense of comfort, security and stability, first, has influence of psychological and emotional state of students, activating their desire to study and work harder. Second, it increases social capital, that, in consequence, strengthens social connections between a student and peers, making him or her increase academic performance in order to socialize with others and raise his/her value in school or university society.

Another factor that researchers are considering being one of the most influential is peer influence [33–35]. Studies have proven that peers have more impact on academic performance of students than their closest relatives. This idea helped to develop different practices, where social networks approach also was used. Wilkinson and Fung [36] suggested that formation of heterogeneous groups, where the characteristics used to group would be student's learning ability, can help to improve learning processes and furthermore the results of the studies. The researchers expected a positive effect that students with a low level of academic performance would get by trying to follow those who were showing better results. Schindler [37] confirmed the positive effect on low-performing students, however, he also noticed a negative one on high-ability ones. To decrease the negative effects of heterogeneous groups, some researchers suggest forming homogeneous collectives, where students will have more or less the same level of performance [33].

2 Methodology

2.1 Objectives and Scheme of the Study

The present study aimed to analyze the factors that can enhance academic performance among students of two universities located in Uzbekistan, where different study methods are used. The *principal objective* is to determine the influence of group formation on students' grades. To accomplish it, the following *specific objectives* were determined:

- To study the changes in group formation that occurred during all four years of studies;
- To identify key actors in last-year students' social networks and their relationship with university group where ego is present;
- To understand better last-year students' social networks composition from different types of relations (formal and informal);
- To identify the problems last-year students might have faced in relation to their group's formation.

Based on the objectives of the study, we divided our work in different parts.

In the first part of the analysis, the data about groups composition of the students who confirmed their participation was collected. The data included not only information on their last year but also on three previous years, starting from their entrance to the University.

Second and third parts focused on the composition on social networks of students, where they were asked to provide information only about those who were currently or who had been previously included in their University group and with whom they were maintaining relations.

The fourth part included analysis of critical incidents that might be related to changes of network composition throughout the years.

2.2 Questionnaire

The main tool used for data collection was an electronic questionnaire created in English, Russian and Uzbek. The original questionnaire was created in Russian, the translations were proofread by native speakers.

The questionnaire contained four parts. The first one was centered on questions related to demographic characteristics of participants. They were asked to provide information about the university they were studying in, their major, age and gender.

The second part asked the participants to provide a list of 45 people who comprise their social network. The number was selected on the basis of experience and previous research, where including 45 network members could provide rich information and content needed to conduct relevant analysis [38]. The participants had to indicate whether an alter is currently a member of a participant's university group and the grade of his/her relationship with this alter, following the scale, where:

- 1 – barely talk to each other;
- 2 – more or less frequently communicate only on study issues;
- 3 – more or less frequently communicate on issues related and not related to studies;
- 4 – are close friends.

The list of alters was followed by a grid where participants had to indicate the grade of relationships between alters, following the scale below:

- 0 – do not know each other;
- 1 – know each other;
- 2 – are friends;
- 3 – are close friends.

The fourth part of the questionnaire was dedicated to describing incidents that potentially could be related to students' social network constitution. The participants were asked to describe the incident providing information on where, when and what exactly happened. There was a possibility to share several experiences that could potentially enrich the study.

2.3 Data Collection and Data Analysis

The data collection was conducted using two different strategies.

First, the researchers contacted last-year bachelor students in two universities located in Tashkent. After receiving the confirmation of interest in participation in the study, electronic questionnaires in English, Russian and Uzbek were distributed. The participants could fill them out in any language they wanted.

Participants spent from 45 to 60 min to fill up the questionnaire, having the possibility to contact the researchers in case of having any problems or misunderstandings of questions. No incidents of this type were registered.

After finishing data collection from students, researchers made a request to administrative staff of the universities, where the participants were studying, asking for information about their groups lists during all four years of the studies. In addition, a list of grades on profiling modules of each participant in the research was requested.

The data on academic performance was collected and the quality of that variable was controlled by ensuring that the modules were taught by international faculty. In two cases the same module was taught by the same people in both universities.

At the data analysis stage, Ucinet 6 [39] and SPSS 22 were used. The analysis included descriptive statistics measures and Independent samples t-test calculation. In addition, basic network measures for each network were calculated (degree, betweenness, closeness, clustering coefficient) and average measures (average degree, average betweenness, average closeness, average clustering coefficient). The visualization of networks was done using Gephi 0.9.2 [40]. Also, the researchers analyzed critical incidents cases to make sure that they were related to changes in group composition and not to other factors.

2.4 Participants

We decided to invite students from two universities located in Tashkent to participate in the study. The egos had to be last-year bachelor degree students from business and management faculties, so we could compare their results on the same or similar modules provided by their universities.

The decision to use data from one local and one international university is related with the academic system applied in those establishments. In the local Uzbek university, it is uncommon to experience changes in groups during all four years of study. Usually the change can be caused by a force majeure, or by student's will. However, in the international university, there a practice to mix the groups every academic year to ensure the students have experience in communication with different people, know how to adapt to a changing environment and learn how to establish new relationships in a short period of time.

Out of 213 students invited to participate in the study from the international university and 164 in the case of the local one, 152 and 149 people respectively took part in the survey (participation rate: 71.36% and 90.85%). After eliminating incomplete questionnaires the researchers have got 136 participants from the international university and 128 from the local one (response rate: 63.85% and 78.05% from the initial list of invitees).

In the case of the international university 58.09% of participants were males, while in the local university this number reached 64.06%.

3 Results

3.1 Networks Composition

Data analysis have shown that personal networks of the participants from the international university are more diverse in comparison with the local one. As you can see from the figure below (Fig. 1), the social network of a local university student tends to be more homogenous, and the majority of nodes included are from the same group as the ego. International university students have shown more diversity by including people from other University groups. In the examples provided in Fig. 1 we can observe that in the case of the local university student, 28.89% of alters are proceeding from other university groups, while the ego from the international university has 48.89% of alters from other groups. In 34.89% of cases the representatives of other groups do not demonstrate strong connection with those proceeding from the university group of ego.

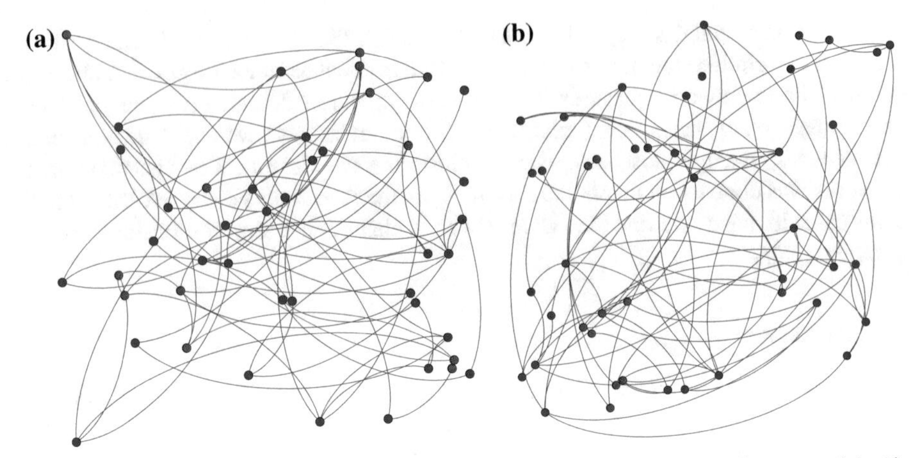

Fig. 1. Personal networks of two students, where **blue color** shows alters who are studying in the same group as ego and **red** – the ones who study in other groups, a – personal network of a student of the local university, b – personal network of a student of the international university.

Analysis of the dynamic of changes in students' lists have provided information on how big was the change experienced by egos in terms of group composition during four years of Bachelor studies. In the case of the local university, the average change experienced after completion of the first course was 0.02% (SD = 1.67), after the second one – 1.35% (SD = 1.14), the third – 1.48 (SD = 1.55). The international university showed different numbers. The change in groups' members made after the foundation year of studies on average was 14.43% (SD = 8.03), after the second year – 12.46% (SD = 7.01), after the third one – 7.11% (SD = 0.33).

The level of close friendships in both types of networks is more or less similar (53.12% in the case of the local university and 57.01% among the students of the international one). In the international university the proportion of those who are only forming part of ego's group and those who are considered a close friend of this one is on average 4.7:1, while in the local university that proportion changes to 3.2:1.

The international university students' personal networks are more equally distributed in terms of gender, while in the case of the local universities the students tend to have more people of their gender included in their networks rather than the opposite one (average proportion: 2.4:1). Students in the local university have more cohesive networks (Average degree = 36.42, SD = 1.44) in comparison to the international university students (Average degree = 23.19, SD = 2.04).

In both cases the majority of connections is formed with those who were part of students' group during their first year of bachelor studies (63.8% among students of the local university and 60.8% among those proceeding from the international one).

3.2 Networks Composition and Academic Performance

As the researchers tried to control the modules where students' academic performance was measured, for analysis four main modules were chosen:

- Economics II;
- Strategic Management;
- Human Resources Management;
- Entrepreneurship.

Economics II and Strategic Management are taught in the third course in both cases, while the other two form part of a curricula of the last year of bachelor studies.

Local university students demonstrated lower grades in comparison with international university representatives in two modules: Strategic Management: 3.87 versus 4.12 (SD = 0.87 and 1.03 respectively); and Entrepreneurship: 3.61 versus 4.29 (SD = 0.92 and 0.88 respectively). Economics II seems to be the most difficult subject from the list for representatives of both universities: the mean of 3.04 (SD = 1.18) was reached in the local university and 3.62 (SD = 1.12) in the international one. Human Resources Management is the module with the highest marks among all four: 4.12 (SD = 0.86) in the local university and 4.85 (SD = 0.60) in the international one.

Analysis shows positive correlation between the heterogeneous composition of students' personal networks and academic performance: the more diverse is the network, the better results a student shows ($r = 0.816$). This factor can explain partially higher grades obtained by participants from the international university.

From 264 questionnaires submitted by participants, 114 critical incidents relevant to the case studied were retrieved: 86 of those were provided by the international university students. In the majority of cases (85.09%) the students described the problems faced after the group's composition change and the decrease in their academic performance in that moment. Search for peers' help in and outside of the group they started to belong to helped to overcome the difficulties experienced.

72.81% of students reported problems after the first group change that usually happened after the foundation year.

4 Limitations and Ideas for Future Research

The present study analyzed the data collected from four-year bachelor students of two universities located in Tashkent. However, the results cannot be extrapolated to all the universities in the country for several reasons.

First, the participants in the study were representing only one of more than 170 specializations that are provided by local and international universities to Uzbekistani students. As Kleijn, van der Floeg & Topman [41] and Fenollar, Román & Cuestas [42] have shown in their works, academic performance can differ depending on the faculty, so it is necessary to conduct an analysis using data of students from different specialties. As the present research aimed to be a pilot one, we can apply the same methods in future analysis.

The study is not taking into consideration the initial motivation of students that are entering a local or an international university. For a more profound analysis in the future, we would like to integrate that variable and determine its effect on academic performance.

However, a longitudinal study can be conducted to obtain data that can enrich the results. Dynamic analysis of students' personal networks and the correlation between them and academic performance will explain the strengths of relation between those indices and potentially provide new insights on other factors that can mediate this correlation.

The study helped to answer the main question asked in the beginning by the researchers and created a basis for future studies in the field.

5 Conclusion

Analysis conducted among four-year bachelor students of two faculties of local and international universities in Tashkent (Uzbekistan) helped to determine the following. The composition of personal networks of students has influence on academic performance, as the participants with more heterogeneous networks showed higher grades levels. On the other hand, the cultural background of Uzbekistan gives more value to diversified networks that, first, help to grow social capital, and second, provide opportunity to use the connections in the network for further enrichment and development.

As the researchers controlled the quality of education provided, having in both cases international faculty, the possible explanation of these phenomena can be related to the diversified social network that influences students' conduct and information exchange between groups, leading to knowledge enrichment.

The conclusions made of the study conducted can prove the effectiveness of the system implemented by the international university and can be a reason to try to apply it in other educational institutions of the country.

References

1. State Committee of the Republic of Uzbekistan on Statistics. https://www.stat.uz/en/. Accessed 30 May 2020
2. Jayanthi, S.V., Balakrishnan, S., Ching, A.L.S., Latiff, N.A.A., Nasirudeen, A.M.A.: Factors contributing to academic performance of students in a tertiary institution in Singapore. American J. Educ. Res. **2**(9), 752–758 (2014)
3. Europedia. http://eacea.ec.europa.eu/education/Eurydice/eurypedia_en.php. Accessed 31 May 2020
4. Latipov, I.A.: Kultura kak factor formirovaniya jiznennih cennostey rossiyskih studentov y studentov iz Uzbekistana. [Culture as a factor influencing formation of life values of students from Russia and Uzbekistan], Vestnik Instituta sotsiologii, **9**(1), 155–170 (2018)
5. Dadabaev, T.: Between state and society: The position of the Mahalla in Uzbekistan. In Social capital construction and governance in Central Asia, 77–95, Palgrave Macmillan, New York (2017)
6. Isaeva, K., Adams, B.G., van de Vijver, F.J.: The kaleidoscope of language, ethnicity, and identity in Uzbekistan. In: Changing Values and Identities in the Post-Communist World, pp. 295–311. Springer, Cham (2018)
7. Tunçer-Kılavuz, İ.: Political and social networks in Tajikistan and Uzbekistan: 'clan', region and beyond. Central Asian Survey **28**(3), 323–334 (2009)
8. Richardson, M., Abraham, C., Bond, R.: Psychological correlates of university students' academic performance: a systematic review and meta-analysis. Psychol. Bulletin **138**(2), 353–387 (2012)

9. Vermunt, J.D.: Relations between student learning patterns and personal and contextual factors and academic performance. Higher Educ. **49**(3), 205–234 (2005)
10. Agus, A., Makhbul, Z.K.: An empirical study on academic achievement of business students in pursuing higher education: an emphasis on the influence of family backgrounds. In: Proceedings of the International Conference on the Challenge of Learning and Teaching in a Brave New World: Issues and Opportunities in Borderless Education (2002)
11. Alva, C., Manuel, L.: Self-Esteem study habits and academic performance among university students. J. Educ. Psychol. Propósitos y Representaciones **5**(1), 101–127 (2017)
12. Bailey, T.H., Phillips, L.J.: The influence of motivation and adaptation on students' subjective well-being, meaning in life and academic performance. Higher Educ. Res. Dev. **35**(2), 201–216 (2016)
13. Bakar, K.A., Tarmizi, R.A., Mahyuddin, R., Elias, H., Luan, W.S., Ayub, A.F.M.: Relationships between university students' achievement motivation, attitude and academic performance in Malaysia. Procedia-Soc. Behav. Sci. **2**(2), 4906–4910 (2010)
14. Thomas, C.L., Cassady, J.C., Heller, M.L.: The influence of emotional intelligence, cognitive test anxiety, and coping strategies on undergraduate academic performance. Learn. Individual Differences **55**, 40–48 (2017)
15. Beblo, M., Lauer, C.: Do family resources matter? Educational attainment during transition in Poland. Econ. Transition **12**(3), 537–558 (2004)
16. Ermisch, J., Francesconi, M.: Family matters: impacts of family background on educational attainments. Economica **68**(270), 137–156 (2001)
17. Collett, P., Gyles, N., Hrasky, S.: Optional formative assessment and class attendance: Their impact on student performance. Global Perspectives on Account. Educ. **4**, 41–59 (2007)
18. Rodgers, J.R.: A panel-data study of the effect of student attendance on university performance. Australian J. Educ. **45**(3), 284–295 (2001)
19. Romer, D.: Do students go to class? should they? J. Econ. Perspectives **7**, 167–174 (1993)
20. Moore, R.: Class attendance: how students' attitudes about attendance relate to their academic performance in introductory science classes. Res. Teach. Dev. Educ. **23**(1), 19–33 (2006)
21. Whitney, G., Kolar, I.: Am I missing something? Universal Access Inf. Soc. **19**(2), 461–469 (2019). https://doi.org/10.1007/s10209-019-00648-z
22. Guest, A., Schneider, B.: Adolescents' extracurricular participation in context: the mediating effects of school, communities and identities. Sociol. Educ. **76**(2), 89–109 (2003)
23. Marsh, H.W., Kleitman, S.: Extracurricular school activities, The good, the bad, and the nonlinear. Harvard Educ. Rev. **72**(4), 464–514 (2002)
24. Sparkes, L.: Academic achievement and academic motivation and its relationship to extracurricular activities and parental involvement in high school students, Senior thesis, San Anselm College, Manchester, New Hampshire (2004)
25. Darling, N., Caldwell, L.L., Smith, R.: Participation in school-based extracurricular activities and adolescent adjustment. J. Leisure Res. **37**(1), 51–76 (2005)
26. Larson, R.W., Kleiber, D.: Daily experience of adolescents, In: Tolan, P.H., Cohler, B.J. (eds.) Wiley Series on Personality Processes. Handbook of clinical research and practice with adolescents, pp. 125–145. Wiley (1993)
27. Eccles, J.S., Barber, B.L.: Student council, volunteering, basketball, or marching band: what kind of extracurricular involvement matters? J. Adolescent Res. **14**(1), 10–43 (1999)
28. Dane, Ş., Can, S., Gürsoy, R., Ezirmik, N.: Sport injuries: relations to sex, sport, injured body region. Perceptual Motor Skills **98**(2), 519–524 (2004)
29. Cantin, S., Boivin, M.: Change and stability in children's social network and self-perceptions during transition from elementary to junior high school. Int. J. Behav. Dev. **28**(6), 561–570 (2004)

30. Gremmen, M.C., Dijkstra, J.K., Steglich, C., Veenstra, R.: First selection, then influence: developmental differences in friendship dynamics regarding academic achievement. Dev. Psychol. **53**(7), 1356–1370 (2017)
31. Greenman, P.S., Schneider, B.H., Tomada, G.: Stability and change in patterns of peer rejection: Implications for children's academic performance over time. School Psychol. Int. **30**(2), 163–183 (2009)
32. Li, L.H.: Impacts of homeownership and residential stability on children's academic performance in Hong Kong. Soc. Indicators Res. **126**(2), 595–616 (2016)
33. Goethals, G.R.: Peer effects, gender, and intellectual performance among students at a highly selective college: a social comparison of abilities analysis, Williams Project on the Economics of Higher Education, DP-61, 1–20 (2001)
34. Gonzales, N.A., Cauce, A.M., Friedman, R.J., Mason, C.A.: Family, peer, and neighborhood influences on academic achievement among African-American adolescents: one-year prospective effects. Am. J. Community Psychol. **24**(3), 365–387 (1996)
35. Hanushek, E.A., Kain, J.F., Markman, J.M., Rivkin, S.G.: Does peer ability affect student achievement? J. Appl. Econ. **18**(5), 527–544 (2003)
36. Wilkinson, I.A., Fung, I.Y.: Small-group composition and peer effects. Int. J. Educ. Res. **37**(5), 425–447 (2002)
37. Schindler B.R.: Educational peer effects quantile regression evidence from Denmark with PISA2000 data, Paper presented at EALE2003 Conference (2003)
38. Alieva, D.: Redes de organizaciones turísticas, itinerarios de viaje y experiencias interculturales de los turistas ruso-hablantes en Andalucía [Ties of touristic organizations, travel itineraries and intercultural experiences of Russian-speaking tourists in Adnalusia] (doctoral thesis) (2018)
39. Borgatti, S.P., Everett, M.G., Freeman, L.C.: Ucinet for Windows: Software for Social Network Analysis. Analytic Technologies, Harvard (2002)
40. Bastian, M., Heymann, S., Jacomy, M.: Gephi: an open source software for exploring and manipulating networks. In: Proceedings of the Third International ICWSM Conference (2009)
41. Kleijn, W.C., van der Ploeg, H.M., Topman, R.M.: Cognition, study habits, test anxiety, and academic performance. Psychol. Rep. **75**(3), 1219–1226 (1994)
42. Fenollar, P., Román, S., Cuestas, P.J.: University students' academic performance: An integrative conceptual framework and empirical analysis. British J. Educ. Psychol. **77**(4), 873–891 (2007)

How Networking and Social Capital Influence Performance: The Role of Long-Term Ties

Mohammed Saqr[1,3](✉), Jalal Nouri[2], Uno Fors[2], Olga Viberg[3], Marya Alsuhaibani[4], Amjad Alharbi[4], Mohammed Alharbi[4], and Abdullah Alamer[4]

[1] University of Eastern Finland, Joensuu, Finland
Mohammed.saqr@uef.fi
[2] Stockholm University, Stockholm, Sweden
{jalal,uno}@dsv.su.se
[3] KTH Royal Institute of Technology, Stockholm, Sweden
oviberg@kth.se
[4] Qassim University, Melida, Saudi Arabia
{Marya.alsuhaibani,Amjad.alharbi,Mohammed.alharbi,
Abdullah.alamer}@qumed.edu.sa

Abstract. Recently, students have become networked in many ways, and evidence is mounting that networking plays a significant role in how students learn, interact, and transfer information. Relationships could translate to opportunities; resources and support that help achieve the pursued goals and objectives. Although students exist, interact, and play different roles within social and information networks, networks have not received the due attention. This research aimed to study medical student's friendship- and information exchange networks as well as assess the correlation between social capital and network position variables and the cumulative Grade Point Average (GPA) which is the average grade obtained over all the years. The relationships considered in our study are the long-term face-to-face and online ties that developed over the full duration of the study in the medical college. More specifically, we have studied face-to-face and information networks. Analysis of student's networks included a combination of visual and social network analysis. The correlation with the performance was performed using resampling permutation correlation coefficient, linear regression, and 10-fold cross-validation of binary logistic regression. The results of correlation and linear regression tests demonstrated that student's social capital was correlated with performance. The most significant factors were the power of close friends regarding connectedness and achievement scores. These findings were evident in the close friends' network and the information network. The results of this study highlight the importance of social capital and networking ties in medical schools and the need to consider peer dynamics in class assignment and support services.

Keywords: Social network analysis · Peer relations · Social capital · Performance · Learning analytics

A. Antonyuk and N. Basov (Eds.): NetGloW 2020, LNNS 181, pp. 335–346, 2021.
https://doi.org/10.1007/978-3-030-64877-0_22

1 Background

Human beings are fundamentally and naturally social creatures who are driven by a pervasive desire to belong and form long-term interpersonal relationships. When relationships are supportive and caring, they positively influence the emotional, cognitive, and the performance of individuals, and the structure they endeavor to belong [1, 2]. Cohesive social structures provide the necessary security and resources in addition to the sense of relatedness, which consequently enhances the motivation of the members and the performance of the group. Therefore, humans strive for connectedness and integration within their social structures and resist the dissolution of social ties [3]. Advocates of the value of the social structure and relationships argue for their importance in addition to personal attributes such as age, gender, socioeconomic in shaping performance [1, 2, 4]. So far, the role of social relationships and how they influence performance remains understudied in competitive health care education [5].

The benefits one gets from the wealth of relationships and the occupying of a particular position in the social structure are conceptually known as the social capital [6]. The social capital theory provides a broad conceptual framework for understanding and studying the individual gains and benefits within the social structure. There are three components for the social capital: relational, structural and cognitive aspects. The relational aspect describes the value one gets from interactions such as relatedness, trust, and loyalty. The structural aspect presents the social interactions and resources associated with it, such as knowledge or opportunities. The cognitive aspect describes the shared values such as interests and norms [7]. In this study, we will investigate the structural and relational aspects and how they influence student performance.

Relationships and social capital in academic environments could provide students with access to informational resources, act as a role model, reinforce positive behavior, and help in executing challenging tasks. Besides, they may also act as a source of happiness, provide necessary emotional support, shield against stressful encounters and promote the sense of relatedness [6, 8–11]. Relatedness has an energizing effect that positively enhances students' motivation and engagement [1, 10]. Moreover, relatedness, through social interactions, teaches students the values and qualities needed to succeed in an academic endeavor [2, 10].

1.1 Related Research

Baldwin et al. [12] examined the effect of three networks of relationships (friends, advisors, and adversaries) on student's performance enrolled in an MBA program. Their results indicated that total number of interactions were positively correlated with performance in the advisors' network. Similar results were reported by Yang et al. [13], who studied an online management course. They found that the strength of interactions with the advisors' network was correlated positively with the course grade and a strong determinant of academic performance. Smith et al.'s research has shown that students who received advice request from peers in the class performed better after controlling for gender and previous grades [14]. A positive correlation between relationships and performance has also been described by others [15–17].

1.2 Social Network Analysis (SNA)

Social network analysis (SNA) has a comprehensive toolset for the study of social relations and patterns of interactions research [18–20], therefore, gained popularity for studying the relational aspects of the social capital [5, 12–17, 21, 22]. A social network is composed of actors and the ties that represent the relationships among the actors. Using SNA, a researcher can visually represent the relationships in the form of a graph, mathematically in the form of metrics and centrality measures, and use simulation methods to understand the network ties [18]. SNA visualization often represents nodes (i.e., actors) as points and the edges (i.e., ties) as lines between the nodes. A node may be called 'ego' if it represents the person of interest, or 'alter' if it represents one of his connections or neighbors. Centrality is an SNA quantitative metric that depicts the node position or importance in the social structure. Since the concept of importance varies according to the context, various centrality measures are computed differently. For instance, in a social network, the individual who has a high number of followers (in-degree centrality) is seen as influential. As this study focuses on relationships and interactions within the social structure, we will study centrality measures that correspond to the extent, value, and strength of relationships. A full account of these measures will be presented in the methods section.

1.3 Rationale Behind This Study

Medical education is competitive, lengthy and requires social support along the demanding studying path. Despite that students have become networked in many ways, and evidence is mounting that networking plays a significant role in how students learn, get support, or transfer information [23, 24], little is known about the role of networks and relationships in student's learning. The problem is even more substantial in the medical education field where we ran our study. A recent review pointed out the paucity of studies in the field of medical education, highlighting an urgent need for the medical education community to cope with the societal shift that is more connected, rather than individualistic [5].

Building personal relationships requires serious investment and effort to develop as well as takes a considerable time to form [1, 2], and evidence suggests that information exchange occurs best among trustable close friends [25]. Nonetheless, most of the studies investigated the short-lasting situational ties over a single course; often the course was online [14, 15, 17, 21, 22], which shows a gap in studying the long-term social and information exchange relationships.

It is against such a background that this research aims to investigate the correlation between interpersonal relationships and network position variables (structural and relational social capital) and medical student's performance as measured by the cumulative GPA. This study takes a departure point from previous research by focusing on long-term relationships and information exchange ties. The relationships considered in our study are the face-to-face ties that were maintained over the duration of study in the college. The research question of this study is:

Is there a relationship between student's social capital (network relations and positions) and final performance?

2 Methods

Data collection: As the study focused on the long-term relationships, we chose the final year students as they have spent a considerable time and this serves our purpose. We have surveyed all students enrolled in the fifth year of the medical school of Qassim University during the last month of the academic year. A total number of 114 students were invited to participate, 110 agreed, and four declined. A survey instrument that included questions about GPA, gender, and the city of residence was distributed. To protect the privacy and encourage students to participate, they were given a list of the class coded as numbers, and students were asked to list their code number. Students listed their friends as follows:

- Close friends: those who were close to the student for most of the college time.
- Information receiving network (advice): peers whom the student resorts to for information.

The survey data were compiled as an aggregate directed network of interaction. An edge was considered when a student nominates another student as a friend.

2.1 Network Analysis

The social capital indicators considered in this study included the popularity measures (in-degree, out-degree, and total degree (henceforth, degree) centrality) as well as position and role in information sharing (betweenness, closeness, and information centrality), measures of strength of relationships (alters): strength of connectedness (eigenvector centrality) and the connectedness and performance (prestige rank) as recommended by previous research [19, 20, 26]. The measures considered are:

Popularity and Extent of Ego Network

- In-degree centrality: the number of students who nominated the student as a friend, a measure of popularity, leadership, and prestige as voted by other students [27].
- Out-degree centrality: the number of the peers whom the student nominated as friends, a measure of student relationships and personal network [27].
- Degree centrality: the sum of the in-degree and out-degree centralities or the total size of student's network and relationships [27].

Role and Information Transfer

- Betweenness centrality: the number of times a student lied in between two otherwise unconnected students, and hence connected them together, or acted as a broker. Brokerage of information is a source of social capital as highlighted previously [27, 28].
- Closeness centrality: the distance of a student to all other students in the network, a measure of reachability and easiness of communication [27, 28].
- Information centrality: the amount of information that flows through an individual, an advantageous position in the friendship information network might help facilitate access to information and resources [29].

Alters Power

- Eigenvector centrality is a measure of student's strength of personal network considering the connectedness and popularity of friends. Thus, connections to popular students translate to higher eigenvector scores [27, 30].
- Prestige rank: similar to eigenvector centrality, however, the power of alters is estimated based on the GPA. Therefore, a connection to well-connected high performing students would translate to higher scores [28].

We also calculated the average mean grade of alters and the difference between a student's grade and his/her alters.

2.2 Statistical Analysis

Statistical analysis was performed using the R environment for statistical computing version 3.5.2. Since classical hypothesis testing with network data carries the risk of interdependencies between the variables and the outcome, permutation methods were selected as they offer a convenient and credible statistical analysis in general and for network data in particular [5, 31–33]. Permutation generates random resampled distributions, which is compared to the observed result to calculate the significance [34]. The association between variables was performed using Pearson's product-moment correlation, the level of significance was set at the level of $p < 0.05$, the p-value was calculated using permutation with 100,000 replicates [34]. For the analysis of variance, we used the permutation version of ANOVA and Levene's test to test for the testing of homogeneity of variance. Permutation-based ANOVA offers better reliability and protection of type I error regardless of the distribution, especially when the sample size is small, and the variance is homogeneous [35].

A stepwise backward multiple linear regression tests were performed to test if SNA centrality measures can be used as predictors of the final grade. Cook's distance was used to test for outliers. Collinearity Statistics (Tolerance and VIF) were performed to check for collinearity among variables. Binary logistic regression was performed to predict high achievers. Model validation was performed using 10-fold cross-validation. The 10-fold cross-validation technique was performed by randomly dividing the dataset into ten equal non-overlapping partitions, one partition was held-out as a validation dataset, and the remaining nine subsamples were used as a training dataset. The process was repeated ten times, in each time a different partition was used as a validation dataset and the remaining data was to produce the model. The final model represents the average of the ten iterations. This method has an advantage of including each observation in the training and validation. Furthermore, the method minimizes the risk of overfitting and offer a method to test the validity of using the data to test future data. The calculation of SNA social capital measures was done by R Statnet package [36], the visual analysis was performed using Gephi version 0.92, Gephi is a cross-platform open source software with a graphical user interface that offers flexibility and control over graph aesthetics [37].

3 Results

There was no statistically significant correlation between centrality indicators of social capital (in-degree, out-degree, closeness centrality, betweenness, and eigenvector centrality) and GPA. Since there is an ample empirical evidence that males and females tend to have distinct friendship patterns, especially in enduring long-term close friendships [38–44], a subgroup analysis of male and female networks in this study have confirmed this idea. Males tended to befriend students with comparable performance and females tended to have diverse friends. The results of the Pearson correlation showed a positive and significant correlation between popularity measures (in-degree) ($r = -0.08$, $p = 0.02$), degree centralities ($r = 0.30$, $p = 0.01$) and performance in the male group. However, the significance was low to moderate. Correlation with prestige rank was also statistically significant and of a moderate degree ($r = 0.4$, $P < 0.01$), which is a strong indication of the importance of whom you know and how strongly connected and well performing your friends are. The alters' performance was also positively and statistically correlated with performance. Full correlation statistics are shown in Table 1.

Table 1. Correlation between SNA parameters and performance in male and female groups

Correlation coefficients			
	Group	r	P
In-degree	F	−0.08	0.61
	M	**0.29**	**0.02**
Out-degree	F	0.14	0.38
	M	0.17	0.17
Degree	F	0.02	0.87
	M	**0.3**	**0.01**
Betweenness C.	F	0.03	0.85
	M	0.13	0.3
Closeness C.	F	−0.27	0.09
	M	0.21	0.08
Information C.	F	0.13	0.44
	M	0.16	0.2
Eigenvector C.	F	0	0.98
	M	**0.31**	**0.01**
Prestige rank	F	−0.01	0.94
	M	**0.4**	**<0.001**

M = male, F = female, C. = centrality

3.1 Do Peer Relationships Predict Grade?

A stepwise backward linear regression test was performed to test if the variables related to the position in the friendship network and relationship to alters would significantly predict the final grade of a student. The predictors included in-degree centrality as it captures the popularity construct, the information centrality as it captures the role in information network, age, gender, the difference between the student's grade and alters as it captures his position to them and the eigenvector centrality as it captures the social power of alters. The results of the final regression indicated that four predictors were statistically significant. Those were age, information centrality, the eigenvector centrality, and grade difference, ($R2 = 0.68$, $F(4,105) = 60.1$, $p < .01$). This indicated the value of one's relationships in shaping the performance. The full regression statistics are presented in Table 2.

Table 2. Stepwise backward regression statistics of SNA variables and performance

Variables	Standardized coefficient	t	P	95.0% Confidence interval for B		Collinearity statistics	
				Lower	Upper	Tolerance	VIF
Constant	–	−3.89	<0.01	−7.82	−2.54	–	–
Inf. C.	0.13	2.30	0.02	2.73	36.78	0.94	1.07
Diff.	0.66	11.80	<0.01	0.55	0.78	0.93	1.07
Age	0.34	6.16	<0.01	0.18	0.34	0.93	1.07
Eigen. C.	0.16	2.89	0.01	0.16	0.86	0.93	1.08

Inf. C. = information centrality, Diff. = difference between a student grade and the average of his alters, Eigen. C. = eigenvector centrality

3.2 The Information (Advice Network)

The advice network was rather different from the friendship network. Ties were distributed among all students regardless of them being close friends. In-degree and degree centralities were both positively correlated with performance; the correlation was relatively higher than the friendship network. Similarly, both measures of strength of connected friends regarding social connections and grade performance were positively and significantly correlated with better performance, and the correlation was the highest with prestige rank ($r = 0.43$, $p < 0.01$) (see Table 3).

Table 3. Correlation between information giving network centrality measures and performance

Centrality	r	P
In-degree	0.39	<0.01
Out-degree	−0.11	0.27
Degree	0.28	<0.01
Betweenness C.	0.10	0.33
Closeness C.	0.12	0.23
Information C.	−0.06	0.52
Eigenvector C.	0.33	<0.01
Prestige rank	0.43	<0.01

C. = centrality

4 Discussion

In the medical education field, research has focused on the individualistic aspect, while often overlooking that students exist, interact, and play different roles within social and information networks [45, 46]. This study was done to evaluate the role that different networks and connections might play in students learning. In order to do that, we studied the friendship and information networks as well as the correlation between student's grades and the various social capital measures. Special attention was given to the enduring long-lasting friendship ties that were developed and maintained over the course of the five years of medical college.

Our results show that friendships ties may differ according to gender: females tended to have a diverse pattern of friendship ties, while males might choose friends who have comparable performance. The results of correlation coefficient, linear and logistic regression demonstrated that a student social capital in the form of social relationships was correlated with better performance. The most significant factors were the number and the power of the student's close friends.

Previous studies used the simple count of close friends as a measure of social capital (e.g., [13, 14]). Our study extended this approach by calculating the power of connected friends regarding social connectedness and academic achievement (eigenvector centrality and prestige rank). Both eigenvector centrality and prestige rank showed a higher correlation with better performance compared to the simple count of friends, an indication that the quality of peers matters more than quantity. The advice network showed comparable results to the friendship network, with slightly higher correlation coefficients. These results emphasize the importance of the social capital and point out which sources of social capital are most important. In our study, that enduring personal relationships and information networks seem to be more effective in bridging information exchange [25].

While this study corroborates some of the previous research findings [14, 26, 47], it sheds light on the value of long-term friendships, the power of connected friends and quality of friendship ties. Previous research has highlighted the role of whom you know, this study has shown that whom your friends know is even more important. A major distinction between the previous research and this study is that their sample included the temporary situational ties over a single course [14, 15, 17, 21], while evidence indicates that efficient transfer of information develops only over time as trust builds up among peers [25].

Another distinction between these results and previous research is the statistical inference model we used for inference. Classical statistical techniques typically assume independence between the outcome and the variables, these assumptions might not be valid using network data, because empirical network data are relational by design, being collected from networked actors who happen to react, interact and influence each other [32, 33]. For the study of influence of the network on individuals or the outcome that results from having relations or occupying a certain position, permutation-based tests are both powerful and credible [31], exponential random graph models (ERGM) is another option that could be used for network statistical inference, we have opted for the former [33].

The findings of this study have implications for the way students are assigned to classes or collaborative groups. There are numerous group assignments in medical education, which occurs in problem-based groups, in team-based learning, in research collaboration settings and in clinical settings, etc. The proper placement of students in diverse groups based on proper stratification with the objective of the dissemination of positive properties would help improve the performance of some students. There is evidence that placing high performing students together would encourage competition and lead to improved performance. Evidence also exists that adding better-performing students to lower performing peers would benefit the low-performing students [48]. As medical educators always receive requests from students who wish to change their group, class, or section. The answer should prioritize the group diversity of perspective and exchange of ideas with new peers. Another implication of this study is that we should do more than simple randomization of group assignments. Proper stratification of a group of collaborators might help the underachievers learn from their peers, motivate the high achievers and offer a platform for diverse perspectives.

Considering the connections and social network position of a student might also help predict his performance using learning analytics methods, or add to the existing algorithms that enable educators to support a student in need for early intervention [49]. On the individual student level, this has implications related to the formation of relationships, having ties to well-connected and informed students might lead to performance gains. On the institutional level, these findings emphasize the need to engage students in social and academic activities that support the development of social skills.

More research in other contexts is needed before concluding that these findings are sufficiently generalizable. The methods implemented here are worth implementing in other schools to understand the networking patterns among students and the networks of information flow. Considering the networked aspect of learning and the study of friendship and course-specific networks would broaden our understanding of learning and allow us to manage groups wisely.

5 Conclusions

This study was performed to assess the role that different networks and connections might play in student's learning and to evaluate the correlation between social capital in the form of long-term social connections and student's performance. The results of correlation tests showed that a student social capital was correlated with better performance; the correlation was highest when the quality of connections was considered. Using student social capital measures, we were able to predict performance with reasonable accuracy. Furthermore, social capital measures were predictive of high achievement. The advice network showed comparable results to the friendship network, with slightly higher correlation coefficients. The medical educators should try to perform proper stratification of groups of collaborators to help motivate the students to learn, exchange ideas with their peers and offer them a platform for diverse perspectives.

Informed Consent
All participants signed an informed consent detailing the purpose of the study, commitment to ethical guidelines regarding data handling and analysis and approval to publish.

Competing Interest
The authors of this manuscript have no conflict of interest to report.

References

1. Baumeister, R.F., Leary, M.R.: The need to belong: Desire for interpersonal attachments as a fundamental human motivation. Psychol. Bull. 117, 497–529 (1995). https://doi.org/10.1037/0033-2909.117.3.497
2. Ne Gagné, M., Deci, E.L.: Self-determination theory and work motivation. J. Organ. Behav. J. Organiz. Behav. **26**, 331–362 (2005). https://doi.org/10.1002/job.322
3. Deci, E.L., Ryan, R.M.: The "what" and "why" of goal pursuits: human needs and the self-determination of behavior. Psychol. Inq. **11**, 227–268 (2000). https://doi.org/10.1207/S15327965PLI1104_01
4. Carolan, B.V.: Social Network Analysis and Education: Theory, Methods & Applications. SAGE (2014). https://doi.org/10.4135/9781452270104
5. Isba, R., Woolf, K., Hanneman, R.: Social network analysis in medical education. Med. Educ. **51**, 81–88 (2017). https://doi.org/10.1111/medu.13152
6. Lin, N.: Building a network theory of social capital. Connections **22**, 28–51 (1999). https://doi.org/10.1108/14691930410550381
7. Tsai, W., Ghoshal, S.: Social capital and value creation: the role of intrafirm networks. Acad. Manag. J. **41**, 464–476 (1998)
8. Kovanovic, V., Joksimovic, S., Gašević, D., Hatala, M.: What is the source of social capital? The association between social network position and social presence in communities of inquiry. Workshop on Graph-Based Educational Data Mining (G-EDM 2014), London, UK, 4–7 July 2014, vol. 1183. CEUR Workshop Proceedings (2014)
9. Dika, S.L., Singh, K.: Applications of social capital in educational literature: a critical synthesis. Rev. Educ. Res. **72**, 31–60 (2002). https://doi.org/10.3102/00346543072001031

10. Martin, A.J., Dowson, M.: Interpersonal relationships, motivation, engagement, and achievement: yields for theory, current issues, and educational practice. Rev. Educ. Res. **79**, 327–365 (2009). https://doi.org/10.3102/0034654308325583
11. Brechwald, W.A., Prinstein, M.J.: Beyond homophily: a decade of advances in understanding peer influence processes. J. Res. Adolesc. **21**, 166–179 (2011). https://doi.org/10.1111/j.1532-7795.2010.00721.x
12. Baldwin, T.T., Bedell, M.D., Johnson, J.L.: The social fabric of a team-based MBA program: network effects on student satisfaction and performance 4. Acad. Manag. J. **40**, 1369–1397 (1997)
13. Yang, H.L., Tang, J.H.: Effects of social network on students' performance: a web-based forum study in Taiwan. J. Asynchron. Learn. Netw. **7**, 93–107 (2003). https://doi.org/10.1016/j.ssresearch.2009.01.003
14. Smith, R.A., Peterson, B.L.: Psst… what do you think?" the relationship between advice prestige, type of advice, and academic performance. Commun. Educ. **56**, 278–291 (2007). https://doi.org/10.1080/03634520701364890
15. Cho, H., Gay, G., Davidson, B., Ingraffea, A.: Social networks, communication styles, and learning performance in a CSCL community. Comput. Educ. **49**, 309–329 (2007). https://doi.org/10.1016/j.compedu.2005.07.003
16. Hommes, J., Rienties, B., de Grave, W., Bos, G., Schuwirth, L., Scherpbier, A.: Visualising the invisible: a network approach to reveal the informal social side of student learning. Adv. Heal. Sci. Educ. **17**, 743–757 (2012). https://doi.org/10.1007/s10459-012-9349-0
17. Reychav, I., Raban, D.R., McHaney, R.: Centrality measures and academic achievement in computerized classroom social networks: an empirical investigation. J. Educ. Comput. Res. **56**, 589–618 (2018). https://doi.org/10.1177/0735633117715749
18. Borgatti, S.P., Halgin, D.S.: On network theory. Organ. Sci. **22**, 1168–1181 (2011). https://doi.org/10.1287/orsc.1100.0641
19. Burt, R.S.: The network structure of social capital. Res. Organ. Behav. **22**, 345–423 (2000). https://doi.org/10.1016/S0191-3085(00)220091
20. Borgatti, S.P., Jones, C.: Network measures of social capital. Connections **21**, 27–36 (1998)
21. Saqr, M., Fors, U., Nouri, J.: Time to focus on the temporal dimension of learning: a learning analytics study of the temporal patterns of students' interactions and self-regulation. Int. J. Technol. Enhanc. Learn. **11**, 398 (2019). https://doi.org/10.1504/ijtel.2019.10020597
22. Saqr, M., Viberg, O., Vartiainen, H.: Capturing the participation and social dimensions of computer-supported collaborative learning through social network analysis: which method and measures matter? Int. J. Comput. Collab. Learn. **15**, 227–248 (2020). https://doi.org/10.1007/s11412-0200-93226
23. Kop, R., Hill, A.: Connectivism: learning theory of the future or vestige of the past? Int. Rev. Res. Open Distance Learn. **9** (2008)
24. Siemens, G.: Connectivism: a learning theory for the digital age. Int. J. Instr. Technol. Distance Learn. **1**, 1–8 (2014). https://doi.org/10.1.1.87.3793
25. Song, D.: The tacit knowledge-sharing strategy analysis in the project work. Int. Bus. Res. **2**, 83–85 (2009)
26. Gašević, D., Zouaq, A., Janzen, R.: Choose your classmates, your GPA is at stake! the association of cross-class social ties and academic performance. Am. Behav. Sci. **57**, 1460–1479 (2013). https://doi.org/10.1177/0002764213479362
27. Stephenson, K., Zelen, M.: Rethinking centrality: methods and examples. Soc. Netw. **11**, 1–37 (1989). https://doi.org/10.1016/0378-8733(89)900166
28. Musiał, K., Kazienko, P., Bródka, P.: User position measures in social networks. In: 3rd SNA-KDD Work 2009, pp. 1–9 (2009). https://doi.org/10.1145/1731011.1731017
29. Marchiori, V.L., Marchiori, M.: A measure of centrality based on network efficiency. New J. Phys. **9**, 188 (2007)

30. Liao, H., Mariani, M.S., Medo, M., Zhang, Y.C., Zhou, M.Y.: Ranking in evolving complex networks. Phys. Rep. **689**, 1–54 (2017). https://doi.org/10.1016/j.physrep.2017.05.001
31. Newman, M.E.J., Watts, D.J., Strogatz, S.H.: Exponential random graph models for social networks. (2011). https://doi.org/10.4135/9781446294413.n32
32. Borgatti, S.P., Everett, M.G., Jhonson, J.C.: Analyzing Social Networks. Sage, Thousand Oaks (2013)
33. Cranmer, S.J., Leifeld, P., McClurg, S.D., Rolfe, M.: Navigating the range of statistical tools for inferential network analysis. Am. J. Pol. Sci. **61**, 237–251 (2017). https://doi.org/10.1111/ajps.12263
34. Hothorn, T., Hornik, K., Wiel, M.A. van de, Zeileis, A.: Implementing a class of permutation tests: the coin package. J. Stat. Softw. **28** (2008). https://doi.org/10.18637/jss.v028.i08
35. Mendeş, M., Akkartal, E.: Comparison of ANOVA F and WELCH tests with their respective permutation versions in terms of type I error rates and test power. Kafkas Univ Vet Fak Derg. **16**, 711–716 (2010)
36. Handcock, M.S., Hunter, D.R., Butts, C.T., Goodreau, S.M., Morris, M.: Statnet : software tools for the representation, visualization, analysis and simulation of network data. J. Stat. Softw. **24**, 1–11 (2008). https://doi.org/10.18637/jss.v024.i01
37. Jacomy, M., Venturini, T., Heymann, S., Bastian, M.: ForceAtlas2, a continuous graph layout algorithm for handy network visualization designed for the Gephi software. PLoS ONE **9**, 1–12 (2014). https://doi.org/10.1371/journal.pone.0098679
38. Mehta, C.M., Strough, J.N.: Sex segregation in friendships and normative contexts across the life span. Dev. Rev. **29**, 201–220 (2009). https://doi.org/10.1016/j.dr.2009.06.001
39. McPherson, M., Smith-Lovin, L., Cook, J.M.: Birds of a feather: homophily in social networks. Annu. Rev. Soc. **27**, 415–444 (2001). https://doi.org/10.1146/annurev.soc.27.1.415
40. Ferris, G.R., Liden, R.C., Munyon, T.P., Summers, J.K., Basik, K.J., Buckley, M.R.: Relationships at work: toward a multidimensional conceptualization of dyadic work relationships. J. Manage. **35**, 1379–1403 (2009). https://doi.org/10.1177/0149206309344741
41. Centola, D., Gonzlez-Avella, J.C., Eguíluz, V.M., San Miguel, M.: Homophily, cultural drift, and the co-evolution of cultural groups. J. Conflict Resolut. **51**, 905–929 (2007). https://doi.org/10.1177/0022002707307632
42. Shrum, W., Cheek, N.H., Hunter, S.M.: Friendship in school: gender and racial homophily. Soc. Educ. **61**, 227–239 (2016). https://doi.org/10.2307/2112441
43. Booth, A., Hess, E., Booth, A.: Cross-Sex Friendship, **36**, 38–47 (2017)
44. Rose, A.J., Rudolph, K.D.: A review of sex differences in peer relationship processes: potential trade-offs for the emotional and behavioral development of girls and boys (2006). https://www-ncbi-nlm-nih-gov.ezp.sub.su.se/pmc/articles/PMC3160171/pdf/nihms314230.pdf. https://doi.org/10.1037/0033-2909.132.1.98
45. Saqr, M.: A literature review of empirical research on learning analytics in medical education. Int. J. Health Sci. (Qassim) **12**, 80–85 (2018)
46. Saqr, M.: Learning analytics and medical education. Int. J. Health Sci. (Qassim). 9 (2015)
47. Joksimović, S., Manataki, A., Gašević, D., Dawson, S., Kovanović, V., de Kereki, I.F.: Translating network position into performance. In: Proceedings of the Sixth International Conference on Learning Analytics & Knowledge - LAK 2016, pp. 314–323. ACM Press, New York, New York, USA (2016). https://doi.org/10.1145/2883851.2883928
48. Sacerdote, B.: Peer effects in education: how might they work, how big are they and how much do we know thus far? In: Handbook of the Economics of Education, 249–277. Elsevier B.V. (2011). https://doi.org/10.1016/B978-0-444-53429-3.00004-1
49. Saqr, M., Fors, U., Tedre, M.: How learning analytics can early predict under-achieving students in a blended medical education course. Med. Teach. (2017). https://doi.org/10.1080/0142159X.2017.1309376

Network Modeling of Blended Communications in the Community of Project Teams of Students

Elena Dudysheva[1](✉) and Olga Solnyshkova[2]

[1] Shukshin Altai State University for Humanities and Pedagogy, Biysk, Russia
dudysheva@yandex.ru

[2] Novosibirsk State University of Architecture and Civil Engineering (Sibstrin), Novosibirsk, Russia
o_sonen@mail.ru

Abstract. It is effective to maintain university communities of interacting project teams to develop the vocational and soft skills of students. Lecturers can organize communities at longitude creative workshops, but monitoring of agile students' teams with high structural dynamics is very time-consuming. The additional complication is blended communications. A significant contribution can be the integration of the following technologies: pedagogical (collaboration based on social constructivism), communicative (hybrid practices in knowledge and information flows), managerial (horizontal and vertical connections in agile teams), informational (cloud services). For solving these problems, social network analysis and modeling can be applied. The paper classified and determined strong ties in the learning community of project teams with blended communications based on students' creative workshop of the university course of Engineering Geodesy. Ties types are Formal or Informal, Vertical or Horizontal, Knowledge or Information-based, Blended or Distant, and Stable or Unstable. Communicating sequential processes is considered as a model of formal joint activities in projects. The issues are the identification and dynamic construction of a didactically stable network, transformation of weak and Unstable ties into Stable strong ones, monitoring, and creation of new Knowledge-based ties in formal and informal communications. We discovered that Formal Distant Information-based ties led to unstable connections and then focused on the formation of Stable ties in triads where they are absent or Unstable. We held the students' survey, compared shares of current and expected types of connections, and analyzed the significance of the organizational events for the stable work of the learning community.

Keywords: Higher education · Learning community of project teams · Communications network analysis · Blended communications network modeling · Formal and informal knowledge flows · Stable strong ties · Agile teamwork

A. Antonyuk and N. Basov (Eds.): NetGloW 2020, LNNS 181, pp. 347–364, 2021.
https://doi.org/10.1007/978-3-030-64877-0_23

1 Introduction

The structure and functions of social, professional, and educational communities transform under the impact of technologies. The example is an active emergence of communities of practice [1] with distant communications as professional virtual groups with limited access for communications, information, and knowledge sharing [2], or informal communities of interest [3] as well as coalitions, virtual teams [4]. Informal internet learning communities attract students with the same educational goals [5]. It may be thesis protection or preparation for tests. Learning communities also can be supported by educational projects, such as Wiki communities [6]. The higher education system has additional socializing functions of young people and potential labor reserves accumulating in universities. The traditional way of educational communications in universities is the classrooms' face-to-face interactions. Established types of communication are inevitably subject to spontaneous pressure from social practices of distant connections using mobile devices and internet services. Contemporary technologies in higher education suggest blended learning. This concept relies on blended educational communications, including face-to-face and distant connections between students and lecturers. Blended learning changes the nature of educational communications: coordination, control, assessment, knowledge access, and formation of educational outcomes. Blended and distant connections replace traditional face-to-face in professional organizations as well, thanks to the processes of diversification and digitalization of the international economy. The potential for developing vocational training may lie in extending the best practices of blended communications to students' learning communities at universities.

The essential skills demanded by employers are solving problems, communications, and teamwork [7]. Some organizations conduct training at workplaces when organizing project teams for the subsequent support of virtual and blended teamwork [8]; educators try to do the same [9]. A significant difference between employees and students is the motivation and application of rigid conditions and methods of project management in firms and organizations. Therefore, it is difficult to transfer the training methods directly to the university learning processes. However, systematic cognitive psychology-based approaches to learning in higher education organizations and their practical methods allow both universal competencies and long-term 'soft' skills [10]. Firms began to introduce agile management of virtual teams' projects [11]. A combination of educational and business communication practices in agile project management can support the development of students' skills in self-development and self-regulation, cooperation, and collaboration, resilience in competition, and uncertainty to professional processes of university graduators in future activities.

Universities encourage the practice of projects teamwork of students. Short-term projects are carried out under the guidance of one lecturer without difficulty. It is more effective to maintain a community of teams that actively interact and communicate to develop whole sets of students' competencies. A lecturer can facilitate the university course communities or a longer-term version, such as students' creative workshops. The problem of management, predicting behavior, increasing the level of knowledge for interacting student teams with high structural dynamics becomes extremely relevant but even more time-consuming. Social network analysis and modeling methods can solve this problem and contribute to the integration of pedagogical, communicative, informational, and managerial technologies.

In general, the work of project teams differs from the community of practice [4]. The first concept brings a more independent organizational character; it has some restrictions set by the rigid logic of project activity. If some students participate in projects in the same organizational conditions repeatedly, the structure and intensity of the connections between them change. In that case, we can talk about a community of project teams. The paper considers the learning community of this kind. We performed the experimental work with analysis of blended communications and projects teamwork outcomes in the community of the student creative workshop 'Geo-S' at the Department of Engineering Geodesy at Novosibirsk State University of Architecture and Civil Engineering. Student project teams developing interactive e-learning tools on engineering geodesy organize a community, which is led by a lecturer – the head of the workshop [12]. As the experimental data source, we analyzed the activities of the last year's teams. Some teams were less stable than the others, up to their breakup. The practical relevance of our research lies in the possibility of agile management and monitoring of blended communications in the learning communities to increase stability. The paper aims to consider structural and dynamic models of learning activities and blended communications in the community of project teams using social network analysis and modeling methods in the transformation of educational, professional, and social practices.

2 Statement of the Problem

The origins of our study lie in the field of practical didactics of higher education. One of the leading paradigms in contemporary learning is constructivism, which implies the creation of educational outcomes by students. Therefore, one of the most popular methods in didactics will continue to be project training. On the one hand, it contributes to the formation of problem-solving skills, and, on the other hand, allows working in a team and promotes the development of communicative and teamwork competencies as well as other 'soft' skills [13]. The social aspects of constructivism in education are related to how activities shape knowledge in a social context [14]. Learners create knowledge and deepen their understanding through observation and interaction with the surrounding reality, as well as discourse and participation in events with other learners. The didactics issues of social constructivism were developed within work activity theory [15], where teams play an essential role. With mobile technology, team members can use remote or blended communications. There were numerous attempts to implement corporate workplace training through computer-supported collaborative project learning [16]. However, the organization of labor actions when replacing traditional communicative practices with distant ones leads to unstable dynamic structures of teams. They need new agile management methods when the participants continue developing necessary knowledge while working on projects. New hybrid communicative practices are being formed, including educational and business, formal and informal, and distant and face-to-face communications.

The transfer of social practices of mobile communication into business and educational activity is one of the leading trends. They are organized through cloud services, where data and e-tools are provided for collective access. It leads to the emergence of virtual (or online) communities. An online community can be defined as a group of users who communicate through Internet-based common interests, resources, and goals [17]. The clarification of this concept includes the following points [18]: community members share common goals and activities or needs; members are involved in active participation characterized by joint activities; access to shared resources is organized according to shared rules; members exchange information and support each other; the community works together based on socially accessible norms and agreements. The professional virtual communities support the exchange of information and knowledge, for example, to solve professional tasks. Open virtual communities are based on voluntary participation and restriction. So, outsourced employees deprived of social support in the workplace can find support in open professional communities. The formalized communities in firms require management efforts to organize structural connections – vertical and horizontal in addition to formal and informal information-based flows.

In vocational education, virtual communities are starting to gain high importance too. They consist of students collaborating in solving educational problems or developing projects. One of such social structures is the learning community of practice [19] in which knowledge is constructed through social interaction in a real-practical context. The most popular tools for interaction between university students are web-services, where learners can jointly discuss topics moderated by a lecturer and work together on common documents [20]. Open virtual learning communities are mostly informal. Our interest lies in the purposeful support of student communities when training both professional knowledge and the formation of personal competencies in university courses. An essential condition is the integration of students' collaboration and competitiveness [21], which is especially useful for the formation of 'soft' skills of resilience and future intra-company entrepreneurship. Student team projects are important for both the cognitive and social aspects of constructivism. In addition to scientific laboratories, professionally oriented creative workshops may be useful in longitude training.

We denote the *didactic problem of ensuring the stable functioning of the community of collaborating and competing student project teams with blended communications in university courses.* The didactically stable functioning does not mean the necessary achievement of a high-performance project result by each team. In contrast to the professional field, it is more relevant to involve students in obtaining educational outcomes in at least one project team, i.e., maintain connections for each member of the community with the achievement of the educational goals according to the project criteria.

There are many studies on virtual project teams [22], but the set of teams does not automatically form the community. Of particular interest are combinations of vertical and horizontal connections in the hierarchical structure of the community supported by university lecturers. A system with a coordinating center with the predominance of vertical connections corresponds to traditional education. It has shortcomings in organizing flexible blended learning. A system without a coordinating center (the existence of only horizontal connections) corresponds to the extreme manifestations of open distance education and is considered unstable for formal education. A system with a hybrid organization and cooperation mechanisms maintains the balance of horizontal and vertical connections. It seems to be the most consistent with the agile management of the hierarchical system of the student team community. It also raises different organizational issues, such as the following. What conditions contribute to the formation of a sustainable community of project teams? Can horizontal connections improve community stability and team performance? What impact do information- and knowledge-based connections have in blended communications?

The flows in communities can be separated according to their dominant functions – organizational (Information-based) or cognitive (Knowledge-based) [23]. Knowledge-based connections in a learning community can involve, for example, solving problems with other students or a joint search of reliable technologies. Application of the knowledge flows model in learning improves students' collaboration, communication skills, and critical evaluation in traditional classroom education [24]. However, we cannot be confident in these facts in blended learning. The hybrid nature of communicative interactions of students in blended learning does not allow to identify the universal method of analysis, for instance, psychological observation of the group's work in the classroom. Thanks to digital literacy, young people in universities are more active in transferring distance practices of social interaction to training and informal knowledge sharing than adult lecturers [25]. One of the research methods for blended activities in the community is social network analysis [26]. It can be applied to different connections: physical, behavioral, or associative [27]: physical connections mean direct face-to-face communication, behavioral connections imply the exchange of information, and associative connections correspond to common educational or personal interests, including the exchange of knowledge. Stability in a social network may depend on knowledge flows between actors [28]. The formalized division into teams introduces an additional hierarchy in the learning community and makes it more manageable. It needs the support of formal connections for organizational interactions. Therefore, we rely mainly on concepts related to the different classification of interpersonal connections: Formal or Informal, Vertical or Horizontal, Knowledge or Information-based, Distant (without face-to-face) or Blended.

Social network analysis may consider different problems of learning communities; for example, students attending university courses build a student-course network that quantitatively describes the links among students and courses [29]. They can be applied to team projects as well. Nevertheless, by educational tasks, we study the concept of community at the individual level.

The connections between members can be inside and outside the team. Such connections do not necessarily appear automatically; they can appear and disappear, sometimes leading to the destruction of teams. Social network analysis, in our case, is based on the identification and description of significant ties and dynamic modeling of network structures to ensure the sustainability of the community network (where vertices mean participants).

Ties can be divided into strong and weak, based on the intensity of communication [30]. Online communication may change the possibilities of tie strength [31]. Therefore, some strong ties in blended communications can further turn (or not) into weak ones; we consider them Unstable. The difference of used concepts is in dynamics: Unstable ties are based on intensive strong connections in the past, while the weak ones are not. If strong ties in the dyads remain intense enough for mutual project joint activities, we consider them Stable. We assume that the learning community differs from the set of teams by the relative balance of Vertical/Horizontal, Formal/Informal, and Knowledge/Information-based Stable ties. The primary purpose is to ensure the didactic stability of the learning community network through monitoring and restructuring, which means eliminating the destruction of student connections with falling out from all teams, i.e., each vertex is embedded in at least one dyad with Stable tie. Prediction of possible tie instability requires not only an ego-centric methodology but also an analysis of the complete network. We intend to answer the following problematic questions.

1) *Does the stability of ties with distant and blended communications depend on other tie types, and what combinations of tie classification types can form Unstable ties?*
2) *How can the ratio of Stable tie types characterize the learning community network of student project teams with blended communications?*
3) *What structural features, metrics, or characteristics can provide didactic stability of the learning community network of student project teams?*

3 Data and Methods

Experimental data was collected by analysis of project teamwork activities in the community of the student creative workshop 'Geo-S' at the Department of Engineering Geodesy (Novosibirsk State University of Architecture and Civil Engineering, Russia). Students are future engineers developing electronic educational resources. Each project takes no more than one year. A project is completed if it passed the testing stages and met the criteria. Some students participate in several projects in different roles at the same time, but more often, they work for several years on different projects. The head of the creative workshop is a lecturer of Engineering Geodesy. The head collects requests for the development of electronic educational resources and resolves formal issues in the workshop. The main coordinator is an experienced participant who attracts new members, coordinates teams, controls the time, prepares events, and discusses projects change with the head. The team's leaders are students who monitor tasks, attract resources, participate as executors, and report. These members perform formal vertical functions in the learning community. An external consultant is a lecturer or senior student or employee.

The 'Geo-S' community is not large: of the more than 200 second-year students studying engineering geodesy, about 20 new members come to the workshop every year. In the year of our experiment, 37 students of different courses took part in the workshop, 15 of them were participants of five project teams. The participants have face-to-face events/meetings, but they actively communicate through web-services. Teamwork is agile; members may work in different teams at the same time or change the team.

In agile project teamwork with high dynamics, the completeness of relevant data of blended communications remains problematic. Structural dynamics of blended interactions in teamwork management can be traced using internet services of collective work, as well as by the visualization of stages of design solutions (mental schemes, Petri nets) [23]. As a promising option, we consider a model of communicating sequential processes [32], which is suitable to model formal ties in the community network, especially for automation in large communities. Some tasks must be performed strictly sequentially; others are random; some are optional, and part of them can be iterative. Parallelism permits performing joint actions. Some tasks have a special communicative meaning: students should take part together, which corresponds to Formal Information-based ties.

Data from interviews and online surveys can be used to apply network analysis to explore the interpersonal knowledge-sharing network with agile teams and vertical and horizontal formal and informal flows [28]. Self-evaluation and interviews are the leading methods for small communities. Self-assessment of the formal and informal flows of information and knowledge in the context of vertical and horizontal connections can be used as a method of data collection in higher education [23]. In our opinion, questions about the information and knowledge flows should be supplemented by examples to avoid misunderstanding in distinguishing them. We believe that for small groups, social network analysis can be supplemented with other methods, in particular, a description of the dynamic structure of the students' community as case study students' connections in interacting project teams.

Our methods were structural network analysis and visualization with identification of tie types using such instruments as surveys, interviews, pedagogical observation, frequency analysis of the content of virtual communications, and case study of activities of project teams. We identified eight types of combinations based on three dichotomic grounds. One is information- or knowledge-based flows. A student survey about teamwork on types of interactions can be conducted without separating questions about information- or knowledge-based flows [23]. We believe it is important to separate these questions and provide examples because it is difficult to identify for learners to identify two types of flows. The second ground is vertical or horizontal connections according to members' functions. The next ground is formal (organizational and projects events and activities) or informal interactions between participants. The nature of blended or predominantly distant communications is determined based on pedagogical observation, the content of web-services, and the results of individual interviews. We got 16 variants of tie type combinations.

We took into account strong ties for the structural analysis of the community network. The tie strength was determined depending on the communication intensity. Weak ties describe confirmed participation in the creative workshop events, including former participants, invited lecturers, and external consultants. The structural analysis obtained from the project case study permits us to find Unstable strong ties and stable structural elements. A Stable tie is a confirmed mutual communication between two members. An Unstable (strong) tie is communication with disrupted interactions. We also studied the impact of homogeneity or homophily (gender, nationality, groups, or campus) on tie strength and stability.

We applied the case study method for five ongoing project teams, analyzing their connections and structural dynamics during six months. We interviewed participants and identified ego networks starting from initial mental schemes for determining the need for additional interviews. Then the network graph was visualized using NodeXL [33] for each project and complete network of the 'Geo-S' community. Overall network metrics or characteristics of vertices – degree centrality, clustering coefficient [26], closeness centrality, betweenness centrality [34], or network structural elements such as clusters and triads [35] are used in studies of dynamic learning groups. For networks of small groups, the network analysis results are well consistent with psychological studies. Examples increase students' interaction (network density) in face-to-face events with tutors, identifying informal team leaders (centrality increases), and so on. We used betweenness centrality as an indicator of the workload coefficient and intensity. The higher the measure of betweenness centrality of a vertex, the more information flows of the community pass through it. Such a community member has a high organizational burden. Therefore, monitoring of the vertex characteristics, in particular, the betweenness centrality of the head, was added with the dynamics of structural constellations. Triads of strong ties were analyzed for the sustainability of the complete network.

We also invited 'Geo-S' members to indicate the preferred distribution of the tie type shares with the distant survey. We proposed to define the shares of preference for combinations of types, which then were transferred to the total share of events for information and knowledge flows. Then we compared the statistics of the existing types of connections with the results of the survey of students' team representatives and identified the differences between current and expected distributions. Finally, we analyzed the significance of the organizational events (by their rank) corresponding to a combination of connections types with insufficient current implementation to improve the efficiency of the workshop work.

4 Results

Five student projects of the 'Geo-S' community contributed data for network modeling and analysis. The network graph includes 16 vertices – the workshop's members. The following is a brief description of project cases.

Project1 included the programmer – leader (Vertex3), the artist (Vertex8), and the assistant (Vertex9); all of them are new members. Communications were mainly through the leader; he also communicated with the Project3 leader. Team members study together and live on the same campus, so the interactions are mostly face-to-face. The leader had constant remote interactions with the head (Vertex1) and the main coordinator (Vertex2). The project was completed on time, with distant communications only in the final stages. The assistant went to Project4 (forming Unstable tie), and then he left the workshop.

Project2 includes members who have previously participated in projects: the leader – the cameraman (Vertex4), the editor (Vertex10), the actress who was a tutor for new teams (Vertex11), the actor was simultaneously the coordinator (Vertex2), and another actor who was the consultant as well (Vertex12). Actors communicated with the cameraman face-to-face, but discussions were distant. The editor (Vertex10) did not communicate inside the team, but he carried out tasks well under outsourcing and communicated with the programmer of another team. This project was the most successful and conflict-free.

We constructed the network graphs for each project two months after the start. Figure 1 represents strong Stable and Unstable ties for Project1 and Project2 inside and outside, including outer triads. Vertices were grouped by projects (visualized using NodeXL [33]) with group 0 designating lecturers. These groups are indicated with shapes: 0 – Diamond,1 – Solid Square, 2 – Circle, 3 – Triangle, 4 – Square, 5 – Solid Triangle.

Figure 2 represents strong Stable and Unstable ties for Project3 (inside and outside, including outer triads) and Project4 & Project5 (both problematic projects together).

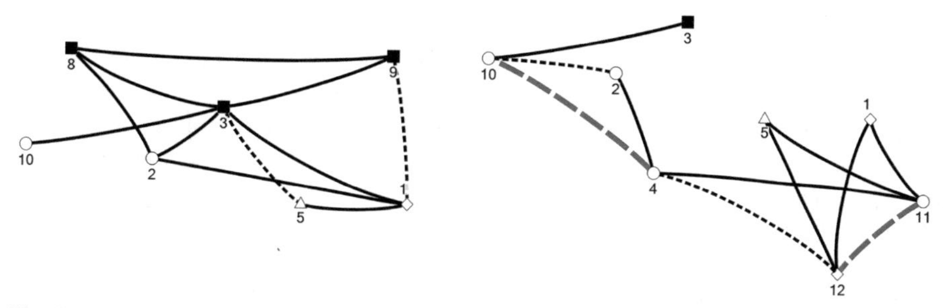

Fig. 1. The communication networks of Project1 (left) and Project2 (right). Vertices are 'Geo-S' members. Bold dash lines are Distant Unstable ties; dot lines are Distant Stable ties; solid lines are other strong Stable ties.

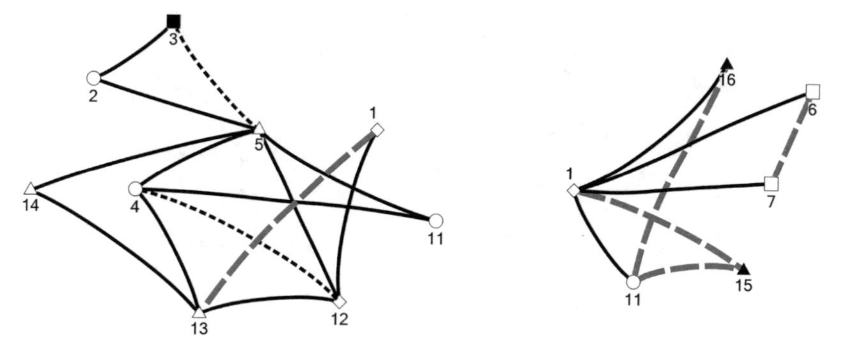

Fig. 2. The communication networks of Project3 (left) and Project4 & Project5 (right). Vertices are 'Geo-S' members. Bold dash lines are Distant Unstable ties; dot lines are Distant Stable ties; solid lines are other strong Stable ties.

Project3 included the leader – programmer (Vertex5), the videographer (Vertex13), later the programmer (Vertex14) joined. The team has a very active leader; she initiatively conducted surveys among students. The project was delivered on time. There were no conflicts. Only during the interview, it turned out that the new participant successfully worked in the team at the leader's initiative.

Project4 included the leader (Vertex6), who was the programmer and the artist as well, and the programmer-animator (Vertex7). The project was held with conflicts, resulting in the retraction of the workshop's head. An experienced workshop participant was involved in tutoring during the month. The project gave an example of the destructive influence of informal communications. Nevertheless, the project was implemented in high quality.

Project5 included new members (Vertex15 & Vertex16), but the workshop head was the actual leader. A tutor (Vertex11) was involved to clarify organizational issues. One participant (Vertex15) provided materials only after reminders; he did not want to support communication with other members and work with anyone except the head. Sometimes he carried out short-term formal correspondence with the main coordinator. Another participant (Vertex16) is the animator and the artist as well. She carried out the project alone, communicating only with the head. Her project's part was completed. The project is scheduled to continue next year.

Further, we constructed the network graph of the 'Geo-S' community (Fig. 3). Vertices were grouped by projects and indicated with shapes.

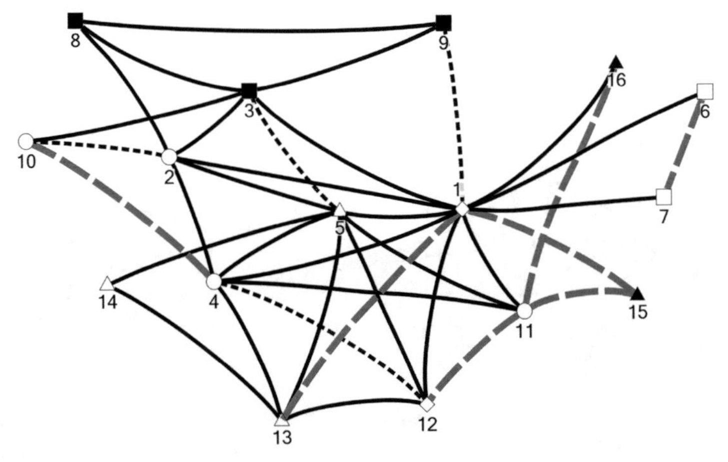

Fig. 3. The communication network of the 'Geo-S' community. Vertices are 'Geo-S' members grouped by project teams. Bold dash lines are Distant Unstable ties; dot lines are Distant Stable ties; solid lines are other strong Stable ties.

The network graph of the learning community can be explored with clustering and cohesion measures [36]. However, cohesion can be a source of rigidity that prevents agile management [37]. There are attempts to avoid these in professional organization management of cross-functional project teams [38], but the effectiveness of the learning community's teamwork remarkably differs from criteria applicable in organizational settings. Besides, the size of our network is insufficient to verify the applicability of such assumptions. Therefore, we focused on the analysis of the dependence of the tie stability on local structural features and characteristics of the vertices. We used betweenness centrality as an indicator of the members' workload (Table 1).

Our finding is that members of students' creative workshop who involve in triads of Stable (strong) ties are more intended to receive project outcome and continue participation in the learning community. Therefore, we can focus on the formation of Stable

Table 1. Characteristics of the 'Geo-S' communication network's vertices

Vertex	Degree	Betweenness centrality	Main group	Formal vertical functions in the group
1	12	52.33	0	Yes (Head)
2	6	8.00	2	Yes (Main Coordinator)
3	6	8.42	1	Yes (Leader)
4	7	7.58	2	Yes (Leader)
5	8	13.25	3	Yes (Leader)
6	2	0.00	4	Yes (Leader)
7	2	0.00	4	No
8	3	0.33	1	No
9	3	2.17	1	No
10	3	0.25	2	No
11	6	4.67	2	At times (Tutor)
12	5	0.25	0	Yes (Expert)
13	5	3.75	3	No
14	2	0.00	3	No
15	2	0.00	5	No
16	2	0.00	5	No

ties in triads for those participants where they are absent, or the tie is Unstable without significantly increasing the workload for the head and coordinators.

Social homogeneity or homophily did not automatically form Stable ties. Nevertheless, informal communication can destroy the emerging organizational ties in teams taking into account the students' characteristics. One of the projects stopped at the start due to the informal relationships of team members. Working in the same team did not necessarily contribute to the formation of a strong tie; there were cases of outsourcing and violation of reciprocity. For example, we found outsourcing execution of tasks by the editor of Project2 (Vertex10) for Project1 when solving programming problems with the programmer of Project1 (Vertex3). One of the examples of violation of reciprocity was systematic attempts of the animator of Project 4 (Vertex7) to interact with the programmer of the same project (Vertex6) to coordinate the results, which often remained without a timely response. These cases affect the strength and stability of ties.

Using Edge filters in NodeXL, we counted confirmed Unstable and Stable strong ties of different types (Table 2). The analysis shows that initially formed Formal and Informal ties (possibly unobservable) turned out to be Unstable based on Information flows and mainly reduced exclusively to Distant (without face-to-face) communication, both Horizontal and Vertical ones (types combination FOD in Table 2). This is not true for blended communications.

We also can characterize the workshop not just as a set of teams but as a learning community due to the relative balance of Vertical/Horizontal (53%/47%)

Table 2. Types of confirmed strong communication ties in the 'Geo-S' community

From vertex no	To vertex no	Vertical or horizontal	Formal or Informal	Knowledge or information	Blended or distant
List of Unstable ties					
1	13	V	F	O	D
4	10	H	F	O	D
6	7	V	F	O	D
11	12	H	F	O	D
11	15	H	F	O	D
11	16	H	F	O	D
Examples of types combinations for Stable ties					
1	9	V	F	K	D
2	10	V	I	K	D
3	5	H	I	O	D
4	13	H	F	K	B
8	9	V	F	O	B
Stable ties, by types: V – 16, H – 14, F – 18, I – 12, K – 15, O – 15, B – 26, D – 4					

and Formal/Informal (60%/40%) Stable ties. It is a community of knowledge since the Knowledge-based ties turned out to be Stable ones, and the share of Stable Knowledge-based ties is half of the total.

At last, we revealed disagreements between the current and expected shares of the community communication types in the creative workshop 'Geo-S' using the survey (for nine members of successful projects). Representatives of the current teams with completed project outcomes and relevant experience of interactions answered two parts of the main survey and eight clarifying ones (4 for each part) voluntarily. Two parts of the main survey concerned the connections based on information and knowledge flows:

Please rate the contribution (in shares from 100%) of the various flows listed below for effective work on the project:

1) *based on Organizational Information (Question1);*
2) *based on Knowledge sharing (Question2).*

We compared the statistics of current communication types with the expected ones with the survey of representatives of students' teams (Table 3).

The comparison shows a low share of Informal/Vertical Information ties as well, and a reduced share of Formal/Horizontal Knowledge-based ties. The first disagreement suggests that there are not enough opportunities in the community for students to communicate with head and experts outside formal events, at the request of students. The second disagreement shows that insufficient attention is paid to organized events for students to discuss subject and technological issues with each other. Equal shares

Table 3. The shares of the types of communication ties developed in the 'Geo-S' community

Ties types	Formal/Vertical	Formal/Horizontal	Informal/Vertical	Informal/Horizontal
Information-based ties (shares, %)				
Expected (mean)	33.8	24.4	23.8	18.1
Current	36.4	31.8	9.1	18.2
Knowledge-based ties (shares, %)				
Expected (mean)	32.5	27.5	21.9	18.1
Current	40.0	20.0	20.0	20.0

for Informal/Horizontal Information- and Knowledge-based ties may be the result of difficulties in their identifying for students (as we mentioned above) in self-study. The higher share of Formal/Horizontal Information-based ties is connected with 'Geo-S' organizational rules, which involve many formal common activities of the participants. The higher share of Formal/Vertical Knowledge-based ties is explained by 'Geo-S' main goals – focus on the in-depth study of engineering geodesy. Equal shares of other types of current Knowledge-based ties may result from empirical attempts at participation of all community members (lecturers and students) in different Knowledge-sharing activities.

We analyzed the significance of the events expected by the participants, calculating total shares. The shares show the average rating among workshop events oriented to Horizontal or Vertical, Formal, or Informal communications separately for Information-based (20 events) and Knowledge-based (16 events) ones. We chose combination with the insufficient current implementation (see Table 3). The events with the highest ratings are Formal Knowledge sharing of materials from current and former participants (16%) and their presentations (9%) and Informal/Vertical Information communication with the workshop management and other lecturers (15%).

We revealed the following insufficient actions with blended communications: informal online meeting with the head and main coordinator; support of the community forum by experts with a good feedback; organization of problem forum subgroups; creation of a virtual knowledge cloud, including examples of current and previous projects.

5 Discussion

When solving research problems, we mainly focused on the following issues: the creating of Stable ties within the learning community based on knowledge-based flows; support of sustainable network constellations arising in blended communications; dynamic monitoring of project teams communications for the transformation of weak or Unstable Distant ties into Stable strong ones.

The results of the experimental work showed that the head of the workshop 'Geo-S' mainly achieved the didactics purpose – the participation of all students in design tasks, excluding one missing participant. Thus, the network of students' communications within the community turned out to be didactically stable. Intensive face-to-face communications with many students achieved this result. Two months after the start of projects, betweenness centrality of the 'Geo-S' head sharply increased when working with unstable teams. It affected the long-term results: insufficient student satisfaction with the intensity of interaction with the head and the refusal of some students to resume participation next year.

Tutoring with experienced members (with past projects) and the involvement of external experts can reduce workload per the lecturer. These conditions correspond to formal horizontal connections and can help in face-to-face events as well as knowledge-sharing. In particular, experienced members should be involved as team leaders for other teams.

Distant ties in the learning community need measures to strengthen them. Knowledge-based communication is the most crucial attribute and method to create Stable ties. Part of knowledge-based connections has developed between different teams of 'Geo-S' members as a result of the knowledge exchange, including experienced and new participants, for example, links (3,10), (5,11) (Fig. 3). Face-to-face events can also turn weak and Unstable ties into Stable ones. Valuable examples of reinforcement of horizontal connections (with reducing the organizational burden) can be events organized by the main coordinator and team leaders: seminars with a simultaneous resolution of vocational issues, professional consultations between teams, as well as problem-solving seminars within the team with participants involved in project tasks. They allowed turning six Unstable ties into Stable ones. Due to the high workload of the head, the idea of pair monitoring of teams to increase their sustainability through triads maintains.

Engaging experts means Vertical connections to support knowledge acquisition; such communications can be Informal or/and Distant. Periodic seminars with other teams on technological and subject issues correspond to Formal Horizontal connections of Knowledge flows in the community; such events can take place online and lead to informal communication on work issues between some participants. Workshop events should involve as many newcomers as possible in a face-to-face (or at least online) format. The main coordinator and team leaders can conduct activities without involving the head or within the team, but the community rules should directly instruct them. On the contrary, some connections with team leaders and coordinators should be weakened by transferring to the distant mode.

Selective construction of triads of Stable strong ties allows community scaling while maintaining stability and increasing the number of successful project teams without a critical increase in the workload of the head. However, the head receives new tasks. The implementation of many projects with structural dynamics requires the timely monitoring and flexible management of blended and distant interactions for solving project tasks by participants. This organizational condition is a possible introduction to outsourcing in the community or team restructuring. A practical solution is to use cloud platforms of common virtual space to support teams to demonstrate the implementation of project tasks by individual participants for coordination and timely completion of the project's stages. Virtual platforms can support forum, include examples of current and previous projects, and provide monitoring.

The undoubted advantage of dynamic network models monitoring with parallel process models is the flexibility of management and visibility since it can show the stage of each participant. The disadvantage of this model is that it is challenging to conduct a global analysis as well as to carry out monitoring of the entire community, rather than a separate project. Performance assessment for the entire community of teams can be performed using asynchronous network models such as Communicating Sequential Processes. It permits unambiguous transformation into nets of parallel processes, which can also be used for teamwork monitoring. Their disadvantage, on the contrary, is the inability to distinguish the tasks of the participants' responsibility. The possibility of combining these two formalisms allows us to effectively monitor and analyze the interaction of participants in joint project activities.

6 Conclusions

The paper considered the didactic problem of ensuring the stable functioning of the community of collaborating and competing student project teams with blended communications in university courses. This can require integrating the following technologies deeply: pedagogical (collaboration based on social constructivism), communicative (hybrid practices in knowledge and information flows), managerial (horizontal and vertical connections in agile teams), informational (cloud services).

We considered actions and conditions contributing to the formation of a sustainable community of effective teams. Social network analysis and modeling methods were applied to solve the problem. Reformulating the problematic questions in the network's terms led to the identification and construction of sustainable structures arising in blended communications both within agile teams and within the whole learning community.

We analyzed connections inside and outside several projects of students' creative workshop 'Geo-S' at the Novosibirsk State University of Architecture and Civil Engineering (Russia) and constructed the network graph of the 'Geo-S' community.

We classified ties based on different types. They are a combination of Formal or Informal, Vertical or Horizontal, Knowledge or Information-based, and Blended or Distant communications.

The tie strength was determined depending on the communication intensity. A Stable strong tie confirmed mutual communication in dyads in joint project activities. We focused on the analysis of the dependence of the tie stability on tie types, structural features, and characteristics of the vertices.

The initially formed Formal and Informal ties turned out to be Unstable based on organizational flows and mainly reduced exclusively to Distant (without face-to-face) communication, both Horizontal and Vertical ones – we answered the first problematic question. So, Distant ties needed measures to strengthen them. The practical relevance lies in the possibility of agile management and monitoring of blended and distant communications in the learning communities to maintain stability. Our paper briefly discussed the possibility of combining the two formalisms. They are communicating sequential processes and Petri nets, which allows to effectively monitor and analyze dynamic structures of formal ties for agile teams' joint activities in large learning communities and can serve the basis for further research.

We discovered that the most crucial attribute and method of creating Stable strong ties are knowledge-based connections and their support. Face-to-face (or online) events can turn weak into Unstable strong ties, and Unstable ones into Stable strong ties within the community. We have not found that homogeneity or homophily automatically form strong ties. Moreover, informal communication can destroy the emerging information-based ties in teams taking into account the personal characteristics of students. Organizational flows of team management, in general, were also not a sufficient condition for stability.

We counted strong ties of different types two months after the start of projects and discovered the relative balance of Vertical/Horizontal, Formal/Informal, and Knowledge/Information-based Stable ties. So, the community is knowledge-based, and it is a real learning community rather than a set of teams – we substantiated our assumption about the difference between communities and sets of teams and answered the second problematic question.

Comparing the network analysis and the results of the pedagogical observation of project cases permit us to find potentially Unstable strong connections and triads as stable structural elements in the community network – we partially could answer the third problematic question. We focused on the formation of Stable strong ties for those participants where they are absent, or the tie is Unstable without significantly increasing the workload for coordinators indicated by betweenness centrality.

Using tutoring with experienced members and the involvement of experts corresponds to Formal Horizontal connections and can help in face-to-face events as well as sharing knowledge. Valuable examples of reinforcement for Horizontal connections reducing the burden on the head can be the following: organizational seminars with a simultaneous resolution of vocational issues, professional consultations between teams, as well as problem-solving seminars within the team with participants involved in project tasks.

We hold the survey, compared shares of current and expected types of connections, and analyzed the significance of events by their rank. Students indicated a lack of Informal/Vertical Information-based ties and Formal/Horizontal Knowledge-based ties. The first disagreement suggests that there are not enough opportunities for students to communicate with the head and experts outside formal events. The second disagreement shows that insufficient attention is paid to organized events for students to discuss learning issues. We selected expected events with insufficient current implementation and suggested suitable actions with blended communications.

Our study also avoided considering cohesion and other overall measures of the network's graph due to a lack of experimental data. These issues can serve as a further broad area of network analysis of blended communications in the learning community of students' project teams.

References

1. Lave, J., Wenger, E.: Situated Learning: Legitimate Peripheral Participation. Cambridge University Press, Cambridge (1991)
2. Lin, M.-J.J., Hung, S.-W., Chen, C.-J.: Fostering the determinants of knowledge sharing in professional virtual communities. Comput. Hum. Behav. **25**(4), 929–939 (2009)

3. Chouchani, N., Abed, M.: Online social network analysis: detection of communities of interest. J. Intell. Inf. Syst. **54**(1), 5–21 (2018). https://doi.org/10.1007/s10844-018-0522-7
4. Kietzmann, J., Plangger, K., Eaton, B., Heilgenberg, K., Pitt, L., Berthon, P.: Mobility at work: a typology of mobile communities of practice and contextual ambidexterity. J. Strategic Inf. Syst. **3**(4), 282–297 (2013)
5. Holland, A.A.: Effective principles of informal online learning design: a theory-building metasynthesis of qualitative research. Comput. Educ. **128**, 214–226 (2019)
6. Hwang, J., Brummans, B.: Learning about media effects by building a Wiki community: students' experiences and satisfaction. In: Wankel C. (ed.) Teaching Arts and Science with the New Social Media, vol. 3, pp. 39–59. Emerald (2011)
7. The Global Skills Gap Report 2019. https://www.qs.com/portfolio-items/the-global-skills-gap-report-2019/. Accessed 01 May 2020
8. Korb, W., Geißler, N., Strauß, G.: Solving challenges in inter- and trans-disciplinary working teams: lessons from the surgical technology field. Artif. Intell. Med. **63**(3), 209–219 (2015)
9. Skelly, M., Bruce, F., Banks, R., Steiner, H.: Designing collaborations at the intersection of academia and industry. In: Berg, A., Bohemia, E., Buck, L., Gulden, T., Kovacevic, A., Pavel, N. (eds.) Building Community: Design Education for a Sustainable Future. Proceedings of E&PDE 2017 (DS 88): The 19th International Conference on Engineering and Product Design Education, pp. 164–169 (2017)
10. Ngang, T.K., Yunus, H.M., Hashim, N.H.: Soft skills integration in teaching professional training: novice teachers' perspectives. In: Procedia - Social and Behavioral Sciences, vol. 186, pp. 835–840 (2014)
11. Santos, V., Goldman, A., de Souza, C.R.B.: Fostering effective inter-team knowledge sharing in agile software development. Emp. Softw. Eng. **20**(4), 1006–1051 (2014). https://doi.org/10.1007/s10664-014-9307-y
12. Solnyshkova, O., Dudysheva, E.: Interactive multimedia educational resources for training of students of architectural and civil engineering university at working with geodetic equipment. In: 2018 IV International Conference on Information Technologies in Engineering Education (Inforino). IEEE (2018). https://doi.org/10.1109/INFORINO.2018.8581861
13. Pillay, J.D.: Selfies 2015: Peer teaching in medical sciences through video clips-a case study. African J. Health Profess. Educ. **9**(4), 164–167 (2017)
14. Van der Veer, R., Yasnitsky, A.: Vygotsky's published works: (an almost) definitive bibliography. In: Yasnitsky, A., van der Veer, R. (eds.) Revisionist Revolution in Vygotsky Studies, pp. 243–260. Routledge, London and New York (2016)
15. Engeström, Y.: The emergence of learning activity as a historical form of human learning. In: Learning by Expanding: An Activity-Theoretical Approach to Developmental Research, pp. 25–108. Cambridge University Press, Cambridge (2014)
16. Romero, M.: Time Awareness Tool for enhancing group times' coordination in the virtual workspace. In: 5th International Conference on Intelligent Environments, vol. 4, pp. 333–337 (2009)
17. Lazar, J., Preece, J.: Social considerations in online communities: usability, sociability, and success factors. In: van Oostendorp, H. Cognition in the Digital World. Lawrence Erlbaum Associates Inc., Mahwah, NJ (2002)
18. Preece, J., Maloney-Krichmar, D.: Online communities: focusing on sociability and usability. Handbook of Human-Computer Interaction, pp. 596–620 (2003)
19. Buckley, S., Strydom, M.: 21st century learning – community of practice for students in higher education. In: 10th International Conference on E-Learning (ICEL), pp. 49–57 (2015)
20. Conrad, M.: Teaching Project Management with Second Life. In: Vincenti, G., Braman, J. (eds.) Multi-user Virtual Environments for the Classroom: Practical Approaches to Teaching in Virtual Worlds, pp. 302–315. IGI GLOBAL, Hersey, USA (2011). https://doi.org/10.4018/978-1-60960-545-2

21. Gorostiaga, A., Aliri, J., Ulacia, I., Soroa, G., Balluerka, N., Aritzeta, A., Muela, A.: Assessment of entrepreneurial orientation in vocational training students: Development of a new scale and relationships with self-efficacy and personal initiative. Front. Psychol. **10**, 1125 (2019)
22. Courtright, S.H., Thurgood, G.R., Stewart, G.L., Pierotti, A.J.: Structural interdependence in teams: An integrative framework and meta-analysis. J. Appl. Psychol. **100**(6), 1825–1846 (2015)
23. Trentin, G.: Graphic knowledge representation as a tool for fostering knowledge flow in informal learning processes. In: Trentin, G. (ed.) Technology and Knowledge Flow: The Power of Networks, pp. 133–156. Chandos Publishing Limited, Cambridge, UK (2011)
24. Oliver, G.R.: A micro intellectual capital knowledge flow model: a critical account of IC inside the classroom. J. Intell. Capital **14**(1), 145–162 (2013)
25. Barrett, B.F.D.: Innovative approach to the formation and sustainability of a learning community connecting students in university classrooms across Asia Pacific. In: Papadopoulos, P.M., Burger, R., Faria, A. (eds.) Innovation and Entrepreneurship in Education, pp. 109–130 (2017)
26. Martínez, A., Dimitriadis, Y., Rubia, B., Gómez, E., de la Fuente, P.: Combining qualitative evaluation and social network analysis for the study of classroom social interactions. Comput. Educ. **41**(4), 353–368 (2003)
27. Carolan, B.V.: Social Network Analysis and Education: Theory, Methods & Applications. SAGE, Thousand Oaks (2013)
28. Panitz, R., Glückler, J.: Network stability in organizational flux: the case of in-house management consulting. Soc. Netw. **61**, 170–180 (2020)
29. Israel, U., Koester, B.P., McKay, T.A.: Campus connections: student and course networks in higher education. Innov. High. Educ. **45**(2), 135–151 (2020). https://doi.org/10.1007/s10755-019-09497-3
30. Granovetter, M.S.: The strength of weak ties. Am. J. Sociol. **78**(6), 1360–1380 (1973)
31. Steffes, E.M., Burgee, L.E.: Social ties and online word of mouth. Internet Res. **19**(1), 42–59 (2009)
32. Hoare, C.A.R.: Communicating sequential processes. Commun. ACM **21**(8), 666–677 (1978)
33. NodeXL: network analysis & insights as easy as pie charts, https://nodexl.com/. Accessed 01 May 2020
34. Saqr, M., Nouri, J., Vartiainen, H., Malmberg, J.: What makes an online problem-based group successful? A learning analytics study using social network analysis. BMC Med. Educ. **20**(1), 1–11 (2020)
35. Krackhardt, D., Handcock, M.S.: Heider vs Simmel: emergent features in dynamic structures. In: Airoldi, E., Blei, D.M., Fienberg, S.E., Goldenberg, A., Xing, E.P., Zheng, A.X. (eds.) Statistical Network Analysis: Models, Issues, and New Directions. ICML 2006, vol. 4503. Springer, Berlin, Heidelberg (2007)
36. Moody, J., Coleman, J.: Clustering and cohesion in networks: concepts and measures. In: Wright, J.D. (ed.) International Encyclopedia of the Social & Behavioral Sciences, 2nd edn., pp. 906–912. Elsevier (2015)
37. Burt, R.S.: Structural Holes. The Social Structure of Competition. Harvard University Press, Cambridge (1992)
38. Gargiulo, M., Benassi, M.: Trapped in your own net? Network cohesion, structural holes, and the adaptation of social capital. Organization Sci. **11**(2), 183–196 (2000). https://doi.org/10.1287/orsc.11.2.183.12514

Author Index

A. Antonyuk and N. Basov (Eds.): NetGloW 2020, LNNS 181, pp. 365–366, 2021.
https://doi.org/10.1007/978-3-030-64877-0

Printed in the United States
by Baker & Taylor Publisher Services